Animals and Science

Animals and Science

A Guide to the Debates

Niall Shanks

ABC CLIO

Santa Barbara, California • Denver, Colorado • Oxford, England

Library of Congress Cataloging-in-Publication Data

Shanks, Niall, 1959–
Animals and science : a guide to the debates / Niall Shanks.
p. cm. —(Controversies in science)
Includes bibliographical references and index.
ISBN 1-57607-246-0 (hardcover: alk. paper); 1-57607-882-5 (e-book)
1. Animal experimentation. 2. Animal models in research. 3. Laboratory animals. 4. Animal welfare. I. Title. II. Series.
HV4915.S495 2002
179.4—dc21
2002001504

06 05 04 03 02 10 9 8 7 6 5 4 3 2 1

This book is also available on the World Wide Web as an e-book.
Visit abc-clio.com for details.

ABC-CLIO, Inc.
130 Cremona Drive, P.O. Box 1911
Santa Barbara, California 93116-1911

This book is printed on acid-free paper ♾.
Manufactured in the United States of America

To James Shanks

Contents

Preface

This is a book about animals—human and nonhuman alike. To be more specific, it is a book about important similarities and differences between humans and nonhuman animals that reflect our relative places in nature and are relevant for a rational discussion of how we humans should treat our fellow creatures. And to be very specific, it is a book about what science has (and has not) told us about these curious matters. So in a way, this is not just a book about animals; it is a book about the nature of science itself and its implications for broader moral questions.

The animal who wrote this book has received help and advice from a number of other animals of his acquaintance here at East Tennessee State University. First and foremost, I have received much valuable help from my colleague Hugh LaFollette (Department of Philosophy), who carefully read parts of the manuscript and who also had many stimulating discussions with me concerning its central theses and claims. Crucial parts of the manuscript were also read by Dan Johnson (Department of Biological Sciences), Jeffrey Gold, Rebecca Hanrahan, and Marie Graves (Department of Philosophy), and George Gale (Department of Philosophy, University of Missouri at Kansas City). My colleague Michael Woodruff (Department of Anatomy and Cell Biology) read parts of the manuscript and provided me with valuable lessons in neuroanatomy and the history of psychology. I have also had many useful discussions with my colleagues Rebecca Pyles, Karl Joplin, and Foster Levy (Department of Biological Sciences). To these worthy and sapient creatures I give my thanks, noting all the while, as is traditional in a preface, that I am responsible for the errors that remain. I also offer thanks to Kevin Downing, Anna Kaltenbach, and Libby Barstow at ABC-CLIO for helping to bring this project to fruition.

—*Niall Shanks*
East Tennessee State University
September 2001

Introduction: Man and Beast in Nature

This is a book about the controversies surrounding our understanding of the similarities and differences between ourselves and nonhuman animals and our relative places in nature. Depending on how these issues are resolved are matters of great contemporary importance. For example, are nonhuman animals—at least some of them—conscious, self-aware, and able to feel pain and experience emotions? If they are like us in these respects, are they like us in other respects? For example, are they then beings worthy of moral respect (so they cannot be treated arbitrarily and capriciously)? Do they even have rights? Do we have obligations and duties toward them? I believe there are very few people who have not given these matters thought, one way or the other. And for good reason, for the ways in which we think about these issues will likely tell us as much about ourselves and what we are as they will about the nature of nonhuman animals themselves.

The basic issues here are as ancient as our ability to think. The earliest cave paintings depict complex interactions between humans and nonhuman animals. As Jim Mason has remarked:

> When we lived as foragers with earthbound religions, animals were the first beings, world-shapers, and teachers and ancestors of people. When we became agriculturalists and looked to the heavens for instruction about the seasons and the elements, we saw animal forms among the stars. Of the forty eight Ptolemaic constellations, all but a few were organic, and twenty five are named for animals. Of the twenty two more that were added in the 17th century, nineteen have animal names. When people built colossal earthworks to appeal to the powers in the heavens, they built them in animal forms. Some in Peru are over a mile long. One in Ohio is in the shape of a giant snake with an egg in its mouth. (1998, p. 38)

It is worth adding that the various religious doctrines and philosophical theories that our species has produced have had numerous and often contradictory implications for our appreciation of the status of nonhuman animals in relation to ourselves. This will be apparent in the first chapter, when I examine influential views emanating from medieval Europe.

So an examination of these issues is a potentially enormous undertaking—it could include, among other things, literature, philosophy, mythology, and religion, as well as the natural and biological sciences. To make the task manageable, I will focus on these issues in the light of science. But the contemporary natural and biological sciences have curious histories, going back to the Renaissance and beyond. And so, though the focus is on science, it will be seen that the issues at hand are nevertheless inextricably intertwined with religious and philosophical matters.

In the course of the analysis it will be seen that the biological sciences have evolved. And our understanding of the place of ourselves and other animals in nature has coevolved with the rise and subsequent development of the sciences. And it is this evolving story, along with the central controversies it has generated, that I am concerned to tell.

But why try to even tell such a complex story? The Duke of Wellington thought some historical narratives should not be told precisely because of the complexity of the causes and effects of events of interest. For example, he did not think that the history of the Battle of Waterloo should be told. He reasoned as follows:

> The history of the battle is not unlike the history of a ball. Some individuals may recollect all the little events of which the great result is the battle won or lost; but no individual can recollect the order in which, or the exact moment at which, they occurred, which makes all the difference as to their value or importance. (Quoted in Keegan 1976, p. 117)

In writing the historical portions of this book, I do not pretend to recollect, even less to present, all the details in the historical evolution of our perceptions of ourselves and other animals in nature, but I do hope, nevertheless, to record some of the great results—the battles won or lost—and to say something about their causation, however incomplete my analysis must be.

Today, our relationships with nonhuman animals betray many contradictions and ambiguities. I suspect that most people think we should refrain from cruelty to animals, yet many of these same people eat meat, wear animal products, and are grateful for the fruits of biomedical research based on animal experimentation. Others believe we need to radically rethink our positions with respect to the status of nonhuman animals. Perhaps the time is ripe to accord them formal rights and to radically extend legal protections.

Our attitudes toward nonhuman animals are undergirded by complex webs of beliefs about similarities and differences between us and other species. These beliefs are shaped by many factors. These include our experiences with animals (perhaps as pet owners, or farmers, or experimenters), religious beliefs, exposure to biological science, philosophical presuppositions, and psychological factors such as our capacity for empathy. As we will see, philosophical, religious, and scientific attitudes about humans and nonhuman animals have a long and complex history of mutual coevolution. In the early parts of the book, I shall try to unravel some of the important threads and strands in this complex and messy fabric of belief.

But at rock bottom, this book is a tale of two themes. On the one hand, it is a book about what science has (and has not) taught us about the nature of nonhuman animals (with implications for how we should treat and otherwise interact with them). On the other hand, it is a book about the nature of science—the sorts of questions science can and cannot answer, the role of theory in shaping the interpretation of evidence, and the very role of experimental methods themselves. In western culture these themes are intertwined. For example, the process of learning how to think systematically about the significance of the results of the dissection of human and nonhuman cadavers and the vivisection of nonhuman animals was a key event in the very rise of modern science itself. Consequently, in the course of my analysis I shall be concerned to display the relevance of the development of modern science for an understanding of the complex, multidimensional relationships between humans and nonhuman animal species.

Today, perhaps more than at any other time in human history, a scientifically informed understanding of the nature of, and relationships between, human and nonhuman animals is not just of biological and ecological importance; it is a matter of crucial moral, social, legal, and political importance. Humans and the nonhuman animals we interact with are similar in some respects, yet different in others. These similarities and differences form the basis for comparative studies in such fields as anatomy, physiology, biochemistry, and genetics, to mention just a few branches of biological inquiry.

But in the long history of human thought about ourselves and nonhuman animals, the important similarities and differences have been with respect to cognitive abilities—for example, the ability to have sensations such as pain; the ability to have a sense of self; and the abilities to reason, to use language, to experience emotional states, and to have hopes, fears, desires, and so on. In the later parts of the book, I will be concerned with animals in the light of the science surrounding these cognitive topics.

As we shall see, the basic scientific issues, although of crucial importance for such questions as whether nonhuman animals have moral standing, transcend the narrow confines of the animal rights debate and are relevant for such issues as the conduct of scientific research itself, especially in biomedicine where there is a heavy reliance on the use of nonhuman research subjects. Thus, as psychologist Clive Wynne has recently remarked:

> Supporters of the Great Ape Project are impressed by the genetic similarity between people and nonhuman apes and by the relatively short period of time since we and our closest relatives (chimpanzees) diverged from a common ancestor. Most advocates of this project point to what they consider the key psychological similarity between nonhuman apes and ourselves: Nonhuman apes, they argue, are self-aware. As a consequence of this self-awareness, gorillas, chimpanzees, bonobos and orangutans must suffer in captivity in ways not so different from what we would experience under similar circumstances. (2001, p. 120)

Clearly, then, there will be much interest in what science has revealed about the cognitive similarities and differences between ourselves and nonhuman animals, in particular chimpanzees.

But scientists have not simply confined their interest in apes to the study of their cognitive abilities, for some 1,600 chimpanzees are currently being used in biomedical research in the United States alone. Thus, discussing some of the types of biomedical research involving chimpanzees, Wynne observes:

> AIDS is another prominent example, because chimpanzees are the only nonhuman species that can be infected with HIV-1, the common form of the virus found throughout the world. The reason became clearer last year, when investigators found proof that HIV-1 spread to humans from chimps early in the 20th century. Now more than 36 million people around the world are infected with this virus, and some 22 million have already died from AIDS. Chimpanzees continue to be immensely valuable in the search for a vaccine. (2001, p. 120)

Here we start to get the first whiffs of some of the serious dilemmas and controversies that are generated by the scientific study of animals. Some of those who study animal cognition are impressed by the apparent cognitive similarities between humans and chimpanzees. These similarities, they allege, confer moral status on the chimpanzees—because they can suffer, they should not be treated arbitrarily or capriciously. But other researchers, noting further biological similarities, allege that these similarities are precisely what make chimpanzees useful for research aimed at ending human suffering.

Very often, in the face of the cognitive and biological *similarities,* it looks as though it is the bare *species difference* between *Homo sapiens* and *Pan troglodytes* or *Pan paniscus* that makes it morally acceptable to perform experimental studies on them that would be grossly immoral if performed on our fellow humans. And indeed some theorists have spoken as if there is indeed morally unwarranted discrimination based on species membership—*speciesism* by analogy with *racism* and *sexism.* And as will be seen later, one does not have to go far to find people within and without the scientific community who have been willing to extend cognitive charity to animal species far more distant from us, evolutionarily speaking, than the apes. Clearly, then, the issues at hand are of enormous importance, both scientifically and morally.

In approaching these matters, I propose to deal with two broad issues. First, I will examine how science has developed through the use of animals as research subjects. Second, I will examine how the scientific study of animals (especially with respect to their cognitive abilities) has transformed our understanding of the place of animals in nature, with serious implications for the ways in which we treat nonhuman animals.

These broad issues will be intertwined with one another in the context of a discussion of their historical development. For to understand the contemporary issues and to locate them in an appropriate cultural context, it will be helpful to examine the historical evolution of modern science and perceptions of its implications for our relationships with nonhuman animals. More specifically, I will discuss issues relating to the nature of science, the rise of the biomedical sciences in the sixteenth and seventeenth centuries, and the consequences of evolutionary biology in the nineteenth and twentieth centuries for our appreciation of the place of animals in nature and our relations to other species.

This book will be of interest to anyone concerned with the relationships between humans and nonhuman species. It will be of interest to people in the animal rights community, people with ecological interests, people in the scientific community (both those who use animals as subjects and those who observe their behavior in the field), and people concerned with the ways in which modern technological innovations are enabling us to exploit nonhuman animal resources (in agriculture, for instance).

In Chapter 1, "The Essence of the Problem: Medieval Legacies," I discuss some of the important philosophical presuppositions of the medieval worldview as it relates to the relationships between humans and nonhuman species. The medieval theorists had a fundamentally theological view of a universe created and designed by God, with a rigid and fixed hierarchical structure. In this scheme there was an absolute difference between humans and nonhuman animals, the latter being viewed as mere *brutes.* There was thus a fixed and unchanging "scale of nature" with humans at the top and other species below. Although this view shaped medieval attitudes toward human and nonhuman animals alike, it did not preclude differential estimates of the nature and worth of nonhuman animals.

In Chapter 2, "Gut Feelings: Anatomy and Physiology in the Renaissance," I show how, impressed by analogies between machines on the one hand and human and animal bodies on the other, the founders of anatomy and physiology in the sixteenth and seventeenth centuries first turned their attention to cadavers (to gain anatomical knowledge of structure). Comparative anatomy showed that there were important structural similarities between human bodies and nonhuman animal bodies. Investigators then turned their attention to live animals (to gain physiological knowledge of vital function). This led to the rise of systematic vivisection of nonhuman animal bodies. The discussion focuses on the work of Andreas Vesalius, William Harvey, and Robert Hooke, all of whom undertook fundamental research using live animals.

A general method emerged from these investigations: the analytic-synthetic method. Here, to understand a complex system (a machine or an animal), you analyze it by taking it to pieces, studying the nature of the parts and the relations (static and dynamic) that obtain between them, and then synthesize this knowledge into an understanding of the whole complex system itself.

This method found its way out of medicine and into physics, culminating in Newtonian mechanics and its view of the universe as a huge machine, obeying deterministic scientific laws, the parts being bound, and the entire system powered, by gravity. Newton's picture of physics (and science generally) will come back and play a role later in the book, in our discussion of nineteenth-century physiology.

In Chapter 3, "Minds, Machines, and Bodies: Intelligent Design in Nature?" I turn to the issue of cognitive differences between humans and nonhuman species. The early investigators saw important structural and functional analogies between humans and nonhuman animals. These observations played a crucial role in establishing important branches of modern scientific inquiry. But even though some observers—Michel Montaigne, for example—argued that there were also cognitive analogies and similarities, others—for example, René Descartes—disagreed. Descartes claimed that only humans had minds and were able to use language. Humans had souls, whereas animals were just mere machines. Later, in Chapter 10, I will examine a recent attempt to revive this argument in modern scientific clothes.

In Chapter 4, "Of Mice and Monkeys: The Moral Relevance of Animal Pain," I turn to the evolution of issues about the moral status of nonhuman animals. Whether or not nonhuman animals are just mere machines, many behave in ways that indicate they can suffer pain. Accompanying the rise of utilitarian moral theory in the early nineteenth century—a theory based on the moral relevance of pleasure and pain—was a growing concern for animal pain. The nineteenth century saw the emergence of the first organized groups whose aims were to promote animal welfare.

In Chapter 4 I will also explain utilitarian moral theory and its relevance for issues surrounding the use of animals in science. Nonhuman animal minds may differ from human minds in many ways, but insofar as they can experience pain, this appears to be a morally relevant cognitive commonality. I will also briefly discuss some contemporary moral theorizing about animals.

In Chapter 5, "The Job of Physiology: Animals in Nineteenth-Century Medicine," I discuss the role played by animal experimentation in the nineteenth century. Even if animals suffer pain, animal experimentation has been crucial to progress in the biomedical sciences. The nineteenth-century physiologist Claude Bernard espoused a deterministic, mechanical view of biological systems inspired by the triumphs of deterministic Newtonian physics.

Today, the American Medical Association recognizes Bernard as one of the theorists who established the methodological foundations of animal experimentation. Bernard produced a thoroughly mechanistic philosophy of biology in terms of which nonhuman animals, though they differ in appearance from humans, are nevertheless so similar in basic design that they can be used as machines in which to test hypotheses about human medicine. So perhaps the enormous medical benefits of animal experimentation outweigh the pain and suffering of animal research subjects.

Lurking in the background is the issue of *species differences*. In his search for universal physiological laws analogous to Newton's laws, Bernard tended to play down the differences and to emphasize the mechanical similarities: In humans and nonhuman animals, same cause is followed by same effect, once differences such as size have been compensated for.

In Chapter 6, "By Accident or Design: Darwin's Theory of Evolution,"I turn at last to a discussion of the implications of evolutionary biology for animals in the light of science. Are animals just machines? The biological world contains many examples of structures and processes that appear to result from deliberate, intelligent design. If animals are machines, machines have makers. The view that God is the maker of humans as well as of the beast-machines science uses to study them emerges from the *natural theology* of William Paley. But natural theology afforded a fundamentally unscientific, supernatural explanation of biological phenomena—in particular, of the way in which organisms seem to be well fitted to their natural environments.

Darwin's theory of evolution challenged the need for supernatural influences in biology by offering a purely natural account of the way in which organisms come to be adapted to their environments. Seeing animals as naturally evolved biological systems, not as the products of supernatural design, challenged the relevance of the mechanical metaphors in terms of which they had been conceptualized prior to the appearance of the theory of evolution.

But the theory of evolution also has profound implications for the biological significance of species differences—they result not from supernatural creation but from natural evolutionary processes. Darwin's theory of evolution, expounded in *The Origin of Species,* had the consequence that species differences were not absolute, categorical differences. The natural processes giving rise to the divergence of new species—the idea of descent from common ancestors with subsequent evolutionary modification—leads one to expect both similarities and differences.

I examine the issue of the cognitive status of nonhuman animals in the light of evolutionary biology. Evolution leads us to expect similarities as well as differences between different species. Charles Darwin himself realized that just as structural similarities could be found between humans and nonhuman animals, so too there was evidence of the existence of cognitive similarities. This idea was pursued by Darwin in both *The Descent of Man* and *The Expression of the Emotions in Man and Animals.*

In these books Darwin presented evidence that nonhuman animals were not just dumb Cartesian machines, or even organisms whose mental lives were perhaps simply restricted to the experience of pleasure and pain. Darwin's work is important precisely because he tried to use scientific evidence to vindicate the claim that nonhuman animals had greater cognitive sophistication than the scientists of his day had allowed. Darwin—a noted opponent of slavery—saw nonhuman animals as sentient beings capable of suffering in ways that were as morally relevant as human suffering. Darwin was an opponent of vivisection. Thus Bernard and Darwin, two giants of nineteenth-century biology, had strikingly different views about animals.

In Chapter 7, "Darwinism Developed: The Ontogeny of an Idea," I discuss how the theory of evolution itself evolved during the course of the twentieth century. This chapter will bring the reader up to date on important recent developments in evolutionary biology. In particular, some basic ideas from genetics will be introduced so that it will be possible to give a gene's eye view of what evolution involves. Doing so will facilitate a discussion of what is involved in the process of speciation and the biological consequences of species differences.

I will also use this chapter as an opportunity to look at some contemporary evolutionary views of organismal development. This will be an occasion to reexamine the issue of how far science has retreated from a literal, mechanical view of organisms, thereby deepening our understanding of the enormous implications of the conceptual revolution initiated by Darwin but continued in the hands of contemporary scientists.

Finally, since the issues about the value and usefulness of animals in the biological sciences have been intertwined with a concern for human health and well-being, I will discuss some of the implications of the new science of *Darwinian medicine*. In the past twenty years there has been a growing realization by evolutionary biologists that the study of human health and disease is something that needs to be seen in the appropriate evolutionary context. Humans, for example, have taken very different evolutionary trajectories from rodents (the most common types of experimental research subjects). The line leading to modern humans diverged from that leading to modern rodents over 60 million years ago, for over 120 million years of independent evolution. Rodents may not simply be men writ small.

And some of the diseases we are prone to—from common types of cancer to heart disease and high cholesterol—reflect the ways in which the modern environment we live in—an environment increasingly shaped by rapid cultural evolution—has diverged from the Stone Age environment in which a lot of our specifically human evolution took place.

The implications of a fully Darwinian view of medicine will become apparent in Chapter 8, "The Mouse as Man Writ Small: Animals in Modern Medicine." I examine in this chapter some of the ways in which animals are actually used in the biomedical sciences—ways that typically do not reflect a deep grasp of the implications of evolutionary biology, since they derive in large measure from the nonevolutionary, mechanistic tradition of nineteenth-century biology.

I will argue that those involved in controversies concerning the use of animals in biomedical research need to confront not just the similarities between humans and nonhuman animals—the mouse as *man writ small*—but also the differences and biological disanalogies consequent upon evolution. Species differences are real differences in nature. Differently adapted biological systems may not respond in the same ways to similar stimuli. Nonhuman animals may not always be good models of human biomedical phenomena. There are obvious similarities between humans and other mammals, for example, but there are important differences too. Aspirin, for example, is safe in humans but causes birth defects in rats and mice.

This side of evolution will thus be relevant for a rational assessment of the alleged great utility of nonhuman animals in biomedical research. The assessment of the consequences of biological evolution for the evaluation of the usefulness of animals in biomedical research is a key factor in controversies about contemporary animal experimentation.

In Chapter 9, "Mice, Mazes, and Minds: Explaining Animal Behavior," I discuss the implications of modern biology and experimental psychology for our understanding of the cognitive abilities of nonhuman animals. I begin with a discussion of the rise of behaviorism, for behaviorist methodologies shaped much of the research on animal (and indeed human) cognition in the first half of the twentieth century. Early behaviorists viewed both human and animal minds as inscrutable black boxes. I discuss the implications of both *classical* and *radical* (Skinnerian) behaviorism for the study of animal cognition. I also discuss reasons for the fall of Skinnerian behaviorism in the late 1950s.

One reason for doing this is to point out that though radical behaviorist ideology fell by the wayside, it nevertheless left an important legacy that continues to shape contemporary debates about animal minds and animal cognition. This residue of behaviorism can be referred to as *evidential behaviorism* (basically the view that all claims about animal cognition and animal minds must be rooted in evidence derived from the study of animal behavior).

Another reason for discussing the fall of behaviorism is that it was a key event leading to the *cognitive revolution* that occurred in the

1960s. As a result of the cognitive revolution, many scientists became willing to countenance hypotheses concerning cognition in nonhuman animals—provided they were examined and tested in accord with the dictates of evidential behaviorism. Still, important controversies exist here between *cognitive minimalists* at one extreme, who maintain that animal behavior can be explained in terms of associative learning, and *cognitive ethologists* at the other extreme, who think that there is evidence to support claims that (some) nonhuman animals have minds with subjective states of awareness not unlike our own. These controversies hinge crucially on what is to count as scientific evidence.

In Chapter 10, "The Evolution of Consciousness: A Question of Animal Pain," I reexamine the issue of consciousness in nonhuman species from the standpoint of modern science. At the dawn of modern science, animals were viewed as mere automata incapable of thought and conscious awareness. These issues recurred at the end of the twentieth century. Do animals feel pain? Do animals feel anger, or sadness, or happiness? The issue in this chapter is what has been termed *feeling-consciousness.* I begin with a discussion of the physiology of pain. This is to provide the reader with the requisite background against which the contemporary controversy over animal pain can be appreciated.

I then examine some contemporary arguments, rooted in evolutionary biology, to the effect that nonhuman animals do not and cannot feel pain. In these arguments, a connection is made between being conscious and being able to use language. Some contemporary theorists think that something like a Cartesian view of animals should get a new lease on life. Humans differ from nonhuman animals not by having a soul but by having a large, evolved brain capable of language use. Ranged against this minimalist estimate of nonhuman animal consciousness are other arguments, also rooted in evolutionary biology, that extend cognitive charity well beyond our own species, to include other mammals, birds, reptiles, and fish.

The differing arguments in this controversy over animal pain, as well as others concerning the emotional lives of nonhuman animals, rest on a mixture of theory and evidence. Scientists disagree over what should count as evidence, how evidence is to be interpreted, and what inferences can be made from behavioral evidence. There are also disagreements over the interpretation of evolutionary theory and its implications for the issues at hand. In this controversy over animal pain, there are many lessons about the nature of science itself—what it can say and what it cannot.

But these issues about whether or not nonhuman animals can feel pain are not merely academic. They are of crucial importance to the moral debate over the use of animals in science, and so I argue that until these matters are satisfactorily resolved, it is reasonable to extend cognitive charity to our phylogenetic kin—they certainly seem to feel pain, and we should treat them as if they do—with all the moral caution that this involves, until it can be shown beyond reasonable doubt that they do not. Mere skepticism about animal pain should not be a license for cruelty.

In Chapter 11, "Animals through the Looking Glass: Language and Self-Consciousness," I examine arguments about the issue of self-consciousness in nonhuman animals, and in chimpanzees in particular. I begin with a discussion of the experiments that appeared to indicate that chimpanzees were capable of learning, comprehending, and using language. I show how the early enthusiasm was later dampened by a careful reexamination of the evidence. Again, the controversies concerning language use by chimpanzees contain lessons about the nature of science itself, as well as lessons about our closest evolutionary relatives and the best explanations of their curious behavior.

I also examine claims about self-consciousness that are divorced from the language issue. Does the behavior of chimpanzees examining their reflections in mirrors, in the context of the so-called mirror self-recognition tests, indicate that they have a sense of self? Do chimpanzees have a *theory of mind?* That is, do they predict the behavior of others (humans or other chimpanzees) by attributing beliefs, hopes, fears, and desires to them? I will argue that the evidence for these claims is ambiguous. Again, by exploring the sources of ambiguity, much can be learned about the nature of science.

My conclusions are conservative. Differential estimates of the cognitive abilities of nonhuman animals, rooted in religious and philosophical theorizing, were present before the rise of science. These differential estimates have survived the rise of science and have evolved with it. But though the evidential and theoretical bases for these differential estimates have changed dramatically with 300 years of science, with a consequent radical transformation of the nature of the basic issues, differences remain, and the scientific and moral controversies arising from these differences are likely to be with us for the foreseeable future. Nevertheless, much of interest can be learned about the nature of science itself from the analysis of these controversies.

Animals and Science

1

The Essence of the Problem: Medieval Legacies

It is a central theme of this book that the rise of modern science in the sixteenth and seventeenth centuries had enormous implications for our understanding of the differences and similarities between ourselves and nonhuman animals. But these events, crucial as they are for this book, did not occur in a cultural vacuum. In this chapter I will present and discuss important features of medieval views of the living world, at least as they had an impact on Western philosophy and science. This will not only provide a theoretical backdrop against which we can see the subsequent rise of science; it will also afford an opportunity to examine some deep cultural presuppositions that not only shaped the rise of modern science but also echo down the ages to our own century and its view of organic nature.

In writing about medieval culture, it is easy to focus on what people in medieval times did not know. Their universe was composed of but four elements: earth, air, fire, and water. They knew nothing of bacteria or cells. They had no biochemistry or molecular biology, they had no understanding of the causes of disease, and importantly for later developments in evolutionary biology, they had no sense of the physical antiquity of the earth and the natural processes whereby new species arise. It is also easy to focus on claims that are, by modern lights, simply false, for example that the earth is at the center of the universe (a basic presupposition of Ptolemaic astronomy) or that diseases are caused by sin, resulting, perhaps, in an imbalance of the vital humors.

Life was undeniably harsh in medieval times, and there was much about which medieval thinkers were ignorant. But this should not blind us to the fact that they had complex and sophisticated theoretical frameworks in terms of which they tried to come to grips with the world around them. Inevitably, in the space of one chapter, I cannot possibly do full justice to the richness, complexity, and sophistication of medieval views of the living world. Nevertheless, I hope to bring out important themes that would come, in the centuries that follow, to have a profound influence on the ways in which we see ourselves in relation to nonhuman animals.

Medieval views of the living world were shaped by religious and philosophical presuppositions. Important for the rise of modern science as a phenomenon in Western civilization was medieval Christianity and its theoretical underpinnings in terms of medieval philosophical theology. Medieval philosophical theology was shaped by influences from ancient Greece. In this regard, the philosophical schemes of Plato and Aristotle are of crucial importance, and their impact on the medieval views of organic nature can be seen in the works of St. Augustine and St. Thomas Aquinas, respectively.

St. Augustine and the Influence of Plato

There is much in the work of the ancient Greek philosopher Plato that can be readily adapted to serve the ends of Christian philosophical theology. In *The Phaedo* (Church 1987), Plato developed several ideas that would come to be important in medieval views about organic nature. First and foremost was the doctrine of *forms.*

Plato made a sharp distinction between an imperfect *visible world,* which contains changeable material objects that can be perceived through the senses, such as tables, chairs, cabbages, and kings, and a perfect, immaterial *intelligible world,* which contains changeless forms—abstract objects—that can only be known through the intellect. From the standpoint of medieval philosophical theology, it is hard to overestimate the importance of the philosophical claim that there were two worlds, one of which was visible and the other invisible.

The impact of this claim becomes clearer when we reflect on the nature of the relationships between the visible world and the invisible world of forms. Plato believed that objects in the visible world exemplified or instantiated various forms in the invisible world. For example, the form *CAT* was what all cats have in common and in virtue of which they were cats and not, for example, dogs! One perceives particular cats with the senses. By contrast, the form *CAT,* which all earthly cats have in common, can only be known, so the story goes, through the operation of the intellect. Since species membership depends on exemplification of a changeless form, it is hard to see how an extant species could give rise to new species—it is hard to see, that is, how the facts of biological evolution could be accommodated in this philosophical scheme.

This latter point can be made slightly differently as follows: For Plato differences between individual cats—intraspecific variation—were the result of imperfections. Such accidental variation was of no great biological significance; what mattered was the exemplification of a species-determining form (that is, *CAT*). By contrast, it will be seen later in this book that Charles Darwin's crucial insight was that where there was such variation between individuals in a population, and where that variation was heritable and had consequences for differential rates of survival and reproduction, then that variation was of crucial importance for it provided the grist for the evolutionary mill of natural selection. It was Darwin's contention that the operation of natural selection over many successive generations could, in favorable circumstances, give rise to new species, to the existence of new forms in organic nature.

Greek philosopher Plato Aristocles (427–347 B.C.) with the philosopher and scientist Aristotle (384–322 B.C.). The philosophical schemes of Plato and Aristotle had tremendous impact on the medieval views of organic nature. (Hulton / Archive)

Equally important, and derivative from this two-worlds view of reality, is Plato's view of human beings. Humans have two parts: a physical body and a separable, nonphysical soul (our moral and intellectual personality) that is capable of surviving the destruction of the body and that properly belongs in the world of forms. It is in virtue of the possession of this soul that humans have the ability to reason, and it is in virtue of the possession and cultivation of the soul that we are fundamentally different from nonhuman animals.

St. Augustine (born 354 A.D.) had an enormous influence on medieval philosophical theology. Augustine's achievement was to Christianize various ideas that had been derived from Platonic philosophy. The physical, material world became the fruit of God's creation *ex nihilo* (literally, from nothing). God, an immaterial being, stood outside the material world but could be known through a spiritual sense derived from our possession of a soul. It was the possession of this soul capable of knowing God that differentiated humans from other animals. Nevertheless, all living things owed their existence to God. To better understand the Augustinian views about living things, it will help to discuss his doctrine of the *rationes seminales* (literally, seminal principles).

Because of his views about creation, Augustine had to confront a version of the problem of the origin of species. For example, in the biblical creation story related in the first book of Genesis, there is a reported sequence of biological events. On the fifth day, God brings forth sea creatures and fowls of the air, whereas on the sixth day he brings forth cattle, things that creepeth, and the primal founder pair for our own species, Adam and Eve. How could all this be reconciled with the view that God created everything—"all things together" (Eccles. 18:1)—in one fell swoop?

Augustine's answer went something like this: From the standpoint of reproduction, mice give rise to mice, humans give rise to humans, and so on. How do mice and men give rise to anything by way of offspring, let alone mice and men respectively? Augustine's answer was that at the time of creation God implanted seminal principles into matter, and this had the effect of building into nature the *potential* for all extant species to emerge. When a species comes to exist—sea creatures on the fifth day, mice and men on the sixth—the potentiality in the species' seminal principle is actualized, and subsequent *seeds* allow for the reproductive continuation of the fixed species.

For Augustine, new biological forms could not be caused or brought about by extant biological forms. Rather, a new biological form, when it appeared, was the result of the actualization of a seminal principle laid down at the moment of creation. Thus Stumpf has observed:

> What the doctrine of *rationes seminales* enabled Augustine to say was that God had created all things at once, meaning by this that he had implanted the seminal principles of all species simultaneously. But, since these germs are principles of potentiality, they are the bearers of things that are to be but that have not yet "flowered." Accordingly, though all species were created at once, they did not all exist fully formed simultaneously. They each achieved their potentiality in a sequence of points in time. (1982, pp. 139–140)

Although Augustine did not pursue evolutionary ideas, his views about *rationes seminales* might be seen as an evolutionary precursor of modern *theistic* evolutionary views, which attempt to reconcile the facts of evolution (in particular, the sequential appearance of various species) with the idea that the whole process is under some kind of divine control. Importantly for Augustine, however, there was no suggestion here of the modern idea that earlier species might be ancestors to later species (Copleston 1961, p. 21).

St. Thomas Aquinas and the Influence of Aristotle

The influence of the ancient Greek philosopher Aristotle on medieval thinking can be seen in the philosophical theology of St. Thomas Aquinas (1224–1274). We will see that in the thought of Aquinas there emerged a number of themes that have had an important shaping influence on subsequent debates about humans and their similarities to, and differences from, nonhuman animals.

To understand the importance of Aquinas's views, it will help to say a little about Aristotle's philosophy of nature. Like Plato before him, Aristotle had to confront the relationship between matter and form. Aristotle rejected Plato's idea that forms could exist separately from individual things, such as mice and men. Aristotle contended that in nature, we never find *matter* on its own or *form* on its own. Everything that exists in nature is a *unity* of matter and form. This unity of matter and form, Aristotle designated as a *substance.* Humans are one type of substance, mice another. Species differences reflect a difference with respect to the form shaping matter.

In order to understand how substances change, Aristotle introduced the idea of the *four causes.* And since this view of causation will turn out to be of importance later, we must examine the basic details here. The doctrine of the four causes was put forward to explain the changes we see in nature. Of any object, be it an inanimate object, an organism, or a human artifact, we can ask four questions: (1) What is it? (2) What is it made of? (3) By what is it made? (4) For what purpose is it made? To answer the first question is to specify the *formal cause,* hence to identify substance and species. To answer the second question is to specify the *material cause* and explain the material composition of the object. To answer the third question is to specify the *efficient cause* and explain by what a thing was made or by what a change was brought about. To answer the fourth question is to specify the *final cause*—the function of the object, the end or purpose for which it was made.

Thus an object might be a clock (formal), made of wood and metal (material), by the clockmaker (efficient), to tell the time (final). But this scheme works for objects that are not human artifacts. An object might be an acorn, made of organic matter, by the parental oak tree, to become an oak tree itself. Importantly for our purposes, Aristotle saw that the form of an object determines its end or function. That is to say, the end or function of an object was determined by its internal nature. This is the sense in which it is the end of an acorn to become an oak tree (Stumpf 1982, pp. 89–92).

For Aristotle, everything in nature, be it organic or inorganic, had a natural end, function, or purpose determined by its form. This is very much a biologist's view of the world as a whole. Yet Aristotle differentiated between organic and inorganic beings through the idea of *souls.* The soul becomes the form of the living, organized body. An organized body has functional parts such that when they attain their end, the organized body as a whole is capable of attaining its end. The parts of the acorn work together that the acorn might become an oak tree.

In fact, the function of anything in nature can be specified by saying what it is there *for the sake of.* This has bearing on our relations with other species, for as Aristotle remarked:

> In like manner we may infer that, after the birth of animals, plants exist for their sake, and that the other animals exist for the sake of man, the tame for use and food, the wild, if not all, at least the greater part of them, for food, and for the provision of clothing and various instruments. Now if nature makes nothing incomplete and nothing in vain, the inference must be that she has made all animals for the sake of man. (Regan and Singer 1989, p. 5)

That humans have a special place in nature, and nature is here for our use, is a theme that will recur in the medieval period duly modified with the trappings of Christianity.

But Aristotle recognized important categorical differences between living things, and so distinguished between three types of soul. *Vegetative souls* conferred upon objects possessing them a simple capacity for life; *sensitive souls* conferred a capacity for life and sensation—nonhuman animals have sensitive souls. Finally, *rational souls* conferred upon their possessors a capacity for life, sensation, and thought. Humans were said to have rational souls, and Aristotle defined humans as rational animals. Right here, in these influential ideas from ancient Greek thought, we see the origins of the philosophical distinctions between humans and nonhuman animals that would influence much subsequent thinking on these matters.

Aristotle himself saw the implications of his views for the relations between humans and nonhuman animals in this way:

> It is clear that the rule of the soul over the body, and of the mind and the rational element over the passionate, is natural and expedient; whereas the equality of the two or the rule of the inferior is always hurtful. The same holds good of animals in relation to men; for tame animals have a better nature than wild, and all tame animals are better off when they are ruled by man. . . . Again, the male is by nature superior, and the female inferior; and the one rules and the other is ruled. . . . Where then there is such a difference as that between soul and body, or between men and animals . . . the lower sort are by nature slaves, and it is better for them as for all inferiors that they should be under the rule of a master. . . . And indeed the use made of slaves and of tame animals is not very different; for both with their bodies minister to the needs of life. . . . (Regan and Singer 1989, pp. 4–5)

The idea that human rationality sets us apart from nonhuman animals, and gives us dominion over them, is a theme that will reappear again and again in the history of Western civilization.

At this juncture, we must also say something about *potentiality* and *actuality* in Aristotle's thought, for there is the seed of an idea here that will mutate and flower in some interesting ways in the medieval thought of Aquinas. The oak tree giving rise to the acorn is an actual tree; the acorn is a potential oak tree. From this Aristotle observed that for a potential thing to become an actual thing, there must be a prior actual thing (the parental oak tree). To explain how there can be a world containing potential things that can become actual things, Aristotle thought that there must be a being that was *pure actuality,* without any *potentiality.* Such a being would be a precondition for the existence of potential beings that could be subsequently actualized. Aristotle called this being the *Unmoved Mover.* Exactly what sort of a being Aristotle was trying to talk about is a little vague. But it was not so for St. Thomas Aquinas.

For Aquinas, the unmoved mover was the Christian God. In many ways, as Augustine Christianized Plato, Aquinas Christianized Aristotle. Aquinas offered five "proofs" for the existence of God, one of which mirrored the pattern of reasoning that led Aristotle to postulate an unmoved mover. But another of Aquinas's "proofs" is much more important for our present concerns. It is the celebrated *argument from design,* an argument that in various mutated forms has worked much mischief on human thinking about the biological world. Although Darwin's theory of evolution can be viewed as a sustained refutation of the argument from design, the argument has been prominent in the writings of contemporary creation "scientists" (see Behe 1996; Johnson 1997), thereby proving that an idea that has been killed stone dead may nevertheless refuse to lie down!

In the hands of Aquinas, the argument proceeded as follows: In the natural world we observe all manner of regularities in the behavior of things that do not possess intelligence. For example, there are regularities in the tides; there are also regularities with respect to the behavior of body parts such as hearts and lungs. For a body part lacking intelligence to nevertheless achieve a function, it must be guided by an intelligence—as the arrow, to use Aquinas's image, is directed by the archer. Aquinas concluded that "some intelligent being exists by whom all natural things are directed to their ends; and this being we call God" (quoted in Stumpf 1982, p. 177). In particular, the functions served by body parts, and indeed the functions served by organisms in the grand scheme of nature, are the result of the purposeful, intelligent design of a benevolent creator. As we will see in later chapters, it took science a long time to dispose of this supernatural way of thinking about natural objects, especially living objects.

But how did Aquinas understand the organisms in the world that are the fruits of the intelligent design of God? In particular, how did he think of differences between humans and nonhuman animals with respect to cognitive capacities? The answer has to do with the relations between form and matter. The cognitive differences between humans and nonhuman animals are ultimately explained in terms of the metaphysics of their composition.

Although Aquinas believed that there were spiritual beings, such as angels, that possessed form alone, he believed that the objects we find here on earth resulted from the unity of matter and form. Aquinas's idea was that when a form was completely *immersed* in matter, the result was a physical object without the capacity for cognition. From where, then, did the capacity for cognition arise?

The answer has to do with the relative degree of independence that a form has from matter. As Hawkins explained it,

> Cognition springs from the possession of form without matter . . . in so far as the form has a certain independence of matter, there arise the various stages of cognitive experience. . . . There are degrees of immateriality. The experiences of a purely sentient being, an animal, have already a certain immateriality; human mind, enjoying a greater degree of immateriality, is capable of an activity of thought which is intrinsically independent of the organism; and beyond human thinking there

are purely spiritual beings, subsistent forms without matter, the angels, whose activity is one of pure thought. (1946, pp. 80–81)

Here, then, is an account of the relative cognitive differences between humans and nonhuman animals. In each case, the soul is the form of the body, but the differences have to do with relative degrees of immersion of form in matter.

In explaining the precise details of the differences, Aquinas argued that the immediate control of nonhuman animal behavior lay in the *concupiscent* and *irascible* appetites. Nonhuman animals are capable of perceiving objects via the senses; the concupiscent appetites then incline the organism toward objects that are pleasurable while the irascible appetites repel the organism from those that are noxious. Humans, by contrast, have sensory abilities and appetites, but they differ from nonhuman animals in the possession of a *free will* and a capacity for *reason,* where it is assumed that the choices made by the will can be guided by reason.

Aquinas arranged the created universe into a hierarchy. This has been referred to as the *chain of being* or *scala naturae.* Beings differ both with respect to species and level of being; some species are *higher,* whereas others are *lower* in nature's hierarchy. This idea of the chain of being is arguably the source of talk of higher and lower organisms. In Aquinas's thought, the idea was cashed out as follows: Below God there is a hierarchy of angels; below angels there are humans, who have a spiritual and a material nature; below humans come animals, plants, and finally the four elements, earth, wind, fire, and water. A human is said to be *brutish* if he behaves like a nonhuman animal, and the latter animal is said to be a *brute.* We have a long history of priding ourselves on the differences between humans on the one hand, and mere brutes on the other.

In this way we see that for Aquinas, there was not merely a hierarchy in created nature, but that humans had a special place, just below angels, and above nonhuman animals—conceived of as mere brutes. The earth, then, was conceived of by medieval thinkers as being at the center of the universe, and according to Aquinas, humans were at the top of the earthly hierarchy of being, thus occupying a special place in the created order.

But if humans have a special place in creation, and are moral beings in terms of their possession of reason and free will, what are their obligations, if any, to nonhuman animals, who, though neither rational nor free, by the present account, are at least sentient? Aquinas's views can be found in his *Summa Contra Gentiles* (chap. 92). Essentially,

Aquinas merged the ancient Greek emphasis on rationality with the Christian emphasis on the special spiritual nature of humans to bring about a sharp differentiation between brutes and ourselves, saying that other creatures were for our use. It is not wrong for us to use them or kill them (see also Lafollette and Shanks 1996, chap. 12). Aquinas thus had a purely instrumental view of nonhuman animals. As Aquinas put it:

> Hereby is refuted the error of those who said it is sinful for a man to kill dumb animals: for by divine providence they are intended for man's use in the natural order. Hence it is not wrong for a man to make use of them, either by killing or in any other way whatever. For this reason the Lord said to Noe (Gen. Xi. 3): *As the green herbs I have delivered all flesh to you.* (Regan and Singer 1989, p. 8)

Does this mean we can treat animals any way we choose? Aquinas thought there were indeed reasons for restraining ourselves from cruelty to animals.

If we should not be cruel to animals, this is not from reasons derived from considerations of charity, for as Aquinas remarked:

> The love of charity extends to none but God and our neighbour. But the word neighbour cannot be extended to irrational creatures, since they have no fellowship with man in the rational life. Therefore charity does not extend to irrational creatures. (Quoted in Rachels 1991, p. 90)

Charity is thus only a consideration for the relationships among rational creatures such as ourselves. But Aquinas pointed out:

> And if any passages of Holy Writ seem to forbid us to be cruel to dumb animals, for instance to kill a bird with its young: this is either to remove man's thoughts from being cruel to other men, and lest through being cruel to animals one become cruel to human beings: or because injury to an animal leads to the temporal hurt of man. . . . (Regan and Singer 1989, pp. 8–9)

Cruelty to animals is thus revealed as not being bad in and of itself, but as being bad for fear that cruel persons will acquire bad habits in the context of their relations to animals that will affect adversely their relations with their fellow rational creatures.

Other Medieval Perspectives

St. Thomas Aquinas is but one medieval thinker, and others had more charitable views about animals. In particular, St. Francis of Assisi

Saint Francis of Assisi (1182– 1226), founder of the Franciscan Order, preaching to the birds. St. Francis of Assisi professed a love for all living things, human and nonhuman alike. Many canonized saints were said to have healed animals and celebrated their life through prayer and preaching. (Hulton / Archive)

(1182–1226)—whom Bertrand Russell referred to as "one of the most lovable men known to history" (1972, p. 449)—professed a love for all living things, human and nonhuman alike. St. Francis was not alone, for as Andrew Linzey has recently observed:

> Something like two-thirds of canonized saints East and West apparently befriended animals, healed them from suffering, assisted them in difficulty, and celebrated their life through prayer and preaching. . . . Despite the negative tradition within Christianity that has frequently downgraded animals, regarding them, at its very worst, as irrational instruments of the Devil, literature on these saints makes clear . . . the common origin of all life in God . . . because of this common origin in God, it necessarily follows that there is a relatedness, a kinship between humans and nonhumans. (Bekoff 1998, pp. 296–297)

That there really is a kinship between humans and nonhuman animals, and that it can be given a good scientific explanation in purely natural terms, shorn of the trappings of Christian theology, is something for which an explanation will be left to Charles Darwin. Like St. Francis, Darwin, too, was an animal lover.

But St. Francis in particular had a dislike of learning and left no enduring body of theoretical work behind him. And so, as James Rachels has pointed out (1991, p. 91), although St. Francis may have preached to the birds and proclaimed his wonder at the living world, it is nevertheless the views of St. Thomas Aquinas that have been most influential in shaping attitudes toward nonhuman animals in Western culture.

This notwithstanding, it should not be forgotten that there were, among the influential religious authorities of medieval Europe, some very different estimates of the nature and worth of nonhuman animals. Echoes of these differences are reflected in Geoffrey Chaucer's *Canterbury Tales.* For there we learn of the Prioress:

> She was so charitable and so pitous
> She wolde wepe, if that she saw a mous
> Kaught in a trappe, if it were deed or bledde.
> Of small houndes hadde she that she fedde
> With rosted flessh, or milk or wastel-breed.
> But soore wepte she if oon of hem were deed,
> Or if men smoot it with a yerde smerte;
> And al was conscience and tendre herte.
> (Winny 1965, p. 59)

By contrast, Chaucer's worldly monk had a purely instrumental view of animals as creatures for his own use. He kept a stable of horses, enjoyed hunting, and did not give a fig for the religious authorities:

> He yaf nat of that text a pulled hen,
> That seith that hunters ben nat hooly men. . . .
> Lat Austin have his swink to him reserved!
> Therefore he was a prikasour aright:
> Geyhoundes he hadde as swift as fowel in flight;
> Of priking and of hunting for the hare
> Was al his lust, for no coste wolde he spare.
> (Winny 1965, pp. 56–57)

The division between those who are tender hearted and those who are not in their relations toward their fellow creatures is probably as old as humanity itself. But in Western culture, these are issues that also came to be informed by theoretical speculation, first of a religious and philosophical nature, and later, with the rise of science, by biological and medical theory.

In time, these issues about the status of nonhuman animals would be transformed into concerns about their cognitive capacities. And the division between the tender hearted and the hard hearted would evolve into a division between those who, having surveyed the evidence in the light of theory, are generous in their estimates of the cognitive capacities—and hence moral worth—of nonhuman animals and those who, having surveyed the same evidence from different theoretical perspectives, see little or nothing by way of cognition in our fellow creatures.

Medicine, Disease, and Sin

Animals, as research subjects, played a very special role in the rise of modern science in the sixteenth and seventeenth centuries, especially in the medical sciences. In order to appreciate the role played by animal experimentation in the emerging medical sciences, it will be useful to pause and examine some salient features of medieval life, especially with respect to public health and medicine. One good reason for doing this is to appreciate the sorry human predicament that medieval medicine was trying to ameliorate. This will provide a background against which the new ideas—and controversies—of the Renaissance can be appreciated.

The best place to begin is with a consideration of water. Waterborne diseases were the scourge of western Europe until the latter part of the nineteenth century, when efforts were made at last to clean up the water supply and to dispose of human waste safely. The medieval world knew little of these public health measures, though they were not devoid of the fruits of sanitary engineering. For example, Roman aqueducts remained in use throughout the Middle Ages, and reservoirs were constructed, as were sophisticated pumping machines. There is also evidence of the use of lead pipe in medieval plumbing (Kirby et al. 1990, p. 117).

But the fact of the matter is that much water was drawn from wells, rivers, and streams; all of it was untreated; and much of it was probably contaminated by pollution and sewage. The result would have been the widespread prevalence of diseases such as typhoid, dysentery, and hepatitis. In medieval towns and cities, human waste was often thrown into the streets, where it would eventually find its way back into the water supply. Even with technological innovation, things were grim, for as Kirby et al. have pointed out, though latrines, sewers, and cesspools were built, they were often too close to water supplies—a serious problem in time of flooding (1990, p. 118). These problems would all be magnified when you bear in mind that some of the waste finding its way into the drinking water was contaminated with microbial, disease-causing agents!

Poor sanitation, in the form of crowded, filthy living conditions, had other influences on the spread of disease. Throughout the Middle Ages, typhus wreaked havoc across Europe. Epidemic typhus is caused by microbes (*Rickettsia prowazekii*) in human body lice. The feces of the lice contain the microbes that get into the body by being scratched into the skin. In medieval Europe, virtually everyone was infested with lice. Murine or endemic typhus is brought to humans by rodent fleas (carrying *Rickettsia typhi*)—the same fleas responsible for bubonic plague (Biddle 1995, p. 154).

Bubonic plague, known as the Black Death in medieval Europe, is caused by microbes (*Yersinia pestis*) carried by rodent fleas. It is estimated that the Black Death reduced the population of China by about 58 million people between 1200 and 1400 A.D. When it spread to Europe, it reduced the population by up to 30 million people—about one-third of the population (Biddle 1995, p. 167).

Other diseases would have taken a grim toll as well. These would include diseases such as leprosy, diphtheria, scarlet fever, whooping cough, mumps, measles, smallpox, and polio, to name but a few. These diseases accounted for a very high level of infant mortality—levels that would not decline significantly until the twentieth century. Insect-borne diseases such as malaria were also prevalent in medieval Europe. But even though there was much disease, there was little that medieval medical science could do to alleviate the suffering.

It is here that medieval ignorance really took its toll. Various theories (not necessarily incompatible with each other) were put forward to explain disease. In the prevailing religious climate, disease was often viewed as having its source in sin, and the cure thus involved fasting, repentance, and prayer. To this end various saints seem to have been associated with various parts of the body: St. Bridget with the eyes, St. Blaise with the throat, St. Erasmus with the stomach, and so on (Gordon 1993, pp. 7–8). Prayer to the appropriate saint was believed to be efficacious. In the absence of statistics, it would be hard to prove the contrary. At a more theoretical level, medieval medicine was characterized by reliance on classical medical authorities such as Hippocrates, Galen, and Avicenna, the practice of medicine being shaped by prayer and the voice of ancient authority.

The classical authorities bequeathed to medieval culture the humoral theory of disease—a theory that had been initially formulated by Empedocles and applied to medicine by no lesser figures than Hippocrates and Galen (Clendening 1960, p. 39). Essentially, the humoral theory maintained that there were four humors, each of which was associated with two qualities: blood was associated with heat and moisture; phlegm with moisture and coldness; black bile with coldness and dryness; and yellow bile with dryness and heat. Health resulted from the four humors being in a state of harmonious balance, whereas specific imbalances in the humors resulted in specific diseases, manifested by excesses or deficiencies in body heat and moisture. (Vestiges of this system of medicine can be found in "Mom medicine"—with Mom's injunction to stay warm and dry on cold, wet days to avoid an excess of phlegm in the form of a nasty cold.) Treatment was aimed at a restoration of a harmonious balance of the humors. Since sin could conceivably cause a humoral imbalance, this theory was not inconsistent with the foregoing remarks about the religious causes of disease.

Actual medical interventions could be both humorous and horrifying. For example, John of Gaddesden (1280–1361) suggested that congestive heart failure could be treated by getting the patient to drink his own urine, or failing that, by a surgical perforation of the bladder. Smallpox, by contrast, could be treated by making everything around the patient's bed red. As for the cure of phthisis (tuberculosis), Gaddesden suggested, "As to food, the best is the milk of a young brunette with her first child, which should be a boy. . . . Failing a wet nurse, the milk of other animals might be used in the following order of choice: the ass, the goat and the cow" (Clendening 1960, p. 84). Sadly, it was only in the late nineteenth century, with Louis Pasteur's germ theory of disease (and measures such as pasteurization), that the correct relationships between milk and tuberculosis were revealed.

Moreover, with no antibiotics, and nothing resembling a modern health service, even minor cuts and scratches could become seriously—perhaps fatally—infected. And understanding little of the need for sanitary measures, and nothing whatsoever about the germ theory of disease, medieval surgical interventions would have carried a very high rate of mortality. Assuming that their ignorance of anatomy did not result in fatal surgical errors, postoperative infection—what used to be known even in the nineteenth century as "surgical fevers," essentially systemic bacterial infection—would have carried away large numbers of patients.

For example, discussing the surgical removal of arrows, fourteenth-century surgeon Guy De Chaulliac suggested that "if the infixed body cannot conveniently be extracted at the first attempt, it ought to

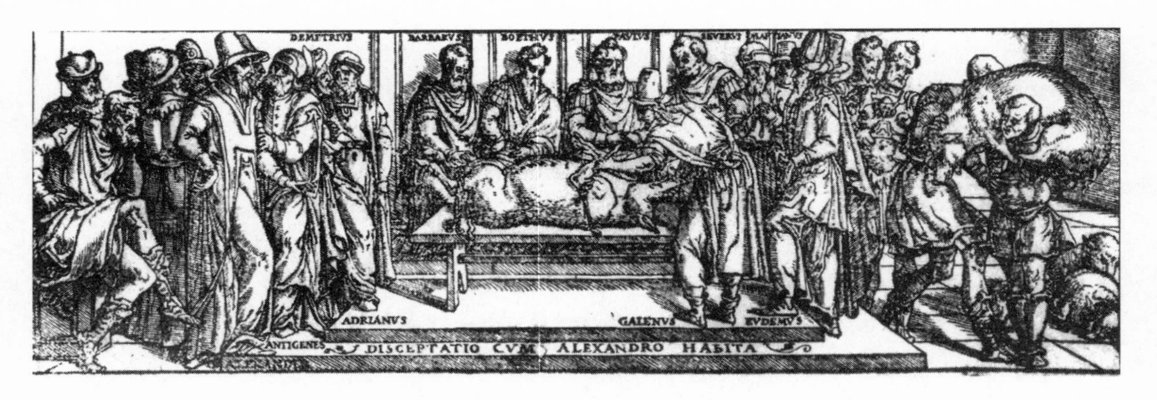

Galen performs a surgical procedure on a pig tied to a table. Besides studying dissected animals, Galen studied human skeletons and conducted experiments on living creatures such as apes. (National Library of Medicine)

be left alone until the flesh withers or corrupts and then by twisting it and moving it here and there the infixed body will be more easily drawn out . . ." (Clendening 1960, p. 91).

The resulting gangrenous wound cannot have boded well for the patient! And when surgery took place, there were no anesthetics, and techniques of hemostasis (control of bleeding) were basically limited to the use of the red-hot poker as a crude cautery.

Amid the horrors of medieval medicine, was there a role for animal experimentation? Animals were certainly hunted, herded, and butchered for food, and along with observations of this and of injuries on the battlefield, it must have been clear to the more astute observers that organs were related to each other in similar ways in different species. Trotula (ca. 1050 A.D.), who was associated with the medical school at Salerno, seems to have dissected pigs, for example, because they looked similar to men on the inside (Gordon 1993, p. 8).

But the fact remains that there was little systematic research on humans to test the results of dissections on animals. Ancient authorities such as Galen, besides studying and dissecting animals, had studied human skeletons and made experiments on living creatures such as apes. These traditions continued into the medieval period:

> [t]here had always been two pitfalls in the practice of ordinary dissection, however. First of all, many conclusions had been reached, not from human but from animal dissection. . . . Secondly—and this was a fault that still had to be pointed out by William Harvey in the seventeenth century—people made wrong inferences when they examined animals which had been bled to death, so that the arteries and the left ventricle of the heart were seen only after the blood had been drained from them. (Butterfield 1957, p. 53)

In the early days of anatomical inquiry, it was thus quite easy to be misled by the results of animal experimentation. But what of systematic dissection of humans?

Partly because of religious belief concerning the sanctity of the human body, there was little systematic human dissection in the medieval period. The body was believed to be made in the image of God and would be needed intact if God's promise of resurrection were to come to pass. This situation began to change in the late medieval period, with postmortems being performed at the great law school of Bologna early in the fourteenth century. But though the students were exposed to dissections as part of their training, the purpose, more often than not, was to illustrate classical authorities such as Galen, not to challenge them (Pledge 1959, p. 25).

Nevertheless, it was Mondino di Lucci (1275–1326), professor of anatomy at the University of Bologna and author of the *Anathomia* (Anatomy) of 1316, who must be credited with having shifted the focus of medical inquiry away from the dissection of pigs to the dissection of human cadavers. According to Hall (1956, p. 40), di Lucci may have been the first investigator since the third century B.C. to perform public demonstrations on the human body. Mondino di Lucci did not escape religious scrutiny and gave up human dissection after 1318 as the result of an edict by Pope Boniface VII (Morley 1961, p. 79).

It is also worth noting that di Lucci performed his own dissections—a practice that was to fall into disrepute until revived by Vesalius. In the period between di Lucci and Vesalius, anatomy, although filthy in practice, was intellectually sterile. The professor sat in a high chair, reading from Galen, while his assistants cut up the body. Dissection simply became theater, with a Galenic script consisting of names and phrases (Hall 1956, p. 40).

But what of di Lucci himself? It is worth noting that di Lucci was also an early exponent of the importance of comparative studies:

> [T]he woman whom I anatomized in the past year, or A.D. 1315, in the month of January had a uterus twice as large as the one whom I anatomized in the month of March in the same year. . . . And because the uterus of a pig which I anatomized in A.D. 1306 was a hundred times larger than it can ever be seen in a human being, there may be another cause, i.e., because it was pregnant and had in the uterus 13 little pigs. In this I showed the anatomy of the foetus or of pregnancy as I shall tell you. (Clendening 1960, p. 123)

But di Lucci's work was crude and contained many errors and was in fact treated by many anatomists as a supplement to Galen.

Moreover, di Lucci's book was basically a practical dissection manual, uninformed by the powerful mechanical metaphors that made possible the explosion of anatomical and physiological inquiry in the Renaissance. As we will see in the next chapter, these metaphors do their work by providing the investigator with a theoretical framework in terms of which structural and functional analogies between animals of different species can be located, organized, and conceptualized.

So it remains the case that the human body, notwithstanding the important work of anatomists such as di Lucci, was not the subject of widespread, systematic anatomical inquiry in medieval Europe. Partly this was for religious reasons: The body, held sacred even in death, was generally considered to be beyond the reach of the dissector's blade. But of perhaps equal importance is the fact that the mechanical metaphors and methods of inquiry that came to inform and indeed fuel the growth of anatomy and physiology in the Renaissance had not yet crystallized in the minds of investigators.

There was thus little knowledge of the structure and function of organs and the relationships between them. The function of the lungs, the nature of the heart, and the circulation of the blood were matters unknown to medieval investigators—investigators for whom air was an element, and who did not even have the word, let alone the concept, of oxygen! This is a situation that would begin to change in the Renaissance—and this change was tied intimately to the rise of modern science and a special role for animals in medical inquiries. But to understand these important changes and innovations, we must look at the ways in which animals were being used in medieval Europe.

Medieval Machine Technologies

When we think of medieval Europe, there is a tendency to think primarily in terms of what is visible today: the buildings and cathedrals, the literature and the scholastic philosophy. Yet what is of equal importance, particularly from the standpoint of the rise of modern science, is the fact that medieval culture was a mechanically sophisticated culture, even if little remains today of their machines. For example, Albertus Magnus (ca. 1200–1280), the teacher of Aquinas, is rumored to have had a robot in his laboratory that could raise a hand and give a greeting. The story is no doubt apocryphal, but there was in any case a great interest in machines, even if science itself was rudimentary and qualitative (Lindeboom 1979, p. 62). What is clear is that the diffusion of mechanical clock technology was underway by the beginning of the fourteenth century. It is also clear that there was a multiplicity of machines at this time with clockwork mechanisms, ranging from monastic alarms, musical machines, and astronomical simulators to striking clocks (Rossum 1996, p. 178).

As Ernst Mayr has pointed out in his discussion of the influences leading to the modern scientific view of the world, we need to look not just at philosophy, but also at "technological developments in late medieval and early Renaissance times. There was great fascination with clocks and other automata—and indeed almost any kind of machine. This eventually culminated in Descartes's claim that all organisms except humans were nothing *but* machines" (1997, p. 3). Indeed, in the late medieval period before the rise of modern science, clockmaking skills and the mechanical fruits of those skills had begun to assume a broader cultural and intellectual significance. Thus, Rossum remarked:

> Parisian natural philosophy at the end of the fourteenth century honored clockmakers by comparing the cosmos or creatures with artful clockworks and the creator-God with a clockmaker. As constructors who designed and built their products, clockmakers thus took their place alongside architects, who were highlighted in these comparisons. (1996, p. 174)

Since these developments have enormous implications for the ways in which later theorists would conceptualize humans and nonhuman animal species, we must trace their development carefully.

Today, with 300 years of modern science, and with science and machine technologies getting firmly coupled in the seventeenth century, and coevolving ever since, we are apt to think of machinery as simply a fruit of scientific knowledge. But this is not correct, and many cultures—including our own in medieval times—have had fairly sophisticated machine technologies, though they had little that resembled modern science (even as it was coming to be understood at the end of the seventeenth century, let alone the beginning of the twenty-first century).

As we will see in the next chapter, machines provided powerful metaphors for the pioneering Renaissance investigators who were trying to understand the anatomy and physiology of humans and nonhuman animals for the first time. Conceptual reliance on the metaphor of *the machine,* coupled with the successes of anatomical and physiological inquiries, helped give rise to *machine thinking,* a new way of thinking about the world, and this contributed greatly to the rise of science in the sixteenth and seventeenth centuries.

The role of animals in all this is initially indirect. Machines need power sources. But although slaves and serfs could be employed as power sources, the medieval period saw a power revolution that took medieval society beyond the exclusive need for human power. Medieval engineers exploited three principal power sources: water power, wind power and animal power. Medieval engineers were not the first to exploit these power sources, but this was a period that saw the systematic development and extension of technologies powered, for example, by water wheels and wind mills. It was also a time when animal power, especially horse power, came to be efficiently harnessed (Kirby et al. 1990, chap. 5).

Animals had been used as power sources since ancient times, but the yoke tended to strangle a horse if it exerted significant effort. And unshod, the horses of ancient cultures also tended to go lame and useless. Things began to change, however:

> By the tenth century, . . . men had developed the horse collar, which rests on the horse's shoulders and does not choke him as he pulls. In addition, the horseshoe was invented, and the tandem harness which allows more than one pair of horses to pull a load. The horse is fast and efficient compared to the ox and proved to be a valuable source of nonhuman power in agriculture and the operation of machines. (Kirby et al. 1990, p. 101)

Animals, horses in particular, would play a prominent role in agriculture, industry, and transportation until the early years of the twentieth century. Then they were gradually phased out, at first by extensions of steam-powered technologies growing out of the eighteenth century, and later by technologies exploiting fossil fuels.

But what of the machines themselves? It is worth dwelling on this matter, because practical experience with machines will help to shape new ways of thinking about the world—with enormous implications for the ways in which we see ourselves, nonhuman animals, and our relative places in nature. Roman engineers knew about pulleys and cranes and how to use gears and cogwheels to transmit motion at any angle from a rotating axle. By 850 A.D. it is clear that the crank had been developed—with the consequence that reciprocal motion could be converted into rotary motion, and vice versa (Kirby et al. 1990, pp. 96–97).

Some of the fruits of medieval machine technologies are preserved in the writings and drawings of Georgius Agricola (1494–1595), whose *De Re Metallica* was first published in 1556 (see Hoover and Hoover [1556] 1950). Agricola's concern was with industrial processes relating to the mining of metal ores and the subsequent extraction of metals from these ores. The drawings preserved in *De*

Peasant farmers plowing, sowing, and harrowing their fields. Horses are used for plowing instead of oxen. In the tenth century, the horse proved to be a valuable source of nonhuman power in agriculture and the operation of machines. (Hulton/ Archive)

Re Metallica afford a glimpse at medieval mechanical sophistication. They also provide examples of some of the ingenious ways in which medieval engineers exploited water, wind, and animal power.

If one looks at one of Agricola's machines—as one might at an old, mechanical clock—one sees that a machine is a complex system, consisting of numerous parts (cogs, gears, levers, cranks, and so on). In short, it has an anatomy. The parts not only themselves have definite characteristics—characteristics that are essential if the machine is to function properly—but they also stand to each other in definite relationships such that the component motions of the parts give rise to motions of the whole machine that enable it to achieve its intended function—it might be a pump in a mine or a clock to tell the time. By the dawn of the Renaissance—Agricola's day—machines with these properties were well known, for medieval culture was a mechanically sophisticated culture.

It should perhaps come as no great surprise to note that people often see the world in terms of metaphors based on their experiences with objects in their immediate environments, and it is arguably no accident that other complex systems—human and nonhuman animal bodies—came to be seen in terms of metaphors based on the complex and rich medieval experience with functional machinery—especially when one bears in mind that the intellectual elites who played a crucial role in this *rethinking of the body* were investigators raised in an academic culture whose metaphysical presuppositions emphasized the ways in which form (formal cause) determined function (final cause).

Further Reading

A good account of ancient Greek thought can be found in T. Irwin, *Classical Thought* (Oxford: Oxford University Press, 1989). For more information on medieval philosophical theology, I would recommend F. C. Copleston, *Medieval Philosophy* (New York: Harper Torchbooks, 1961). Readable accounts of these matters can also be found in B. Russell, *A History of Western Philosophy* (New York: Simon and Schuster, 1972), or S. E. Stumpf, *Socrates to Sartre: A History of Philosophy* (New York: McGraw-Hill, 1982).

Some interesting perspectives concerning medieval science can be found in A. C. Crombie, *Medieval and Early Modern Science,* Volume 1. (New York: Doubleday Anchor Books, 1959). An informative discussion of medieval engineering technologies can be found in R. Kirby et al., *Engineering in History* (New York: Dover, 1990). An excellent discussion of medieval clock-making technologies can be found in G. Rossum, *History of the Hour: Clocks and Modern Temporal Orders* (Chicago: University of Chicago Press, 1996).

R. Gordon, *The Alarming History of Medicine: Amusing Anecdotes from Hippocrates to Heart Transplants* (New York: St. Martin's Press, 1993), provides an entertaining perspective on the history of medicine in the medieval period. L. Clendening, ed., *Sourcebook of Medical History* (New York: Dover, 1960), provides excerpts from the writings of medieval medical theorists. W. Biddle, *A Field Guide to Germs* (New York: Doubleday, 1995), provides a readable discussion of a variety of infectious diseases, including the great scourges of medieval Europe.

2

Gut Feelings: Anatomy and Physiology in the Renaissance

In the preceding chapter it was seen that late medieval investigators undertook anatomical inquiries involving human and nonhuman animal cadavers. Nevertheless, though human and animal dissection took place in the medieval period, the explosion of interest in anatomical and physiological inquiry—involving widespread, systematic studies of human cadavers and live, nonhuman animals—occurred in the Renaissance.

It would be a mistake to ascribe all this growth of interest in human anatomy to a radical decline in religious opposition to the dissection of the dead. The great Renaissance anatomist Andreas Vesalius (1514–1564) encountered significant opposition from the ecclesiastical authorities, and Michael Servetus (1509–1553) was even burned at the stake by John Calvin for work that had led to the uncovering of the details of human pulmonary circulation (Gordon 1993, p. 10). Moreover, it would be a mistake to suppose that those who undertook these early anatomical and physiological investigations had abandoned their religious beliefs.

It will emerge in this chapter and the next that early modern science, far from displacing religious belief, readily adapted to it, was also accommodated by it, and indeed appeared to provide justification for it. And science did so in ways that are relevant for the present discussion of controversies concerning the cognitive abilities of nonhuman animals, for as was seen in the preceding chapter, important strands of theological theory required that there be fundamental, metaphysical differences between humans and nonhuman animals.

An important part of the explanation for this blossoming of anatomical and physiological inquiry lies in the way that Renaissance investigators became increasingly reliant on mechanical metaphors to conceptualize the objects of their inquiries—*bodies*—in mechanical terms. In this chapter we will see how the metaphor of *body-as-machine* not only came to guide anatomical and physiological studies but also reflected a systematic method of inquiry, known as the *resolutio-compositive method* (method of analysis and synthesis). We will see that the combination of metaphor and method resulted in an intellectual whole greater than the sum of its parts.

The metaphors we use to understand new aspects of the world around us typically reflect prior experience with other, more familiar features of the world. So we can say, for example, it is as though *X* (the body revealed by dissection) is like a machine *Y* (perhaps a mechanical clock with the back removed to display the interacting cogs, springs, and other components). The metaphor of body-as-machine did not appear all at once, but rather it evolved from crude mechanical analogies early in the Renaissance to a fully crystallized and articulated mechanical picture of human and nonhuman animal bodies by the middle of the seventeenth century.

We will see that the metaphor of body-as-machine had enormous implications for medical inquiries, and especially for the use of nonhuman animals as research subjects. We will also see that the mechanical metaphors that fueled the growth of anatomical and physiological inquiry had broader implications, culminating in the metaphor of *nature-as-machine.* It is arguably no accident that a method that had proved so fruitful in the domain of *physik* should come to shape early modern inquiries in *physics* as well.

And to anticipate later developments, especially in the next chapter, machines need designers and makers. God, as the designer of the natural machine, was just one of the ways in which science and religion came to enjoy a cooperative relationship—a relationship that would only be soured by events, forced in large measure (but by no means exclusively) by a growing understanding of the consequences of Darwin's theory of evolution.

But this is to look ahead. Here the central task is to understand how methods so valuable in anatomy and physiology should also find application in broader inquiries into what might be termed the nature of nature. First, the administrative and conceptual compartmentalization of inquiry that separates a department of physics from a department of physiology in a contemporary university did not exist in the Renaissance. So an interest in physiology did not preclude an interest in physics. And, moreover, the physiologist and the physicist were often one and the same person.

Second, institutions such as the University of Padua, in what is now Italy, were intellectual centers that maintained traditions of scientific inquiry and provided the academic contexts in which interdisciplinary cross-fertilization could occur. For example, at one time Padua could boast of Vesalius as professor of anatomy; at a later time it could list among its students the great English physiologist William Harvey and the great physicist Galileo Galilei.

To understand the changes in the Renaissance with respect to our understanding of the internal structure and functioning of human and nonhuman animal bodies, it will help, following one of the themes in the preceding chapter, to discuss briefly the ways in which disease, and in particular the process of *infection,* were being reconceptualized.

Fracastorius and the Nature of Contagious Disease

Girolamo Fracastorius (1478–1553) is arguably the founder of scientific epidemiology. While at the University of Padua, he was a classmate of Nicolaus Copernicus and a pupil of the anatomist Alessandro Achillini. His great work, *De Contagionibus et Contagiosis Morbis et Eorum Curatione* (On Contagion, Contagious Diseases, and Their Cure), published in 1546, broke with medieval tradition and the idea that disease is caused by sin and offered instead a mechanical account of the spread of disease. This is as much of a Copernican revolution in medicine as the one initiated in astronomy by his classmate.

Writing before the discovery of the microscope, and long before the modern germ theory of disease, Fracastorius came up with three causal pathways through which disease could spread. Disease could spread through direct, physical contact with an infected person, through direct contact with *fomites* (items such as blankets that have themselves been in physical contact with an infected person), or indirectly, at a distance, with no contact with either patient or fomites.

To explain what occurs during infection, Fracastorius introduced the idea of *essential seeds of contagion* (*seminaria*)—physical particles that can be transferred from one body to another, contact with which can induce states of disease. Although these seeds of contagion are not to be confused with modern talk of germs, Fracastorius does offer a natural, mechanical account of infection. Thus, explaining how this idea can be used to explain disease transmitted at a distance, Arturo Castiglioni commented:

> In such cases he imagined that the germs were propagated by selecting the humors for which they have the greatest affinity, entering the organism by means of the respiration. The germs . . . were then thought to be absorbed from the breath and adhered to those humors which carried them to the heart. These germs . . . had the power of multiplying rapidly. (Palter 1961b, p. 60)

Fracastorius, ever an astute observer, was aware that different diseases exhibited different organ specificities. He knew that there were gender differences and age differences with respect to disease.

But he also pointed out that there were *species differences.* There are not just cognitive and morphological differences between animal species; there are also differences with respect to pathological susceptibility. Fracastorius thus introduced another dimension along which the similarities and differences between humans and nonhuman animals can be assessed. The relevance of this, and related biological dimensions of assessment, would assume greater importance as later biomedical investigators came to place more and more reliance on the use of nonhuman animals as models of the human medical condition.

But why should later investigators come to see nonhuman animals as humans writ small? It is to this issue that I now turn. In my discussion of the use of human and nonhuman animal bodies in medicine, I will begin with an examination of the science of anatomy (the study of the structure of organisms) and will then discuss physiology (the study of organic function).

The Art of Anatomy: Leonardo da Vinci

The study of anatomy requires a mixture of skills. One must be well steeped in the biological sciences, one must be well steeped in the practical art of dissection, and even today, in our visually sophisticated culture that affords all the benefits of photography, one must be an artist. Leonardo da Vinci (1452–1519) is an early and notable example of this very mixture of skills, and the fruits of his anatomical inquiries are preserved in a magnificent series of drawings.

The study of da Vinci's drawings reveals a dual interest in machines on the one hand and the anatomy of humans and nonhuman animals on the other. Da Vinci's mechanical drawings reveal a good acquaintance with the sophisticated engineering technologies of his day. Yet it would be a mistake to categorize him as a mere practical engineer, for his drawings show that his mechanical ideas, though rooted in the technology of his time, are highly innovative. They afford many examples of how extant technology can be evolved and adapted to serve new mechanical ends:

> Besides his machine guns, breech-loading cannons, tanks, a submarine, and a flying machine, Leonardo's sketches included lathes, pumps, cranes, jacks, water wheels, a canal lock, drawbridges, wheelbarrows, a diver's helmet with air hose, roller bearings, a self-propelled carriage, a double decked city street,

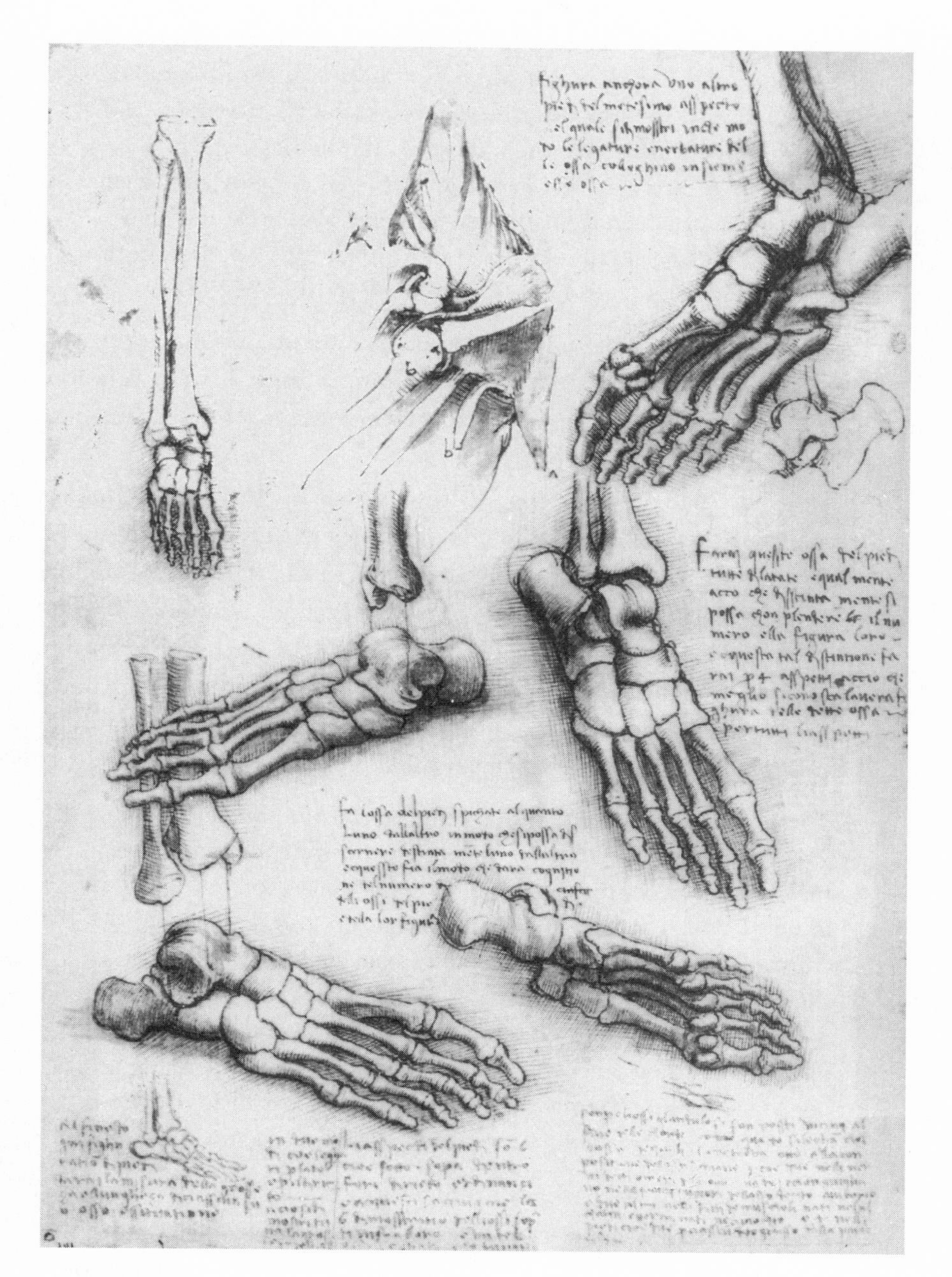

Leonardo da Vinci's drawings revealed an interest in machines and in human and nonhuman animal anatomy. (Hulton/Archive)

> sprocket chains, an automatic printing press, a universal joint, a helicopter, and a wooden truss bridge. (Kirby et al. 1990, p. 124)

But what of the anatomical drawings? The significance of these drawing for da Vinci was intertwined with his interests in the structure and functioning of machines.

Antonio de Beatis, who visited da Vinci in 1517, two years before the great artist's death, observed at the time:

> This gentleman [Leonardo] has compiled a particular treatise of anatomy, with the demonstration in draft not only of the members, but also the muscles, nerves, veins, joints, intestines, and of whatever can be reasoned about in the bodies both of men and women, in a way that has never yet been done by any other person. All of which we have seen with our own eyes; and he has said that he has already dissected more than thirty bodies, both men and women of all ages. (Quoted in Popham 1952, p. 69)

In addition to the drawings of relevance to human anatomy, there are drawings reflecting the dissection of nonhuman animals such as horses and bears, and in at least one case there is a drawing suggestive of an interest in comparative anatomy.

The fact that da Vinci made drawings of both the structure of machines and the anatomical details of bodies is suggestive, though it does not demonstrate, that he saw the latter in the light of metaphors drawn from his study of the former. Certainly, the metaphor of body-as-machine had not fully crystallized in da Vinci's day, yet lurking in his anatomical writings are hints of a role being played by mechanical metaphors and analogies. Thus, in discussing the action of the muscles in respiration, da Vinci himself observed:

> These muscles have a voluntary and an involuntary movement seeing that they are those which open and shut the lung. When they open they suspend their function which is to contract, for the ribs which at first were drawn up and compressed by the contraction of these muscles then remain at liberty and resume their natural distance as the breast expands. And since there is no vacuum in nature the lung which touches the ribs from within must necessarily follow their expansion; and the lung therefore *opening like a pair of bellows* draws in the air in order to fill the space so formed. (Clendening 1960, p. 124; my italics)

The analogy of lung-as-bellows is elegantly simple and explanatory of the phenomenon of respiration that da Vinci was trying to describe, and it illustrates how, in the early development of scientific anatomy, comparisons with mechanical devices could help illuminate anatomical puzzles.

But anatomy needs more than metaphors; it also needs a method. And the very practice of anatomical dissection in the sixteenth century, first as a means to an understanding of the structural properties of organisms and later as a means to an understanding of physiological function, came to both reflect and promote a new and emerging, yet powerful, method of study: the *resolutio-compositive method,* or *method of analysis and synthesis.*

A pig laid out on a dissecting table, illustrated in Andreas Vesalius's De Fabrica Humani Corporis, *published in 1543. The topic of vivisection, the dissection of live animals, has been controversial ever since the publication of this work. (National Library of Medicine)*

Vesalius: Dissection, Vivisection, and Method

Andreas Vesalius was perhaps the greatest of the Renaissance anatomists, and his book, *De Fabrica Humani Corporis* (The Structure of the Human Body) was published in 1543—the same year that saw the publication of Copernicus's *De Revolutionibus Orbium Caelestium* (Of the Rotation of Celestial Bodies). The *Fabrica* deserves attention partly because it corrected errors in the Galenical tradition of anatomy—Vesalius showed that the jaw was one bone, not two; that the breastbone consisted of three bones, not seven; and that men and women had the same number of ribs.

But more importantly, the text also included a discussion of the importance of *vivisection,* the dissection of live animals. This topic has been a source of much controversy ever since. The *Fabrica* thus takes us from a static study of organic structure (something that could be accomplished through the study of cadavers) to the study of organic function—a dynamical study that depends on observations of the working bodily machine and thereby moves the focus of the discussion from anatomy to physiology.

In Vesalius's day there was a rigid division of labor in the teaching of anatomy. Various assistants would perform dissections while the professor of anatomy would cite ancient authorities to explain the significance of the dissection to the students. Vesalius, who clearly thought that the professor of anatomy should have firsthand knowledge based on skillful dissection, observed of this pedagogical practice:

> The latter . . . from a lofty chair arrogantly cackle like jackdaws about things they never have tried, but which they commit to memory from the books of others or which they place in written form before their eyes. The former, however, are so unskilled in languages that they cannot explain the dissections to the spectators. They merely chop up the things which are to be

> shown on the instructions of the physician, who, never having put his hand to cutting, simply steers the boat from the commentary. . . . Fewer facts are placed before the tumult than a butcher could teach a doctor in his meat market. I shall not mention those schools where they hardly ever think of dissecting the structure of the human body. (Clendening 1960, p. 133)

By contrast, Vesalius sought to unite theory and practice: Anatomical theory must be based on careful observation of, and experience in, dissection. The anatomist must literally have a feel for the very texture of the organs and structures examined. Their properties must be known manually, and not merely visually. Anatomy, for Vesalius, was very much a hands-on science.

But there is also a method at work. Anatomical knowledge does not proceed by mere cutting and observation. It is a systematic inquiry whose nature is reflected in the very structure of the *Fabrica*. Thus Vesalius, having paid tribute to the University of Padua, described his work as follows:

> And so, in the first book I have narrated the nature of all the bones and cartilages, which, because the other parts are supported by them and are marked off by them, come to be known first by students of anatomy. The second book dwells upon the ligaments, by aid of which the bones and the cartilages are in turn connected; and then the muscles, the workmen of motion, depending upon our will. The third includes the commonest series of veins. . . . The fourth teaches not only the layers of nerves which bear the animal spirit to the muscles, but all the rest of the nerves. The fifth treats the construction of the organs serving for nutrition . . . and furthermore, because of proximity of position, it contains also the instruments fashioned by the Supreme Creator of things for the succession of the species. (Clendening 1960, pp. 137–138)

There are several points worth noting here.

First, although the division of the work in part reflects anatomical tradition, it also reflects the idea that in order to understand the structure of a complex system, you must first understand the structural properties of the most basic parts of the system (in Vesalius's day, organs and tissues). The second stage of inquiry consists of an analysis of the structural relationships among the parts so identified, with the synthesis of this information of the parts and their relationships yielding knowledge of the fabric of the body—an interwoven, intertwined system of parts. This fabric was described by Vesalius as "a fabric not built of ten or twelve parts, as it appears to the causal observer, but of several thousand diverse parts"(Clendening 1960, p. 139), but a fabric that could nevertheless be systematically unraveled to reveal its secrets.

Second, Vesalius's anatomical method accords well with the resolutio-compositive method, a systematic method of inquiry developed by medical logicians at the medical school at Padua, beginning in the early fourteenth century with the work of Pietro d'Abano and culminating in the early sixteenth century with the work of Jacopo Zabarella (1533–1589). The resolutio-compositive method is a method designed to enable the investigator to understand a complex phenomenon, such as disease, or a complex system, such as a body.

Suppose one wants to understand a complex system such as a body. The first step involves the *resolution* or *analysis* of the body into its component parts. Having resolved the system into its parts, the properties of the parts must then be studied, so that their causes, and what in turn they cause, may be understood. Having understood the properties of the parts, one must go on to discern the static, structural relationships between the parts—how they are connected or stand to each other. In short, one must discern the anatomy of the system.

From knowledge of the anatomy of the system, one must then study the dynamical relationships between the parts, not only to see how changes in the parts are brought about, and themselves bring about further changes in the system, but also to discern the *functions* of the parts—in short, their physiology. This understanding of the parts and their mutual relationships can then be *composed* or *synthesized* into an understanding of the complex system of interest.

As we shall see shortly, this method was employed by William Harvey in his investigations of the motions of the heart, and a variant of the method was employed by Galileo in investigations of the motions of falling bodies and also by Thomas Hobbes in the formulation of his mechanical conception of humans and nonhuman animals. It should be obvious that the resolutio-compositive method is ideally suited for the study of complex machines with many interrelated parts (for example, clocks), and for other complex systems conceptualized as being mechanical in nature.

But the method is also evident in the foundations of Vesalius's anatomy, and especially in his physiology, to which I now turn. Anatomy can teach us much about the structure of the parts of the body and their static relationships, but to understand vital function, one must look not just at the parts and their static relationships, one must also understand the dynamical interactions between the parts. This in turn means going beyond cadaver studies on human and nonhuman animals. It involves the dissection of living organisms—vivisection—to see the parts in action. Moral and legal concerns limited the possibilities for human vivisection (though there are records of experiments on prisoners), and one effect of this was that nonhuman animals became the veritable subjects of such harrowing trials that nineteenth-century physiologist Claude Bernard would refer to them as the Jobs of physiology.

The methodological importance of vivisection of nonhuman animals was explained by Vesalius as follows:

> Just as the dissection of the dead teaches well the number, position and shape of each part, and most accurately the nature and composition of its material substance, thus also dissection of a living animal clearly demonstrates at once the function itself, at another time it shows very clearly the reasons for the existence of the parts. Therefore, even though students deservedly first come to be skilled in the study of dead animals, afterward when about to investigate the action and use of the parts of the body they must become acquainted with the living animal. (Clendening 1960, p. 142)

But why should studies of nonhuman animals be relevant for human physiology?

The answer to this question is to be found in the observation that by the time Vesalius was writing, there was an extensive body of comparative anatomical knowledge. Vesalius, like da Vinci before him, had noticed all manner of structural similarities and analogies between humans and nonhuman animals (including, but not restricted to, dogs, cats, sheep, pigs, and horses). For example, in the study of the arrangements of the nerves in man and beast Vesalius wrote:

> They will be arranged almost in this manner: In man one of them is the third, and is carried into the forearm along the anterior side of the elbow joint; another, in fact, the fifth, runs to the elbow next to the posterior portion of the internal tuberosity of the humerus. For in this manner the nerves are also observed in the dog. And these nerves having been tied somewhere before they reach the elbow joint, the motion of the muscles flexing the digits and arm will be abolished, and if thou wilt intercept with a band the nerve which in man is reckoned by me the fourth and is extended along the humerus to its external tuberosity, then the motion of the muscles extending the foreleg and digits will be abolished. (Clendening 1960, p. 143)

It is these structural similarities that permit the testing of hypotheses about the function of structures found in human cadavers by experiments on live animals.

Vesalius's experiments on live animals were performed without the benefit of anesthesia, but it is clear that he had found ways to control the screams and shrieks that must have pervaded his laboratory:

> I note the recurrent nerves lying on the sides of the rough artery [trachea] which I sometimes intercept with ligatures, at other times I cut. And first I do the same on the other side, in order that it may be clearly seen when one nerve has been tied or cut how half the voice disappears and is totally lost when both nerves are cut. And if I loosen the ligatures the voice will return again . . . and it is clearly proved how the animal struggles for deep breaths without its voice when the recurrent nerves have been divided with a sharp knife. (Clendening 1960, pp. 149–150)

This is all in a manner reminiscent of the way in which someone studying the function and dysfunction of a machine such as a clock might systematically study the parts in motion and then remove or impede the action of selected parts to assess specific kinds of defect. But it is also part and parcel of the new methods that are both shaping, and being reflected in, medical inquiry in the Renaissance.

Harvey: The Heart of the Matter

After graduating from Cambridge, William Harvey (1578- 1657) spent five years at the University of Padua working with Fabrizio di Aquapendente (1537–1619), who was Galileo's personal physician. It was at Padua that Harvey came to understand and appreciate the importance of comparative anatomy.

Harvey's use of the resolutio-compositive method, along with the use of explicit mechanical metaphors, can be found in the context of work on the motions of the heart—published in 1628 as *De Motu Cordis et Sanguinis* (Of the Motions of the Heart and Blood)—work that was made possible only through extensive experimentation on a wide variety of animal subjects. The problem confronting Harvey was the problem of understanding the complex motions of the heart. This was a daunting task, for as Harvey himself noted, "I found the task so truly arduous, so full of difficulties, that I was almost tempted to think with Fracastorius, that the motion of the heart was only to be comprehended by God" (Clendening 1960, p. 155).

The problem was generated by the speed with which the heart's motions occur, especially in mammals whose hearts had been exposed to public view without benefit of anesthesia and who consequently were in great physical distress:

> For I could neither rightly perceive at first when the systole and when the diastole took place, nor when and where dilatation and contraction occurred, by reason of the rapidity of the motion, which in many animals is accomplished in the twinkling of an eye, coming and going like a flash of lightening. . . . (Clendening 1960, p. 155)

The problem can be solved through an application of the resolutio-compositive method if there is some way to analyze these fast, complex motions into component motions and to understand the causes of the component motions.

This could be achieved through experimentation on appropriate animal subjects—subjects in whom the component motions would be visible:

> These things are more obvious in the colder animals, such as toads, frogs, serpents, small fishes. . . . They also become more distinct in warm-blooded animals, such as the dog and the hog, if they be attentively noted when the heart begins to flag, to move more slowly, and, as it were, to die: the movements then become slower and rarer, the pauses longer, by which it is made much more easy to perceive and unravel what the motions really are, and how they are performed. (Clendening 1960, p. 156)

So with the help of appropriate subjects—from the standpoint of species and physiological condition—the component motions of the heart can be resolved.

Harvey analyzed the complex motion into component motions associated with structures discernible in the heart (*ventricles* and *auricles,* the latter being the old word for *atria*). Harvey was then able to synthesize his understanding of the properties of the parts into an understanding of the complex motion of the whole system:

> These two motions, one of the ventricles, another of the auricles, take place consecutively, but in such a manner that there is a kind of harmony or rhythm preserved between them, the two concurring in such a wise that but one motion is apparent. . . . *Nor is this for any other reason than it is in a piece of machinery, in which, though one wheel gives motion to another, yet all the wheels seem to move simultaneously;* or in that mechanical contrivance which is adapted to firearms, where the trigger being touched, down comes the flint, strikes against the steel, elicits a spark, which falling among the powder, it is ignited, upon which the flame extends, enters the barrel, causes the explosion, propels the ball, and the mark is attained—all of which incidents, by reason of the celerity with which they happen,

> seem to take place in the twinkling of an eye. (Clendening 1960, p. 161; my italics)

In this passage, we see how the fruits of animal experimentation, united with the resolutio-compositive method and the explicit use of mechanical metaphors, could yield natural resolutions of problems that had hitherto been viewed as mysteries beyond the reach of human ken.

Thinking of the operation of the heart in mechanical terms—and hence as a system admitting of a quantitative mathematical description—yielded further fruits. Even granting a large margin of error, Harvey estimated that in an hour the heart could pump more blood by weight than its human owner. Where was all this blood coming from, and where did it go? Harvey had a radical solution. There is a mystery:

> [U]nless the blood should somehow find its way from the arteries into the veins, and so return to the right side of the heart; I began to think whether there might not be A MOTION, AS IT WERE, IN A CIRCLE. Now this I afterwards found to be true; and I finally saw that the blood, forced by the action of the left ventricle into the arteries, was distributed to the body at large, and its several parts, in the same manner as it is sent through the lungs, impelled by the right ventricle into the pulmonary artery, and it then passed through the veins and along the vena cava, and so round to the left ventricle in the manner already indicated. (Clendening 1960, p. 164)

Harvey thereby united his own research on the structure and function of the heart, with earlier work on pulmonary circulation, to conceptualize the conjoined system of heart and blood vessels as a closed, mechanical circulatory system.

It should not be forgotten in all this that Harvey himself was an Aristotelian and that like Aristotle he set special store on circular motions. Indeed, Harvey made the heart the principle organ of the body; it is not just a pump, it is the source of life itself:

> All the parts may be fed, warmed and quickened by the warmer and more perfect vaporous, spirituous and, as I may say, nutritive blood; and this, on the contrary, may become, in contact with the parts, cooled, thickened, and so to speak effete, so that it returns to its origin, the heart, as to its source, the inmost temple of the body, to recover its perfection and virtue. (Quoted in Crombie 1959b, p. 236)

We will see in the next chapter how Hobbes built on Harvey's estimate of the special nature of the heart.

Hooke: Machines and Medicine

In the course of the development of the biomedical sciences in the sixteenth and seventeenth centuries, humans and nonhuman animals gradually came to be seen as machines. The mechanical metaphors lost their original heuristic purpose and came to take on a more literal significance. The significance of this for controversies concerning the relative places of humans and nonhuman animals in nature will be examined in the next chapter. Here I wish to point out, however, that the evolution of a thoroughly mechanistic view of organisms was accompanied by a view that if organisms were *natural* machines resulting from providential design, they could nevertheless be studied with the aid of machines—artifacts—of our own design. This is a fragment of a more general phenomenon in the seventeenth century, where the emerging conception of *nature-as-machine* was accompanied by the view that it could be understood with the aid of machines (such as telescopes, microscopes, barometers, vacuum pumps, and so on). In biomedical investigations, this had some interesting consequences for the direction of research.

First, investigators began designing machines to gain quantitative data about medical phenomena. An early example is provided by the work of Santorio Santorii (1561–1636), a professor of medicine at Padua and a colleague of Galileo. He is credited with the invention of a number of devices such as the pulsilogium, which was a device for measuring pulse rate, as well as a clinical thermometer. In addition, "he described an experiment which laid the foundation for the modern study of metabolism. He spent days on an enormous balance, weighing food and excrement, and estimated that the body lost weight through invisible perspiration" (Crombie 1959b, p. 221).

Second, nonhuman animals could also be mechanically coupled to machines to resolve more radical physiological puzzles. In this regard there were some issues about the function of the motions of the lungs. With the aid of an air pump, Leonardo da Vinci had shown that air did not travel from the lungs to the heart. After the work of Servetus and Harvey, the details of pulmonary circulation and its coupling to the general circulatory system had become much clearer.

Yet there were still puzzles concerning the function of the motion of the lungs. Some physicians maintained that the motions of the lungs served to promote the circulation of the blood. Writing after Harvey, Robert Hooke (1635–1703) performed an experiment in 1667 to settle this matter—an experiment that involved the coupling of an *animal machine* to machines of human design. The report of the experiment was originally published in the *Philosophical Transactions of the Royal Society of London* under the title "Preserving Animals Alive by Blowing through Their Lungs with Bellows."

Hooke's experiment was performed before members of the Royal Society. He described his experimental setup as follows:

> I caus'd at the last meeting the same experiment to be shewn in the presence of this *Noble Company,* and that with the same success, as it had been made by me at first; the dog being kept alive by the Reciprocal blowing up of his lungs with *Bellowes,* and they suffered to subside, for the space of an hour or more, after his *Thorax* had been so display'd, and his *Aspera arteria* cut off just below the *Epiglotis* and bound on upon the nose of the Bellows. (Zucker 1996, p. 22)

In other words, the experimental subject, the dog, has had his windpipe cut and then tied to the nose of a pair of bellows, and has been kept alive. It is clear from other remarks Hooke makes that the dog's lungs were exposed in the course of the experiment, so that they could be observed and manipulated. This experiment gives us an opportunity to examine an early example of the use of an artificial, mechanical respirator.

The purpose of this part of the experiment is to give the investigator control over the motions of the lungs through the use of the bellows. Thus Hooke commented:

> The Dog having been kept alive . . . for above an hour, in which time the Tryal hath often been repeated, in suffering the dog to fall into *Convulsive* motions by ceasing to blow the Bellows, and permitting the Lungs to subside and lye still, and of suddenly reviving him again by renewing the blast, and consequently the motion of the Lungs. (Zucker 1996, p. 22)

This part of the experiment shows at least three things. First, it shows the dog can be kept alive with the aid of the respirator (there had apparently been some earlier failures to replicate Hooke's first results with this technique). Second, it shows that there is a connection between vital function and the motion of the lungs. Third, it sets us up for the crucial question to be investigated: Is the immediate cause of vital function the motion of the lungs in and of itself, or is it something else (something that might be *supported by* the motion of the lungs in a natural setting)?

To investigate this question, Hooke modified the structure of his respirator in an ingenious manner:

> I caused another pair of Bellows to be immediately joyn'd to the first, by a contrivance, I had prepar'd, and pricking all the outer-coat of the Lungs with the slender point of a very sharp pen-knive, the second pair of Bellows was mov'd very quick, whereby the first pair was always kept full and was always blowing into the Lungs; by which means the Lungs also were always kept very full, and without any motion, there being a continual blast of Air forc'd into the Lungs. (Zucker 1996, p. 22)

Here the experimental modification of both the respirator and the lungs of the animal subject permitted the introduction of continuous airflow through the lungs.

This in turn allowed the observation of airflow with no motion of the lungs, as well as the cessation of airflow, again with no motion of the lungs. It permitted the investigator to discern whether the motion of the lungs was the immediate cause of vital function, or whether it was something else—the airflow—which was the immediate cause of vital function (with the motion of the lungs supporting such airflow in the natural context of an unmodified experimental subject). Hooke reported his results as follows:

> This being continued for a pretty while, the dog, as I expected, lay still, as before, his eyes being all the time very quick, and his heart beating very regularly. But upon ceasing the blast, and suffering the Lungs to fall and lye still, the Dog would immediately fall into Dying convulsive fits; but be as soon reviv'd again by the renewing the fullness of his Lungs with the constant blast of fresh Air. (Zucker 1996, p. 22)

In this way Hooke was able to ascertain that the immediate cause of vital function was the airflow, not the motion of the lungs.

It is important to note that the experiment was performed without the benefit of anesthesia—something that would not find its way into medical use until the middle of the nineteenth century. The dog would have been fixed to the experimenter's table, so that the observations could be performed. The dog lies still because it has been immobilized. The motions of its eyes are used to indicate the presence of vital function. In the very nature of this experiment, it would have been unable to yelp or shriek, except in the very earliest stages of surgical preparation.

Hooke was able to cross-check his result by some further observations of his experimental subject:

> Towards the latter end of the Experiment a piece of the Lungs was cut quite off; where 'twas observable, that the blood did

> freely circulate, and pass thorow the Lungs, not only when the Lungs were kept thus constantly extended, but also when they were suffered to subside and lye still. (Zucker 1996, p. 22)

In this way, Hooke was able to establish that the motion of the lungs did not support circulation—that occurs whether the lungs move or are stationary. What matters for vital function is a flow of fresh air. Normally this is achieved through the motion of the lungs, but it can be achieved without such motion. But what it was about fresh air that supported life was unknown at this time—since it would be more than a hundred years before oxygen was isolated, let alone named.

Conclusion

Correlative with the rise of modern science is the dual phenomenon of nature's being conceptualized with the aid of mechanical metaphors and nature's being studied with the aid of machines. It was the incredible success of this new way of thinking and this new way of exploring nature that cemented the union between science and technology—a union that owes its existence in no small measure to the work of investigators in anatomy and physiology.

But there is a method too. The resolutio-compositive method predated a fully articulated conception of nature as a machine, growing out of medical inquiries at Padua. Nevertheless, the resolutio-compositive method is ideally suited for an understanding of mechanical objects such as clocks or of objects conceptualized in mechanical terms, for example, cats and dogs. Thus it was not just useful in the study of bodily motions in humans and nonhuman animals; it found applications in the study of motion generally. For example, a variant of the method was used by Galileo to study the motions of falling bodies. In particular he was able to show that the complex motion of a projectile describing a parabolic trajectory can be resolved into the combined effects of two independent motions (Cohen 1985, p. 107).

More importantly, in the course of the seventeenth century nature itself came to be seen as a complex system of interacting bodies in motion that could be understood in mechanical terms. Arguably, the crowning achievement of seventeenth-century physics is to be found in Sir Isaac Newton's (1642–1727) great work, *Principia Mathematica,* published in 1687. The resulting system of physics—Newtonian mechanics—provides a vision of the universe itself as a giant machine whose parts are held together, and whose motions are interrelated, through gravitational interactions. The law of gravity itself reflects the resolutio-compositive method, for to understand the complex motions of a system of bodies in a state of mutual gravitational attraction, one must understand the properties of the parts, that is, the masses of the bodies, and how they are related to each other, in particular the (squares of) the distances between them.

In 1687, Sir Isaac Newton (1642–1727) published Principia Mathematica, *a crowning achievement of the seventeenth century. (National Library of Medicine)*

In Newton's England, the emergence of modern science in the seventeenth century occurred at a time of tumultuous social and political transformation. This was the century that saw the execution of Charles I and English civil war. It was also the century that saw the birth of the idea of *progress*—of development and improvement through processes of transformation. In a sense, then, it was the century that saw the emergence of the modern world. Science played a part in this shift in mind-set, as witnessed by John Aubrey:

> Till about the yeare 1649 when Experimental Philosophy was first cultivated by a Club at Oxford, 'twas held a strange presumption for a Man to attempt an Innovation in Learning; and not to be good Manners, to be more knowing than his Neighbours and Forefathers; even to attempt an improvement in Husbandry (though it succeeded with profit) was look'd upon with an ill Eie. Their Neighbours did scorne to follow it, though not to doe it, was to their own Detriment. 'Twas held a Sin to make a Scrutinie into the Waies of Nature; Whereas it is certainly a profound part of Religion to glorify God in his Workes: and to take no notice at all of what is dayly offered before our Eyes is grosse Stupidity. (Dick 1978, pp. 50–51)

The conflict between science and religion that is evident in the period following the publication of Darwin's *Origin of Species* in 1859 had yet to crystallize. Though the scientists of Aubrey's day would have found atheism unthinkable, it will be shown in the next chapter that the seeds of future discontents had by this time been sown.

So what is the picture that emerges of humans and nonhuman

animals, and their relative places in nature, as a result of the rise of modern science? In particular, are all organisms just cogs in nature's machine, or are some organisms special? This will be the subject of the next chapter.

Further Reading

R. Gordon, *The Alarming History of Medicine: Amusing Anecdotes from Hippocrates to Heart Transplants* (New York: St. Martin's Press, 1993), provides an entertaining perspective on the history of medicine. L. Clendening, ed., *Sourcebook of Medical History* (New York: Dover, 1960), provides excerpts from the writings of Renaissance anatomists and physiologists. Valuable essays about medieval and Renaissance medicine can be found in R. M. Palter, ed., *Toward Modern Science,* volumes 1 and 2 (New York: Farrar, Strauss and Cudahy, 1961).

The reader interested in events surrounding the origins of modern science may find some of the chapters in A. C. Crombie, *Medieval and Early Modern Science,* volumes 1 and 2 (New York: Doubleday Anchor Books, 1959), to be helpful. I also recommend H. F. Kearney, ed., *Origins of the Scientific Revolution* (London: Longmans, 1966), and S. F. Mason: *A History of the Sciences* (New York: Collier Books, 1962).

3

Minds, Machines, and Bodies: Intelligent Design in Nature?

In the preceding chapter we saw how the emergence of the biomedical sciences of anatomy and physiology in the sixteenth and seventeenth centuries succeeded, in no small measure, through the use of mechanical metaphors and analogies, to conceptualize the animal objects of inquiry. In the seventeenth century, the fecundity of mechanical metaphors was such that nature itself came to be seen in mechanical terms, as a giant machine. This chapter is concerned with the consequences and controversies generated by attempts to locate humans and nonhuman animals in a mechanical universe.

Importantly, this mechanical view of nature was quite consistent with the medieval idea that nature, in particular organic nature, was the result of providential design. Indeed, it is very natural, if one sees nature in mechanical terms, to see the universe and its organic parts as intelligently designed artifacts. These connections between early modern science and religious belief will be examined later in this chapter, for the controversies they generated reverberate down the generations to our own day.

So what exactly were the implications of this mechanical view of nature for humans, nonhuman animals, and their relative places in the grand scheme of the universe? Are organisms themselves just machines? Are humans special machines—perhaps thinking machines? What are the relevant differences between humans and nonhuman animals? In this chapter we will examine some important figures who helped to shape views of ourselves and nonhuman animals that are of a great deal of contemporary interest.

The broader implications of this mechanistic scheme for the place of humans and nonhuman animals in nature were worked out by Thomas Hobbes, who was a great admirer of Harvey, and by the French philosopher René Descartes. Since Descartes in particular was responding in part to views expressed by his countryman, Montaigne, however, it will not go amiss to review some of the latter's views on the cognitive abilities of animals.

Montaigne: Talking like an Animal

Michel Montaigne (1533–1592), writing about humans and nonhuman animals at the very dawn of modern science, represents the spirit of skepticism. Montaigne was much less certain than his contemporaries that there were fundamental differences with respect to cognitive ability and behavioral sophistication between humans and nonhuman animals. Why do people suppose there are fundamental differences between us and nonhuman animals? Montaigne suggested:

> It is by the vanity of this same imagination that he equals himself to God, attributes to himself divine characteristics, picks himself out and separates himself from the horde of other creatures . . . and distributes among them such portions of faculties and powers as he sees fit. How does he know, by the force of his intelligence, the secret internal stirrings of animals? By what comparison between them and us does he infer the stupidity that he attributes to them? (Frame 1968, p. 331)

It was Montaigne's contention that a closer look at animal behavior would point in the direction of a revision of the lowly estimate we give to the cognitive abilities of nonhuman species.

Montaigne was an early advocate of the idea that nonhuman species not merely had elaborate communication systems but also were genuine language users. Since this belief will be a considerable bone of contention in what follows later in this book, it is worth examining his views. Pointing out that we failed to understand the language used by some human groups (he mentions the Basques and the Troglodytes), he asked whether it was not possible that we had not attended carefully enough to animal language and had jumped too hastily to the conclusion that there was no language there at all. Perhaps we needed to look and see, rather than to rely on the results of philosophical theorization. After all:

> [W]e discover very evidently that there is full communication between them and that they understand each other, not only those of the same species, but those of different species. . . . In a certain bark of the dog the horse knows there is anger; at a certain other sound of his he is not frightened. Even in the beasts that have no voice, from the mutual services we see between them we easily infer some other means of communication. . . . (Frame 1968, pp. 331–332)

And Montaigne rightly realized that the ability to use language does not depend on speech, for humans had gestural communication systems long before the advent of American Sign Language:

> Why not; just as well as our mutes dispute, argue, and tell stories by signs? I have seen some so supple and versed in this, that in truth they lacked nothing of perfection in being able to make themselves understood. Lovers grow angry, are reconciled, entreat, thank, make assignations, and in fine say everything, with their eyes. . . . (p. 332)

Montaigne's critics would have to deal with two related issues: the question of animal language and the issue of the sophistication of nonverbal communication systems. As we will see in later chapters, both of these issues are alive and well in contemporary science—though not, perhaps, as Montaigne imagined them.

There are other issues about animal behavior that go beyond the language issue. For example, Montaigne asked:

> Do the swallows that we see on the return of spring ferreting in all the corners of our houses search without judgement, and choose without discrimination, out of a thousand places, the one which is most suitable for them to dwell in? . . . Do they take now water, now clay, without judging that hardness is softened by moistening? Do they floor their palace with moss or with down, without foreseeing that the tender limbs of their little ones will lie softer and more comfortably upon it? (Frame 1968, p. 333)

In other words, Montaigne was asking whether in the complexity and apparent purposiveness of animal behavior there was not evidence of intelligence—independently of the question of language. This issue too, in the mutated form of the issue of nonpropositional knowledge (the possession of problem-solving skills in the absence of the linguistic tools to express the practical knowledge), is alive and well in contemporary science—which is to say it is an issue concerning which there is much contemporary dispute.

Montaigne was an astute observer of animal behavior. The issue of intelligence in nonhuman animals was not one to be simply dismissed out of hand. Thus he had observed the use of dogs to guide the blind:

Montaigne (1533–1592), an astute observer of animal behavior, advocated the idea that nonhuman species not only had elaborate communication systems but were genuine language users. (National Library of Medicine)

> I have seen one, along a town ditch, leave a smooth flat path and take a worse one, to keep his master away from the ditch. How can this dog have been made to understand that it was his responsibility to consider solely the safety of his master and to despise his own comforts in order to serve him? And how did he know that a given road was quite broad enough for himself, which would not be so for a blind man? Can all this be understood without reasoning and intelligence? (Frame 1968, p. 340)

Again, it is an issue of much contemporary interest as to exactly what can be inferred from such apparent problem-solving animal behavior.

Montaigne had one further line of argumentation that is worthy of mention. He observed commonalities between us and nonhuman animals where others had sought to emphasize differences:

> Our weeping is common to most of the other animals; and there is scarcely any who are not observed to complain and wail long after birth, since it is a demeanor most appropriate to the helplessness that they feel. As to the habit of eating, it is, in us as in them, natural and needing no instruction. . . . (Frame 1968, p. 334)

Given the observation of such commonalities and given the apparent purposiveness and intelligence of animal behavior:

> [T]here is no apparent reason to judge that the beasts do by natural and obligatory instinct the same things that we do by choice and cleverness. We must infer from like results, like faculties, and consequently confess that this same reason, this same method we have for working, is also that of the animals. Why do we imagine in them this compulsion of nature, we who feel no similar effect? (p. 336–337)

The inference here is not trivial or stupid. In fact, the pattern of inference would become enshrined in science a century or so later by Sir Isaac Newton in his second rule of reasoning: "*Therefore to the same natural effects we must, as far as possible, assign the same causes.* As to respiration in a man and in a beast, the descent of stones in Europe and America. . . ." (Thayer 1953, p. 3). Montaigne had merely raised the issue, *as to cognition in man and in a beast.* The issue is not as simple, perhaps, as respiration, but it cannot be dismissed on that account alone.

Montaigne was apparently little influenced by the emerging mechanical view of nature that was already underway as he wrote. He is worthy of attention primarily for his willingness to extend cognitive charity to nonhuman species. But others were not so charitable, and

Descartes in particular responded to Montaigne's challenges. Descartes was not alone in tracing out the implications of the new way of thinking. His contemporary, Hobbes, was as much a stimulus to thought on these matters as Montaigne himself, and it is to Hobbes that I now turn, since his views lie somewhere between the extremes of Montaigne and Descartes.

Thomas Hobbes (1588–1679) saw points of cognitive similarity, rooted in our common biology, between humans and nonhuman species. (Hulton / Archive)

Hobbes's Mechanical Man

Thomas Hobbes (1588–1679) is an important figure for the purposes of this book because his philosophical writings reflect important effects of the rise of modern science on the intellectual climate of the seventeenth century and the way in which animals were coming to be located in nature. Hobbes was an admirer of both Harvey and Galileo (he had met the latter in Florence in 1636), and their influences can be seen in his writings, both from the standpoint of his interest in matter in motion and from his reliance on that great Paduan legacy, the resolutio-compositive method. Hobbes attempted to articulate a unified philosophy of nature in which humans and nonhuman animals are seen as machines made of matter. He wrote of physical nature in *De Corpore* (Concerning Body, 1655), of human nature in *De Homine* (Concerning Man,1658), and the nature of human society in *De Cive* (Concerning the Citizen, 1642).

Hobbes was first and foremost a *materialist*. In his *Leviathan,* published in 1651, he made the following observations about the nature of the universe:

> The World . . . is Corporeal, that is to say, Body; and hath the dimensions of Magnitude, namely, Length, Breadth, and Depth: also every part of Body, is likewise Body, and hath the like dimensions; and consequently every part of the Universe, is Body; and that which is not Body, is no part of the Universe. And because the universe is all, that which is no part of it, is Nothing. (Minogue 1973, pp. 367–368)

Hobbes explicitly rejected the idea that had dominated much medieval thought, that there are noncorporeal, nonmaterial substances.

He found such talk unintelligible. The universe that science described was a universe of matter in motion.

But though Hobbes's universe was made of matter, he was nevertheless not an atheist (a matter that was well-nigh unthinkable in the seventeenth century). In fact, he conceded that reasoning about cause and effect could easily lead to the idea of God:

> But the acknowledging of one God, Eternal, Infinite, and Omnipotent, may be more easily be derived, from the desire of men to know the causes of natural bodies and their severall vertues, and operations. . . . For he that from any effect hee seeth come to pass, should reason to the next and immediate cause thereof, and from thence to the cause of that cause, and plunge himself profoundly into the pursuit of causes; shall at last come to this, that there must be . . . one First Mover; that is, a First, and an Eternal cause of all things; which is that which men mean by the name of God. . . . (Minogue 1973, p. 55)

For Hobbes, then, *materialism*—the view that the world consists of matter in motion—did not imply *atheism.* In fact, Hobbes had a materialist theology. He thought that God had a corporeal existence: "God is Body." Although there is nothing to contradict this in the Bible or creeds—nothing there claims that God must be made of nonmaterial substance—his materialism was at odds with the religious orthodoxy of his day. For example, the thirty-nine articles of the Anglican Church, promulgated in 1652, begin: "There is but one living and true God, everlasting, without *body, parts* or *passions* . . ."(my italics).

Hobbes's materialism, however, does have some important consequences for the similarities and differences between human and nonhuman animals. In his *Leviathan,* the grand mechanical scheme was explained as follows:

> Nature (the art whereby God hath made and governs the World) is by the *Art* of man, as in many other things, so in this also imitated, that it can make an Artificial Animal. For seeing life is but a motion of Limbs, the beginning whereof is in some principall part within; why may we not say, that all *Automata* (Engines that move themselves by springs and wheels as doth a watch) have an artificial life? For what is the *Heart,* but a *Spring*; and the *Nerves,* but so many *Strings;* and the *Joynts,* but so many *Wheeles,* giving motion to the whole Body, such as was intended by the artificer? (Minogue 1973, p. 1)

For Hobbes, human beings were mechanical artifacts resulting from providential design and creation. The human machine may even be understood by analogy with mechanical devices, such as pocket watches, that are the fruits of human design and creation. In this scheme, life itself is to be understood in terms of matter in motion. Humans (and nonhuman animals) are far more complicated contrivances than mechanical pocket watches of human design, but humans are no less mechanical for all that.

How then did Hobbes analyze human beings and human nature, and what differences and similarities did he see with respect to humans and nonhuman animals? Living things, for Hobbes, consisted of matter in motion. Hobbes's inquiries into these matters were shaped by his adherence to the resolutio-compositive method—the same method Harvey had used to unravel the mysteries of the complex motions of the heart.

First of all, Hobbes was clearly impressed that some objects (humans) but not others (trees or rocks) were conscious, had sensations, and even had internal representations of external objects. As Hobbes put it: "Of all the phenomena or appearances which are near us, the most admirable is apparition itself . . . namely, that some natural bodies have in themselves the patterns almost of all things, and others of none at all" (quoted in Woolhouse, 1988, p. 38).

Some things, then, have minds capable of perception, but where other theorists were to think of this phenomenon in nonphysical, even spiritual, terms, Hobbes was convinced that the mind was nothing but the motions in certain parts of an organic, material body. Hobbes had learned from Harvey how fundamental was the study of biological motion, such as the circulation of the blood. What else could be explained in terms of the motions in organic bodies?

Hobbes analyzed the motions that are common to humans and nonhuman animals into *vital motions* and *voluntary motions.* In terms of some special features of the latter, he was able to differentiate humans from nonhuman animals in terms of *cognitive abilities.* Concerning the nature of vital and voluntary motions, Hobbes offered the following explanation:

> There be in Animals, two sorts of *Motions* peculiar to them: One called *Vitall;* begun in generation, and continued without interruption through their whole life; such as are the *course* of the *Bloud,* the *Pulse,* the *Breathing,* the *Concoction, Nutrition, Excretion,* &c; to which motions there needs be no help of Imagination: The other is *Animal motion,* otherwise called *Voluntary motion;* as to *go,* to *speak,* to *move* any of our limbs, in such a manner as is first fancied in our minds. (Minogue 1973, p. 23)

Vital motions thus refer to physiological processes, whereas voluntary motions refer to what we now call *behavioral characteristics* of organisms. It was with respect to his account of voluntary motions that Hobbes saw similarities and differences between humans and nonhuman animals.

Hobbes's account of voluntary motions explained how organisms interact with objects in their environment. He told a rich, causal story about internal processes to explain how sensory inputs bring about behavioral outputs. Though modern science may disagree with Hobbes on the details of his causal story, the issue of whether there is a story to be told (for example, an issue to be confronted in the discussion of *behaviorism* later in this book), and if there is such a story, how it is to be told, is one of the themes explored in this book.

Hobbes began with *sensation:* External objects produce motions within us called sensations. Images remain with us after the objects causing them have been removed. This retention of images Hobbes called *imagination;* as images fade, we have *memory.* Hobbes took from Harvey the idea that the heart was not just a pump but was the principle organ in the body. In the thought of Hobbes, the vital motions produced by the heart do not just sustain life but govern the way we perceive think, feel, and desire. They are crucial, that is, to an understanding of the causes of human and animal behavior. Woolhouse summarized Hobbes's view succinctly as follows:

> Motions from outside the body often affect this "vital motion," either helping or hindering it, and our awareness of these changes constitutes pleasure and pain. Learning from experience which things help or hinder the blood's motion, we come to develop appetites and aversions; we come to hope for what produces pleasure, and to fear what produces pain. In Hobbes' view, humans always act so as to produce the increase in vital motion, which is pleasure. (1988, p. 41)

In these terms there are similarities between humans and animals. Both have sense and memory; both can learn from experience; both hope for pleasure and are averse to pain. Indeed, Hobbes was aware, as was Aristotle before him, that nonhuman animals were often capable of exhibiting remarkable degrees of behavioral sophistication—he mentioned the sociality exhibited by bees and ants, for example (Minogue 1973, p. 88).

Yet there are important cognitive differences between humans and nonhuman animals too, and these differences are reflected in his account of animal behavior. Humans and some nonhuman animal species are capable of social behavior and of being cooperative. Yet between us and such species Hobbes saw some fundamental differences. It will not go amiss to record them (Minogue 1973, pp. 88–89):

1. Humans, unlike nonhuman animals, are in competition for honor and dignity. In our species this is a source of envy, hatred, and ultimately war.
2. Unlike humans, in nonhuman animals the common good does not differ from the private good.
3. Unlike humans, nonhuman animals do not have the use of reason; such creatures do not reflect upon their social arrangements. In humans, by contrast, "there are very many, that thinke themselves wiser, and abler to govern the Publique, better than the rest . . . and thereby bring [society] into Distraction and Civill warre."
4. Nonhuman animals "have some use of voice"—they have communication systems by means of which they can communicate their desires and affections—but they lack the human capacity for language. Humans can form signs and names to discuss their sensations and ideas and can thus "represent to others, that which is Good, in the likeness of Evill; and Evill, in the likeness of Good; and augment, or diminish the apparent greatnesse of Good and Evill; discontenting men, and troubling their peace at their pleasure." Linguistic sophistication, then, is an important difference that comes with a price.
5. Nonhuman animals, lacking reason, cannot distinguish between injury and damage, and by this Hobbes means that as long as they are not in competition for resources—they are "at ease"—they are not offended by their fellows, "whereas Man is then most troublesome, when he is at ease."
6. Agreement, sociality, and cooperation, where they are found among nonhuman animals, are a natural part of their behavioral repertoire. Humans are not by nature social or cooperative, and human sociality and cooperation result from the selfish use of reason leading to artificial, social conventions and pacts. Humans without these social conventions—living in a *state of nature*—would find themselves in a condition

   > where every man is Enemy to every man. . . . In such a condition, there is no place for Industry; because the fruit thereof is uncertain: and consequently no culture of the Earth, no Navigation, nor use of the commodities that may be

> imported by Sea; no commodious building . . . no Arts; no Letters; no Society; and which is worst of all, continuall feare, and danger of violent death; And the life of man, solitary, poore, nasty, brutish, and short. (Minogue 1973, pp. 64–65)

For Hobbes, these cognitive differences between humans and nonhuman animals—especially differences with respect to linguistic ability—were important, for they meant that nonhuman species were forever cut off from the *verbal* social conventions that underlie human sociality and cooperation and that undergird the moral aspects of human existence. They could not be participants in human society. As Hobbes puts it, "To make covenants with bruit Beasts, is impossible; because not understanding our speech, they understand not, nor accept of any translation of Right; nor can translate any Right to another; and without mutual acceptation, there is no covenant" (Minogue 1973, p. 71). Humans thus have obligations to each other in society that none have to nonhuman species, and language is the basis of this difference.

Hobbes's view was that humans can only live socially and cooperatively if there is a strong civil power to hold the citizens in awe. This power imposes law to regulate behavior for the common good. Though Hobbes had no interest in animal rights, it is not inconsistent with his views that there be laws regulating our treatment of nonhuman animals, especially if such legislation were necessary to preserve the structure of the social fabric. But given his bleak view of human nature, it is likely that the life of animals in contact with humans, like the life of humans in the state of nature, would be solitary, poor, nasty, brutish, and short!

Hobbes's importance lies in the fact that he saw points of cognitive similarity, rooted in our common biology, between us and nonhuman species. His political and moral theory was not rich enough to ascribe moral significance to these cognitive insights. But perhaps Hobbes was wrong. Perhaps there are no cognitive similarities between humans and nonhuman animals. This is a view to which I now turn, because it will afford us an opportunity to see what happens when differences between humans and nonhuman animals are exaggerated.

Descartes: Physics and Cosmology

René Descartes (1596–1650), philosopher, mathematician, physiologist, and amateur anatomist, was writing at the same time as Hobbes—indeed he and Hobbes were acquainted with each other's work. But Descartes is important in the present context because he articulated, against the same intellectual background as Hobbes, a very different estimate of the nature of humans and nonhuman animals and of their relative places in nature. Descartes's views in this regard have been enormously influential, and one can, even today, find their distant echoes resonating in debates about animal welfare and animal experimentation. I will examine these echoes of Cartesianism in the last four chapters of this book.

Descartes's interests were far ranging, and I must of necessity focus on the fragments of his theories that are relevant to the concerns of this book. It will be useful to begin with a brief overview of the salient characteristics of Cartesian physics and cosmology: first, because the distinctions between the *physical* and the *mental* will be very crucial in what follows, and second, because such an overview affords us a glance at the type of universe in which Descartes was trying to locate humans and nonhuman animals. It will also be useful to examine Descartes's cosmology, to see how it differs from the medieval view of the universe but also how it is in turn different from modern cosmology. Although we do not need to review all the features of Cartesian physics (which was supplanted by Newtonian mechanics), the following are relevant to the present inquiries.

To the question of what the physical world was made of, Descartes had the following answer: matter. But how did he think of matter? The answer is that he thought matter was literally *res extensa*—extended substance, literally that with spatial extension. If this were so, it was the hope of Cartesian physics to explain all physical phenomena without needing to postulate anything in matter other than movement, size, shape, and arrangement of its parts. Or as Descartes himself put it:

> Who has ever doubted that bodies move and have various sizes and shapes, and that their various different motions correspond to these differences in size and shape, or who doubts that when bodies collide bigger bodies are divided into smaller one and change their shape? . . . This is much better than explaining matters by inventing all sorts of strange objects which have no resemblance to what is perceived by the senses, such as "prime matter," "substantial forms" and the whole range of qualities that people habitually introduce, all of which are harder to understand than the things they are supposed to explain. (Quoted in Cottingham 1986, p. 95)

Like that of Hobbes, Descartes's scheme stood in contrast to medieval views about the nature of substance, as could be found, for instance, in the writings of St. Thomas Aquinas.

Though Descartes is classified as a *rationalist* philosopher, as one who places special emphasis on the role of reason, his scheme for physics had an important place for sensory experience, for though reason might tell us that the universe consists of matter in motion,

> we cannot determine by reason alone how big these pieces of matter are, or how fast they move, or what kinds of circle they describe. Since there are countless different configurations which God might have instituted here, experience alone must teach us which configuration he actually selected in preference to the rest. We are thus free to make any assumption on these matters, with the sole proviso that all the consequences of our assumption must agree with experience. (Quoted in Cottingham, 1986, p. 92)

There is good sense, then, to be given to the claim that the program in Cartesian physics was an empirical program richly informed by theory.

For all this, there are questions about the place of God in Cartesian cosmology, for Descartes was convinced that there was a God—his theory of knowledge rested on the idea of God, and he produced various arguments to "prove" that God existed. Moreover, God created the world and set its parts in motion, and according to Descartes, was needed to keep it all in existence from moment to moment (Cottingham 1986, p. 99). Once again, science and theology are seen to be richly intertwined.

Questions about the role of God are important because the relative places of humans and nonhuman animals in nature often hinge on how God is introduced into the overall cosmological scheme of things. Descartes's philosophy is no exception. Yet his views were somewhat unique. As pointed out by Cottingham (1986, pp. 96–97), he accepted the Copernican revolution in astronomy and indeed went beyond it, stating that the earth is a planet that moves around the sun, and the sun is but one star among many. The earth has no special cosmological status as the center of the universe. Descartes himself observed:

> It is a common habit of men to suppose that they are the dearest of God's creatures, and that all things are made for their benefit. They think that their own dwelling place, the Earth, is of supreme importance, that it contains everything that exists, and that for its sake everything was created. But what do we know of what God may have created outside the earth, on the stars and so on? How do we know that he has not placed there other crea-

> tures, other lives, and other "men" or at least beings analogous to men . . . ? (Quoted in Cottingham 1986, p. 97)

The earth, and life upon it, has no special place or status in the grand scheme of things. The possibility of alien life elsewhere in the cosmos cannot be excluded.

And this is a strong rejection of medieval views according to which everything was designed by a creator for the sake of humans, the pinnacle of organic creation, inhabiting the planet at the center of the universe. For as Descartes put it:

> In the study of physics it would be utterly ridiculous to suppose that all things were made for our benefit since there is no doubt that many things exist or once existed, though they are here no longer, which have never been seen or thought of by any man, and have never been of use to anyone. (Quoted in Cottingham 1986, p. 98)

There is a clear tension here with interpretations of the Bible according to which the created world was for our benefit. This is a clear departure from medieval cosmology. And it is a useful caution to contemporary advocates of "fine-tuning" theology based on the anthropic principle, according to which the basic constants of the universe were selected at the dawn of time by the creator in such a way as to be favorable to the appearance of human life (see Miller 1999, p. 228).

For Descartes, a scientific understanding of the world and its contents was not something that would proceed most fruitfully from a consideration of the details of divine creation:

> Nevertheless, if we want to understand the nature of plants or of men, it is much better to consider how they can gradually grow from seeds than to consider how they were created by God in the beginning. . . . Thus we may be able to think up certain very simple and easily known principles which can serve, as it were, as seeds from which we can demonstrate that the stars, the Earth and indeed everything we observe in this visible world could have sprung. . . . (Quoted in Cottingham 1986, p. 99)

This emphasis on the study of natural processes is laudable, yet even if we focus on the "simple and easily known principles," rather than religious dogma, we will discover that Descartes is cut off from the very possibility of anything like a modern, fully evolutionary perspective of humans and nonhuman animals.

Importantly for what follows, Descartes was unable to accept that conscious beings such as ourselves could descend with modification from unconscious beings. In Cartesian philosophy, consciousness is not explicable as arising from a modification of matter. It is with respect to consciousness that Descartes drew the line in the sand separating humans and nonhuman animals—a line far more radical than that drawn by Hobbes, and a line that would have dismayed his countryman Montaigne.

Descartes: Of Mice and Men

Against this background of physics and cosmology, we can now begin to unravel the Cartesian view of the nature of humans and nonhuman animals. Like Hobbes, Descartes thought that (involuntary) physiological processes can be explained in purely mechanical terms. His work, though fanciful by modern lights, helped to focus attention on the nervous system (Descartes believed that the nerves were hollow pipes, flowing in which, like hydraulic fluid, was a substance known as *animal spirits,* and these spirits could exert mechanical control over the muscles). There are passages where it seems he anticipated something like the reflex arc:

> When people take a fall and stick out their hands so as to protect their head, it is not reason that instructs them to do this; it is simply that the sight of the impending fall reaches the brain and sends the animal spirits into the nerves in the manner necessary to produce this movement even without any mental volition, just as it would be produced in a machine. (Quoted in Cottingham 1986, p. 108)

We would have to wait for the work of Marshall Hall in the 1830s to provide an experimental vindication of the phenomenon of the reflex arc.

In the *Discourse on Method,* having discussed Harvey and the circulation of the blood, Descartes commented on the existence of complex, involuntary motions in organisms as follows:

> This will hardly seem strange to those who know how many motions can be produced in automata or machines which can be made by human industry, although these automata employ very few wheels and other parts in comparison to the large number of bones, muscles, nerves, arteries, veins, and all other component parts of each animal. Such persons will therefore think of this body as a machine created by the hand of God, and in consequence incomparably better designed and with more admirable movements than any machine that can be invented by man. (Lafleur 1987, p. 41)

For Descartes, the motions of nonhuman animals were purely involuntary. They are machines, and nothing more than mechanical principles are needed to account for their behavior. We have obviously not told the full story yet, for as we saw earlier in this chapter, Hobbes saw animals as machines, but he saw them also as having sensations. So what are the differences between Hobbes's machines and the machines of Descartes?

To answer this question, it is necessary to examine Descartes's views about humans and what makes us fundamentally, indeed metaphysically, different from nonhuman animals. For Descartes, humans have rich mental lives. We are conscious. We have bodily sensations, including phantom limb sensations. We have memories. We think and reason. We experience emotions such as anger. We have ideas, and we dream. In virtue of what is all this possible? Descartes, in the process of defining what sort of a being he is, said, "I think therefore I am" (*Cogito ergo sum* in Latin and *je pense donc je suis* in French) (Lafleur 1987, p. 24). He concluded that he was a being that thought (*Sum res cogitans*). This was all possible because, in addition to his physical body made of extended substance (*res extensa*), he had a mind or soul made up of spatially unextended thinking substance (*res cogitans*). Descartes summarized his view thus:

> I then described the rational soul, and showed that it could not possibly be derived from the powers of matter . . . but must have been specially created . . . next to the error of those who deny God, which I think I have sufficiently refuted, there is none which is so apt to make weak characters stray from the path of virtue as the idea that the souls of animals are of the same nature as our own, and that in consequence we have no more to fear or to hope for after this life than have flies and ants. Actually when we know how different they are, we understand more fully the reasons which prove that our soul is by nature independent of the body, and consequently does not die with it . . . we are naturally led to conclude that it is immortal. (Lafleur 1987, pp. 43–44)

Humans are thus characterized as being very different, not merely morphologically, but metaphysically, from nonhuman animals.

In Descartes's philosophy, our mental lives in all their richness are possible because we, unlike nonhuman animals, have nonphysical minds—minds that do not depend for their existence, moreover, on the physical brain or other arrangements of matter. Because of this dualism of substance, no purely material being can have any mental life whatsoever, nor could such a being be modified by purely material, evolutionary processes to possess a mind capable of supporting a mental life. But this in turn means that animals are purely mechanical entities with no mental lives whatsoever—they lack the substance necessary for a mental life of any kind.

To see this a bit more clearly, it will help to reflect on the meanings of verbs such as *cogitare* in Latin and *penser* in French. These have been rendered in English translations of Descartes in terms of the verb *think* and the noun *thought.* But this suggests an intellectualistic rendering of the original words, which is not really supportable. As Anscombe and Geach have pointed out:

> In everyday XVIIth-century French, *penseé* had a rather wider application than in modern French; it was then natural, as it would not now be, to call an emotion *une penseé.* Similarly, *cogitare* and its derivatives had long been used in a very wide sense in philosophical Latin; for example, *cogitationes cordium* in Aquinas covers all internal states of mind. Descartes himself defines the words as applying not only to intellectual processes but also to acts of will, passions, mental images, and even sensations. (1976, p. xlvii)

When Descartes denied that animals can think, it is possible that he was not denying with Hobbes that they can reason; he was rather denying them mental lives in this very broad sense. They have neither reason, nor passions, nor memories, nor sensations.

Descartes did not merely rest his case with a simple metaphysical appeal to a difference in substantial constitution. He had additional arguments to establish a fundamental divide between us and nonhuman animals. Descartes imagined that someone might make a machine with the "organs and appearance of a monkey"—perhaps such an artifact could result from an experiment in bioengineering. He suggested that a really good machine could be constructed to be behaviorally indistinguishable from an actual monkey.

Not so for us human beings. Someone might make a machine in our own image—similar to Mr. Data on *Star Trek: The Next Generation*—but it would be differentiable from a real human in at least two ways:

> The first is that it could never use words or other signs for the purpose of communicating its thoughts to others, as we do. It is indeed conceivable that a machine could be so made that it would utter words, and even words appropriate to the presence of physical acts of objects which cause some change in its organs; as, for example, if it was touched in some spot that it

> would ask what you wanted to say to it. . . . But it could never modify its phrases to reply to the sense of whatever was said in its presence, as even the most stupid men can do. (Lafleur 1987, pp. 41–42)

Descartes thus believed that machines cannot be constructed to use language in flexible and creative ways to respond appropriately to arbitrary stimuli in the environment.

> The second method of recognition is that, although such machines could do many things as well as, or perhaps even better than, men, they would infallibly fail in certain others, by which we would discover that they did not act by understanding or reason, but only by the disposition of their organs. For while reason is a universal instrument which can be used in all sorts of situations, the organs have to be arranged in a particular way for each action. From this it follows that it is . . . clearly incredible that there should be enough devices in a machine to make it behave in all occurrences of life as our reason makes us behave. (Lafleur 1987, p. 43)

Descartes's view here was that a machine's behavioral output, though it might be complex, reflected a finite set of instructions and once again would lack the flexibility and adaptability of genuine human behavior.

These defects in machine behavior enable us to differentiate humans from nonhuman animals:

> For it is a very remarkable thing that there are no men, not even the insane, so dull and stupid that they cannot put words together in a manner to convey their thoughts. On the contrary, there is no other animal . . . that can do the same. And this is not because they lack the organs, for we see that magpies and parrots can pronounce words as well as we can. . . . On the other hand, even those men born deaf and dumb . . . usually invent for themselves some signs by which they can make themselves understood by those who are with them enough to learn their language. And this proves not merely that animals have less reason than men, but that they have none at all, for we see that very little is needed in order to talk. (Lafleur 1987, p. 42)

In later chapters, I shall return to both the issue of language use in nonhuman animals and the issue of behavioral flexibility. For the present, it suffices to note that for Descartes, these differences between human and nonhuman animals did not reflect differences with respect to the complexity of our brains or the fact that we have taken distinct evolutionary trajectories; rather they reflected fundamental metaphysical differences between us and them.

That there are differences between humans and nonhuman animals, no one denies, but there is a great deal of debate about the nature of the differences, especially with respect to the issues of behavioral flexibility and cognitive sophistication. An understanding of these issues will require that we get beyond the model of animal-as-machine, fruitful as it may have been in the seventeenth century. In particular, it remains to be seen whether being a *dumb* animal (unable to use language like "one of the most stupid children, or at least a child of infirm mind") implies that nonhuman animals are behaviorally dumb (completely lacking behavioral flexibility) and cognitively vacant.

René Descartes (1596–1650) believed that the physical and behavioral differences between human and nonhuman animals reflect fundamental metaphysical differences between the two. His position was that if animals feel pain, they do so with no accompanying thoughts. (National Library of Medicine)

For the present, Cottingham has pointed out that the Cartesian idea that animals are nothing more than machines

> gained a distinctly sinister reputation in the century following Descartes: if animals were mere mechanical automata, then Cartesian anatomists could blandly claim that the screams of vivisected animals nailed to their work benches were of no more significance than the chimes of a clock, or the piping of a church organ when certain keys were pressed. Not surprisingly, Descartes is, for this reason, apt to be regarded as something of an ogre by modern animal liberationists. (1986, p. 108)

And there are passages where Descartes compared animals to clocks and where he suggested that their remarkable behavioral abilities were not more surprising than those of a clock that can purely mechanically keep time better than we, with all of our intelligence. Not everyone agreed, however, and Descartes's contemporary, the physiologist Niels Stensen (Steno) (1638–1686), could not accept the received Cartesian view that animals feel no pain—and eventually gave up the practice of vivisection (Lindeboom 1979, p. 64).

But what of Descartes himself? Cottingham contends that his views about animals were not quite as straightforward as I have so far suggested and that the issue has to do with *sentience*. For there are passages where Descartes did not simply deny that animals have souls but said that the differences between us and them would not arise, "if their soul were not of a wholly different nature from ours" (1987, p. 43). That this is no mere *façon de parler* is a matter to which I now turn.

In 1646, Descartes wrote a lengthy letter to the Marquis of Newcastle in which he responded to criticisms and explained why he did not agree with the views of his countryman, Montaigne, that animals could think:

> If you teach a magpie to say good-day to its mistress, when it sees her approach, this can only be by making the utterance of the word the expression of one of its passions. For instance, it will be the expression of the hope of eating, if it has always been given a tidbit when it says the word. Similarly, all the things which dogs, horses and monkeys are taught to perform are only the expression of their fear, their hope or their joy. . . . But the use of words, so defined, is peculiar to human beings. (Quoted in Cottingham 1986, p. 110)

This represented something of a retreat from his earlier views, for it meant that he was now willing to countenance a difference between nonhuman animals and clocks: Animal behavior is a little different from the behavior of clocks in that hopes, fears, and desires can play a causal role in the production of animal behavior, even if no *thoughts* accompany them.

Descartes also seems to backtrack on the issue of pain (and sensations), again in ways suggestive of a breakdown of the machine analogy. Animals certainly act as if they feel pain, and Descartes made references to the souls of animals, albeit souls different from our own. His position was that if animals felt pain, they did so with no accompanying thoughts. He seemed to suggest that though nonhuman animals have no *rational* soul, they did have a *sensitive* soul and experienced bodily sensations (*sens organique*). Descartes referred to this animal soul as a bodily soul (*âme corporelle*) and he located it in the animal spirits (Lindeboom 1979, pp. 64–65).

So notwithstanding the extensive use of mechanical imagery and metaphor, even Descartes, toward the end of his life, was compelled to recognize that nonhuman animals had minimal mental lives in the form of passions and sensations, even if there were no accompanying thoughts. What is the moral significance of this observation that animals might feel pain? This will be the subject of the next chapter. For the present, I will conclude by noting that Descartes's mechanical scheme for animals was very influential—and the idea that mechanical principles could be used in biological explanation found successful expression in Giovanni Borelli's *On the Motion of Animals* (1680).

As for observations of the adapted nature of animal behavior—for example, the nest building of birds and the return of swallows in the spring, as well as observations of physiological, morphological, and anatomical adaptation—these were evidences of providential machine design. For the scientist at the end of the seventeenth century, these features of the organic world were captured by the title of John Ray's (1627–1704) book, *The Wisdom of God Manifested in the Works of Creation* (1693).

The Intelligent Design of the World

The picture of organisms that emerges from seventeenth-century science is filled with mechanical metaphors: stomach as retort, veins and arteries as hydraulic tubes, the heart as pump, the viscera as sieves, lungs as bellows, and muscles and bones as a system of cords, struts, and pulleys (Crombie 1959b, pp. 243–244). The metaphors bolstered a picture of organisms as special machines made by God. As the philosopher Gottfried Leibniz put it in his *Monadology* (1714):

> Thus each organic body of a living thing is a kind of divine machine, or natural automaton, which infinitely surpasses all artificial automata. Because a machine which is made by the art of man is not a machine in each of its parts; for example, the tooth of a metal wheel has parts or fragments which as far as we are concerned are not artificial and which have about them nothing of the character of a machine, in relation to the use for which the wheel was intended. But the machines of nature, that is to say living bodies, are still machines in the least of their parts *ad infinitum*. This it is which makes the difference between nature and art, that is to say, between Divine art and ours. (Parkinson 1977, p. 189)

Thus, organisms, unlike watches, are machines *all the way down,* and this is what differentiates God's handicraft from that of mere mortal mechanics.

But inorganic nature, too, was seen in mechanical terms. Newton's universe was a clockwork universe—a giant machine with many interacting, moving parts. And wheels within wheels could be seen everywhere. Not only did the organism have its mechanical parts each *adapted* for specific functions necessary for life, but different organisms had distinct places in nature. Specialized in distinct and unique ways, they, like the parts within them, had proper places in the natural machine.

But how literally were these mechanical metaphors to be taken? The answer is that in many scientific circles they were taken quite literally. The universe was a grand machine, with many finely crafted moving parts. It was, in fact, the result of deliberate, intelligent design on a grand scale. God was the cosmic engineer, and the date of creation was set by Archbishop James Ussher (1581–1656) as 4004 B.C., 23 October at midday (Gould 1993, p. 181).

The intellectual tradition of studying nature—the mechanical fruit of God's providential design—in order to make discoveries about the creator (both his very existence, as well as particular properties, such as benevolence and so on) is known as *natural theology*. We have already seen that a version of the *argument from design* was formulated in the medieval period—in particular in the work of St. Thomas Aquinas. But the argument, far from being dispelled by the rise of modern science, was in fact bolstered by it. That God had set everything in motion was generally agreed. But was the world of today essentially the same as it had been at the time of creation? Or had there been dramatic changes? Some natural theologians, such as John Ray, saw the existing earth as both the fruit of intelligent design and the best of all possible worlds. Others disagreed.

Natural theologian Thomas Burnet (1635? –1715), who corresponded with his friend Newton about these matters, was puzzled by the current observation of the relatively short life spans of humans and nonhuman animals, especially since the Bible reported longer life spans prior to the Noachian flood:

> [S]uppose a Mill, where the Water may represent the nourishment and humours in our Body, and the frame of Wood and Stone, the solid parts; If we could suppose this Mill to have the power of nourishing itself by the Water it receiv'd, and of repairing all the parts that were worn away, whether of the Woodwork or of the Stone, feed it but with a constant stream, and it would subsist and grind forever. And 'tis the same for all other Artificial Machines of this nature, if they had a faculty of nourishing themselves, and repairing their parts. And seeing those natural Machines we are speaking of, the Body of Man, and of other Animals, have and enjoy this faculty, why should they not be able to preserve themselves beyond that short period of time which is now the measure of their life. (Burnet [1691] 1965, p. 153)

Burnet's answer was that though God had designed a perfect world, the earth as we know it today is a ruined paradise. The ruination was brought on by sin and by the Noachian flood and was evidence of

God's anger. The ruination of the planetary system had adverse consequences for the longevity of the machines crawling on its surface!

Prior to Darwin, *natural science* and *natural theology* were coupled enterprises, with figures prominent in one of these intellectual enterprises often being prominent in the other. My purpose here is to examine the idea of the intelligent design of nature, to see how it shaped the eighteenth- and early-nineteenth-century view of humans, nonhuman animals, and their relative places in nature.

Today, one can still find advocates of natural theology, though this formerly respectable discipline has mutated into a grim parody of itself under the heading of *creation science* (or *intelligent design theory,* as one version is currently being peddled to the public—see Shanks 2000; Shanks and Joplin 1999; 2000; 2001). But where the natural theologians of old were in awe of the grandeur of nature, reveled in the discoveries of natural science, and saw the *book of nature* as a supplementary volume to the *book of God,* the contemporary creation scientist feels compelled to substitute for the book of nature as we now know it, a grotesque work of science fiction and fantasy, so that consistency may be maintained between preferred interpretations of the two books.

Concerning contemporary creation science, I shall give Thomas Burnet the last words: "'Tis a dangerous thing to ingage the authority of Scripture in disputes about the Natural World, in opposition to reason lest Time, which brings all things to light, should discover that to be evidently false which *we had made Scripture to assert . . .*" ([1691] 1965, p. 16; my italics). But what of the state of affairs in the seventeenth and eighteenth centuries?

Newton and Design in Nature

It will come as a shock to some to realize that no lesser personage than Sir Isaac Newton, the champion of the mechanical view of the universe, was also a natural theologian. Yet here he is, setting the tone of the debate for the next hundred and fifty years:

> The six primary planets are revolved about the sun in circles concentric with the sun, and with motions directed toward the same parts and almost in the same plane . . . but it is not to be conceived that mere mechanical causes could give birth to so many regular motions, since the comets range over all parts of the heavens in very eccentric orbits. This most beautiful system of the sun, planets and comets could only proceed from the counsel and dominion of an intelligent and powerful Being . . .

> and lest the systems of the fixed stars should, by their gravity, fall on each other, he hath placed those systems at immense distances from one another. (Thayer 1953, p. 53)

Like Aquinas before him, Newton was impressed with the natural motions observed in the heavens and saw in them evidences of providential design.

Importantly for the present purposes, Newton saw evidence of intelligent design in the living world too:

> Opposite to godliness is atheism in profession and idolatry in practice. Atheism is so senseless and odious to mankind that it never had many professors. Can it be by accident that all birds, beasts and men have their right side and left side alike shaped (except in their bowels); and just two eyes, and no more, on either side of the face . . . and either two forelegs or two wings or two arms on the shoulders, and two legs on the hips, and no more? Whence arises this uniformity in all their outward shapes but from the counsel and contrivance of an Author? (Thayer 1953, p. 65)

For Newton, morphological similarities were evidence of deliberate intelligent design. Atheism was odious because it could offer no good account of the similarities, save that they were, perhaps, fortuitous accidents.

But Newton did not rest his case simply with the observation of morphological similarities. There was also evidence of adapted complexity:

> Whence is it that the eyes of all sorts of living creatures are transparent to the very bottom, and the only transparent members in the body, having on the outside a hard transparent skin and within transparent humors, with a crystalline lens in the middle and a pupil before the lens, all of them so finely shaped and fitted for vision that no artist can mend them? Did blind chance know that there was light and what was its refraction, and fit the eyes of all creatures after the most curious manner to make use of it? These and suchlike considerations always have and ever will prevail with mankind to believe that there is a Being who made all things and has all things in his power, and who is therefore to be feared. . . . (Thayer 1953, pp. 65–66)

For Newton, such adapted complexity had two possible explanations. First, that it was the result of intelligent design, or second, that it all came about by chance and happenstance. Newton was inclined to the former, as the latter is—and everyone will admit this—so implausible as to be silly and beyond belief. Part of Darwin's achievement, as we shall see, is to offer a third possibility—one that Newton never considered—to explain the same appearances: the morphological similarities and the existence of adapted complexity. Darwin's views will be examined in Chapter 6.

Newton, though clearly a believer in both God and creation, was no Biblical literalist, and this sets him apart from many contemporary advocates of creation science. As Newton himself put it in a letter to Thomas Burnet, "As to Moses, I do not think his description of the creation either philosophical or feigned, but that he described realities in a language artificially adapted to the sense of the vulgar" (Thayer 1953, p. 60), adding:

> If it be said that the expression of making and setting two great lights in the firmament is more poetical than natural, so also are some other expressions of Moses, as when he tells us the windows or floodgates of heaven were opened. . . . For Moses, accommodating his words to the gross conceptions of the vulgar, describes things much after the manner as one of the vulgar would have been inclined to do had he lived and seen the whole series of what Moses describes. (Thayer 1953, pp. 63–64)

For Newton, there was no conflict between science and religion, and his own account of nature, especially organic nature, was thoroughly intertwined with his religious beliefs. In fact, the natural theologians of the seventeenth and eighteenth centuries were daring, sophisticated thinkers and a veritable far cry from the creation scientists of our own day, who at their very best are but poor mimics of this earlier intellectual tradition.

Paley and the Evidences of Design

William Paley's (1743–1805) great work, *Natural Theology, or Evidence of the Existence and Attributes of the Deity, Collected from the Appearances of Nature,* was published in 1801. It was a book that Darwin had read and admired. Moreover, it is of interest to us, since, along with the mechanical view of nature and its contents that emerged with the rise of science, consideration of it will help provide a backdrop against which we can better appreciate the significance of Darwinian evolutionary thought.

Paley was impressed by observations of the way organisms show adaptation to their natural surroundings. Organisms contain structures serving specific functions that enable them to fit into their allotted places in nature. In the grand tradition of thinking in terms of mechanical metaphors and analogies, Paley reasoned as follows:

> In crossing a heath, suppose I pitched my foot against a *stone,* and were asked how the stone came to be there, I might possibly answer, that for anything I knew to the contrary it had lain there for ever. . . . But suppose I had found a *watch* upon the ground. . . . I should hardly think of the answer which I had before given, that for anything I knew the watch might always have been there. ([1801] 1850, p. 1).

Watches are machines with many finely crafted, moving parts adjusted so as to produce motions enabling the whole device to keep time. It would make sense to infer, in the case of such a functional, complex piece of machinery, that

> we think it inevitable, that the watch must have had a maker—that there must have existed at some place or other, an artificer or artificers who formed it for the purpose which we find it actually to answer, who comprehended its construction and designed its use. (p. 10)

Even if the watch contained the odd property of being able, in the course of its movement, to produce another watch, an initial watch with these curious reproductive capabilities would still have to have been the fruit of design and manufacture (p. 19).

The next step in the argument was to consider the eye, which, like the watch, appears to be a complex piece of machinery with many finely crafted, moving parts, all enabling the organ to achieve its function. Eyes are compared to telescopes, and Paley was led to the conclusion that the eye, like the watch and the telescope, must have had a designer ([1801] 1850, chap. 3). More than this, Paley compared the eyes of birds and fishes and concluded, "But this, though much, is not the whole: by different species of animals, the faculty we are describing is possessed in degrees suited to the different range of vision which their mode of life and of procuring their food requires" (p. 27). Different species occupy different places in nature, and for each species, the machinery of the eye has been adapted to suit the needs consequent upon their allotted place.

In his discussion of the fruits of comparative anatomy, Paley explained these similarities and differences with the aid of mechanical metaphors:

> Arkwright's mill was invented for the spinning of cotton. We see it employed for the spinning of wool, flax, and hemp, with such

modifications of the original principle, such variety in the same plan, as the texture of those different materials rendered necessary. Of the machine's being put together with design . . . we could not refuse any longer our assent to the proposition, "that intelligence . . . had been employed, as well in the primitive plan as in the several changes and accommodations which it is made to undergo." ([1801] 1850, p. 143)

Comparative anatomy, then, yielded, as it did for Newton, further evidence of intelligent design in the natural world, with mechanical metaphors carrying much explanatory weight.

Consistent with the biological theories current at the end of the eighteenth century, Paley saw nature arranged in a hierarchy. Thus in a discussion of the causes of reproduction, he remarked:

> We may advance from animals which bring forth eggs, to animals which bring forth their young alive; and from this latter class, from the lowest to the highest—from irrational to rational life, from brutes to the human species. . . . The rational animal does not produce its offspring with more certainty or success than the irrational animal; a man than a quadruped, a quadruped than a bird; nor—for we may follow the gradation through its whole scale—a bird than a plant. . . . ([1801] 1850, p. 43)

The distinction between humans and nonhuman species was drawn on the basis of our possession of the cognitive trait of rationality, and we see that Paley conceived of the natural world as being arranged in a hierarchy. In this regard, little has changed since medieval times.

Though Paley drew a sharp distinction between humans and nonhuman animals, it is clear that he had a more charitable estimate of their mental lives than that to be found in the official Cartesian position, for it is part of the evidence of benevolent design that animals experience *pleasure* as well as *pain:*

> Assuming the necessity of food for the support of animal life, it is requisite that the animal be provided with organs fitted for the procuring, receiving and digesting of its food. It may also be necessary, that the animal be impelled by its sensations to exert its organs. But the pain of hunger would do all this. Why add pleasure to the act of eating? . . . This is a constitution which, so far as appears to me, can be resolved into nothing but the pure benevolence of the Creator. ([1801] 1850, p. 313)

In the next chapter we will see that the admission that animals are capable of experiencing sensations such as pleasure and pain turns out to be crucial, quite independently of further scientific developments, for their moral status. In this way we will see that even minimal admissions of cognitive sophistication turn out to have dramatic consequences for our relations to nonhuman species.

Could chance or natural causes be behind the adapted complexity we see in nature? Paley was uncompromising on this topic: "In the human body, for instance, *chance, that is, the operation of causes without design,* may produce a wen, a wart, a mole, a pimple, but never an eye. . . . In no assignable instance has such a thing existed without intention somewhere" ([1801] 1850, p. 49; my italics). Notice that Paley equated *chance* not with *uncaused events* but with events that may have natural causes but that are unguided by intelligence. This is a claim we will have to reexamine in Chapter 6, when consideration is given to the Darwinian revolution. For the present, what other explanations could there be of such complex, adapted structures other than deliberate design?

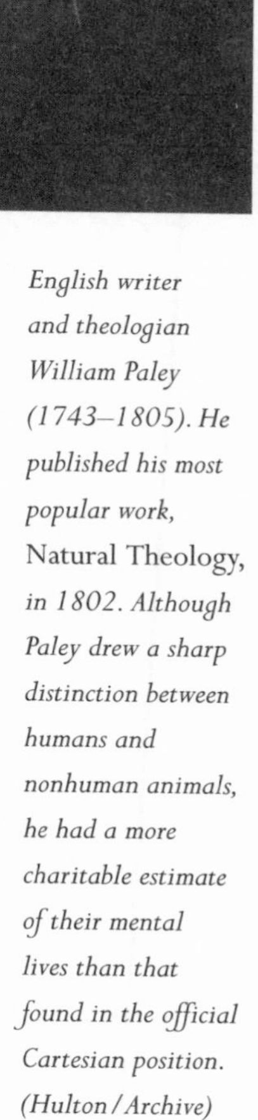
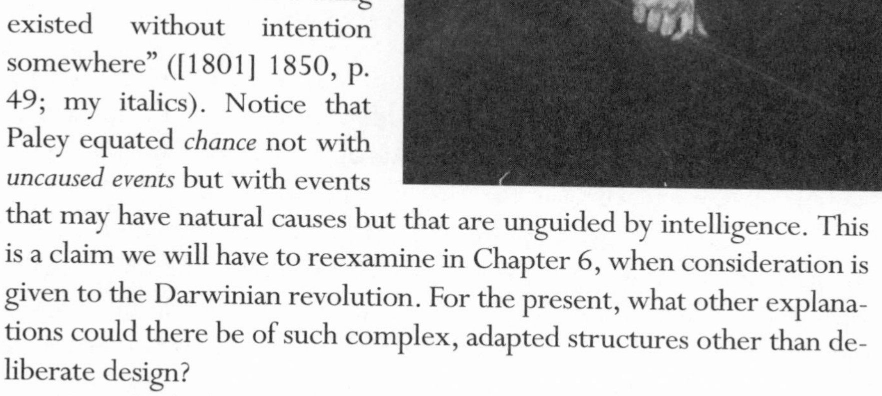

English writer and theologian William Paley (1743–1805). He published his most popular work, Natural Theology, *in 1802. Although Paley drew a sharp distinction between humans and nonhuman animals, he had a more charitable estimate of their mental lives than that found in the official Cartesian position. (Hulton/Archive)*

In Paley's day, nearly sixty years before the publication of Darwin's *Origin of Species,* there had been speculations about the possibilities for evolution. Georges Buffon (1707–1788), for example, had suggested that extant organisms in all their variety might have descended from a much smaller group of earlier organisms. Mason has remarked:

> Maupertuis in 1751 revived the view of the Atomists that the various organic species arose from different chance combinations of organic units or atoms endowed with life. Diderot followed in 1754 with a version of Empedocles' theory, according to which various animal organs, heads, limbs and so on, had come together fortuitously, giving some monstrous and grotesque creatures which did not survive, and present day forms which proved to be more viable. (1970, pp. 338–339)

I cannot prove that Paley was acquainted with these specific theories. But it is clear that he had some acquaintance with naturalistic, evolutionary hypotheses, however fanciful they may have been, that attempted to explain the appearance of adapted complexity without the existence of a supernatural designer.

Paley, as Gould has noted (1993, chap. 9), had sufficient courage of his convictions that he was prepared to seriously consider alternatives to his proposed scheme of intelligent design. Among these are evolutionary alternatives:

> There is another answer which has the same effect as the resolving of things into chance; which answer would persuade us to believe that the eye, the animal to which it belongs, every other animal, every plant, indeed every organized body which we can see, are only so many out of the possible varieties and combinations of being which the lapse of infinite ages has brought into existence; that the present world is the relic of that variety; millions of other bodily forms and other species having perished, being, by the defect of their constitution, incapable of preservation, or to continuance by generation. (Paley [1801] 1850, pp. 49–50)

In this passage we see a role for variation and for differential reproductive success. Darwin, who had studied Paley carefully, must have noticed this passage. But Paley did not see how to develop the ideas, and in the same discussion, the insights were lost.

Paley lost evolutionary insights for at least three reasons. First, he had no real appreciation for the extent of the extinction of earlier species, owing no doubt, to the fact that paleontology in his day was essentially an unborn fetus and the idea of extinction was as much an offense to God's plan as was the origination of new species:

> We may modify any one species many different ways, all consistent with life, and with the actions necessary to preservation. . . . And if we carry these modifications through the different species which are known to subsist, their number would be incalculable. No reason can be given why, if these deperdits ever existed, they have now disappeared. Yet if all possible existences have been tried, they must have formed part of the catalogue. ([1801] 1850, p. 50)

Second, he had no mechanism to drive the process he described. The third reason that Paley missed the evolutionary insight had to do with the state of systematics in his day, which, unlike modern, evolutionary approaches to systematics, had no historical component (because none was deemed necessary):

> The hypothesis teaches, that every possible variety of being hath, at one time or another, found its way into existence—*by what cause or in what manner is not said*—and that those which were badly formed perished; but how or why those which survived should be cast, as we see the plants and animals are cast, into regular classes, the hypothesis does not explain; or rather the hypothesis is inconsistent with this phenomenon. (p. 51; my italics)

For Paley, regularity in the form of the taxonomic order seen in nature—the division of organic beings into plants and animals, and subdivisions of each into genera and subspecies—was not a convenience imposed by systematists—"an arbitrary act of mind"(p. 51)—but reflected an underlying intentional order and plan.

Undergirding Paley's grand scheme of argument was his intellectual inheritance of the conception of nature-as-machine composed in part of organisms-as-machines. Paley, far from bucking the science of his day, was entirely consistent with it:

> What should we think of a man who, because we had never ourselves seen watches, telescopes, stocking-mills, steam-engines, etc., made, knew not how they were made, nor could prove by testimony when they were made, or by whom, would have us believe that these machines, instead of deriving their curious structures from the thought and design of their inventors and contrivers, in truth derive them from no other origin than this, namely, that a mass of metals and other materials having run, when melted, into all possible figures, and combined themselves into all possible forms. . . . These things which we see are what were left from the incident, as best worth preserving, and as such are become the remaining stock of a magazine which, at one time or other, has by this means contained every mechanism, useful and useless, convenient and inconvenient, into which such like materials could be thrown? (1850, p. 51)

But the possibility remains that organisms are not like machines at all, and if so, that the processes by which they originate and change are nothing like the fruits of intentional design and engineering processes. If organisms are not machines, it is no longer absurd to deny design. But that will involve a scientific revolution in the truest sense.

In Chapter 6, I turn to examine Darwin's theory of evolution and its implications for humans, nonhuman animals, and their relative places in nature. There it will be seen that Darwin, in getting away from the idea that organisms are deliberately designed machines, fitting their niches like cogs in nature's grand mechanism, saw a need for a radical reappraisal of what we are and how we stand to other organisms. Darwin's response to Paley was in fact a response to a whole way of thinking about organic nature that goes back to the origins of modern science itself. In a way, his work was far more revolutionary than that of Newton, for where Newton was a champion for a preexisting mechanical tradition, Darwin was the initiator of a radical new way of viewing organic nature.

Further Reading

In addition to the primary source materials cited in the text, readable accounts of the views of Hobbes and Descartes can be found in B. Russell, *A History of Western Philosophy* (New York: Simon and Schuster, 1972), or S. E. Stumpf, *Socrates to Sartre: A History of Philosophy* (New York: McGraw-Hill, 1982). Stumpf also contains a helpful discussion of the work of Montaigne.

A useful discussion of Hobbes's place in the empiricist tradition of thought can be found in R. S. Woolhouse, *The Empiricists* (Oxford: Oxford University Press, 1988), whereas a similar discussion of the place of Descartes in rationalist thinking can be found in J. Cottingham, *The Rationalists* (Oxford: Oxford University Press, 1988).

Though William Paley was not the first theorist to put forward the argument from design, his version has been spectacularly influential. Indeed, the writings of central figures in the contemporary creation "science" movement contain little more than footnotes to Paley—and often neither as literate nor insightful. Consequently, to understand the argument from design as it figures in the creation/evolution debates, one should consult Paley himself. For a good, readable discussion of the contrasting views of Paley and Darwin, I would recommend S. J. Gould's essay, "Darwin and Paley Meet the Invisible Hand," in *Eight Little Piggies: Reflections on Natural History* (New York: W. W. Norton, 1993).

4

Of Mice and Monkeys: The Moral Relevance of Animal Pain

Many nonhuman animals behave in ways that indicate that they feel pain and experience a wide range of bodily sensations. In this chapter, we will examine the ways in which moral theorists of the Enlightenment saw the place of animals with respect to pain, suffering, and cruelty. We will then consider these same matters in the light of contemporary moral philosophy. By understanding the roles played by considerations of pain, suffering, and cruelty in the context of moral philosophy, it will be possible to appreciate the importance of contemporary controversies concerning animal consciousness—controversies that will be examined later in this book. The issue of animal consciousness is inextricably intertwined with issues surrounding the moral status of nonhuman animals.

First, a word of warning. Philosophers tell us that we cannot infer an *ought* from an *is*—that we cannot logically infer what we ought to do, for example, with respect to the treatment of animals, by looking at the facts about animals uncovered by science. Nevertheless, what science tells us about humans and nonhuman animals can be *relevant* to a discussion about what we ought to do, even if it does not logically determine what we ought to do. And it is a fact that the picture of humans and nonhuman animals that emerges from the rise of modern science has had some interesting shaping influences upon the moral debate about the status of nonhuman animals.

We will begin by briefly examining some of the ways in which ideas about humans and nonhuman animals developed in the century after Descartes. We will then examine two very different approaches to moral philosophy that emerge from the Enlightenment. We will examine the moral philosophy of Immanuel Kant, a theorist who saw fundamental moral differences between humans and nonhuman animals derived from the observation that the former, unlike the latter, were rational and in virtue of this constituted a moral community from which the latter were in the nature of the case excluded.

Then we shall examine the work of Jeremy Bentham and the rise of utilitarian moral theory. Accompanying the rise of utilitarian moral theory in the early nineteenth century—a theory based on the moral relevance of pleasure and pain—is a growing concern for animal pain. Nonhuman animal minds might be different from human minds in many important respects, but insofar as they can experience pain, this seems to be a relevant cognitive commonality. If they are not mere machines whose yelps and shrieks have no more significance that the twitterings of a cuckoo clock, the issue of animal pain must certainly be addressed.

Partly as a result of Bentham's groundbreaking work, the nineteenth century saw the emergence of the first organized groups whose aims were to promote animal welfare. The influence of Bentham and Kant can be seen in the works of contemporary moral philosophers. It is in this context that we will discuss the emergence of *animal rights* and *animal welfare* movements in the late twentieth century.

Man and Beast after Descartes

We saw in the preceding chapter how for Descartes the human mind or soul was made of a special, created, immaterial, unextended, nonphysical substance. It was the possession of this mind that separated humans from nonhuman animals. We also saw that Descartes's official position not merely denied thoughts to animals; it even denied them the experience of sensations such as pain. Having said this, we saw that Descartes did seem, toward the end of his life, to come round to the idea that animals could have bodily sensations such as pain. Notwithstanding this admission, many vivisectionists took up the idea that their animal subjects could feel no pain—that their yelps and shrieks were simply mechanical noise.

One response to Descartes was the development of a fully materialistic theory of the human mind. Philosophers such as Julien La Mettrie (1709–1751), Claude-Adrien Helvétius (1715–1771), and Paul Holbach (1723–1789) attempted to reduce the Cartesian account of mind to an account of matter. This group of thinkers became known as the *Cartesian materialists* (Reé 1974, p. 103). According to these theorists, humans may have special mental abilities, but these do not imply the existence of nonphysical substances or souls, or indeed of there being a fundamental metaphysical divide between humans and nonhuman animals, even if their mental abilities should turn out to be very different.

Philosopher La Mettrie (1709–1751) saw little difference in general organization and composition between human brains and animal brains. He even thought that it might be possible to teach an orangutan to use language. (National Library of Medicine)

But perhaps not that different after all. La Mettrie, for example, engaged in comparative anatomical studies of the brain. He saw little difference in general organization and composition between human brains and animal brains—especially the brains of apes. This observation suggested to La Mettrie that there may be no absolute, qualitative difference between the mental lives of humans and nonhuman animals. La Mettrie even thought that it might be possible to teach an orangutan to use language (Macphail 1998, p. 45). This idea that humans and nonhuman primates (in this case) differ only by degrees is an idea we shall examine again in connection with Darwin's theory of evolution.

Be this as it may, the idea began to crystallize that humans were part of nature and could be understood in naturalistic terms devoid of appeals to supernatural entities and properties. And if the human animal was part of nature, it was part of a nature that could be understood in terms of Newton's physics, for it was the eighteenth century that saw the blossoming of Newton's mechanical program for physics. This program for physics, now known as *classical mechanics,* was fully deterministic in the sense that the complete physical state of the universe at any time determines, through the laws of physics, a complete state of the universe at any other time. The great eighteenth-century physicist Pierre Simon Laplace (1749–1827) described the idea as follows:

> We may regard the present state of the universe as the effect of its past and the cause of its future. An intellect which at any given moment knew all the forces that animate nature and the mutual positions of the beings that compose it, if this intellect were vast enough to submit its data to analysis, could condense into a single formula the movement of the greatest bodies of the universe and that of the lightest atom: for such an intellect nothing could be uncertain; and the future just like the past would be before its eyes. (Quoted in Kline 1953, p. 291)

In the century after Descartes, humans became viewed—at least in important scientific circles—as fully natural parts of nature subject to the same scientific laws that govern the behavior of other parts of nature. Capturing the idea that humans had lost the special place in nature they had claimed since the dawn of Western civilization, François-Marie Voltaire (1694–1778) observed:

> It would be very singular that all nature and all the stars should obey eternal laws and that there should be one little animal, five feet tall, which, despite these laws, could always act as suited its caprice. (Quoted in Reé 1974, p. 102)

But if the place of humans in nature was being rethought, so was the place of nonhuman animals, both from the standpoint of the comparative study of animal behavior and from the standpoint of comparative anatomy.

Thus Voltaire could echo Montaigne by comparing animal behavior and human behavior and suggesting that we

> [b]ring the same judgement to bear on this dog which has lost its master, which has sought him on every road with sorrowful cries, which enters the house agitated, uneasy, which goes down the stairs, up the stairs, from room to room, which at last finds in his study the master it loves, and which shows him its joy by its cries of delight, by its leaps, by its caresses. (Regan and Singer 1989, p. 21)

But Voltaire continued:

> Barbarians seize this dog, which in friendship surpasses man so prodigiously; they nail it on a table, and they dissect it alive in order to show the mesenteric veins. You discover in it all the same organs of feeling that are in yourself. Answer me, machinist, has nature arranged all the means of feeling in this animal, so that it may not feel? Has it nerves in order to be impassible? Do not suppose this impertinent contradiction in nature. (p. 21)

But even if we suppose that both we and the nonhuman animals we interact with are parts of nature, we are nevertheless different. In particular, we humans are still rational beings. How then should we use our rationality to conceptualize our duties and obligations to each other and (if at all) to nonhuman animals? To deal with this issue, we must go beyond the physics and physiology of the Enlightenment to examine the moral philosophies it prompted.

Kant: Rationality and Duty

Immanuel Kant (1724–1804) saw nonhuman animals as being inferior to humans. Most thinkers prior to Kant had seen some connection or other between *morality* and *rationality,* but arguably none had welded these two notions together as firmly as Kant did. According to Kant, rationality is not a mere tool for discerning what is moral and what is not. Rather, morality can only make sense for a rational, autonomous agent. Morality emerges from considerations of the nature of rationality itself. The concern of morality is primarily with rational beings and their relationships to, and interactions with, other rational beings. It concerns other things (nonhuman animals, viewed as nonrational beings, and inanimate objects) only insofar as these things have bearing on rational beings and their interactions. In modern language, the *moral community* is the community of rational beings, and nonrational creatures are not members of that community.

Although Immanuel Kant (1724–1804) believed nonhuman animals were inferior to humans, he did not believe that humans should mistreat animals or treat them arbitrarily. (National Library of Medicine)

Our theoretical reason provides us with basic principles—such as the causal principle that *all effects have causes*—in terms of which we can structure and organize our particular sensory experiences of the world around us. (Thus, gravity causes the apple to fall, the beating of the heart causes the blood to circulate, and so on.) In a similar way, our practical reason yields moral principles to structure and organize the behavior of rational beings. Just as the laws of physics are the same for all observers, so the moral law is the same for all rational beings and results in behavior that is universally good, as opposed to being good just for this particular rational being in this particular context. The moral law that forms the basis of particular actions is this: *Act as if the maxim of your action were to become a universal law of nature* (a principle to be followed by all beings in the moral community). Behaving so as to tell the truth will thus commend itself to rational beings, whereas behaving so as to steal from one's neighbor will not pass the test: The advantage you hope from your neighbor would be negated by his following the same maxim, thereby stealing from you.

Related to this is Kant's view that rational beings are *persons,* not mere *things.* Things are used as means to other ends, whereas persons are ends-in-themselves. Your sense of intrinsic moral worth, for example, is your sense of personhood, and this is the same for all rational beings. Telling the truth respects the recipient of the information as a person, as an end-in-herself. Lying to a rational being treats that person as a means to another end. The moral law thus requires that we act so as to respect the persons in the moral community. But where does this leave nonhuman animals?

Kant was consistent to his principles. Nonhuman animals, in virtue of being nonrational, are not persons and are therefore not directly subject to the moral law—they are not members of the moral community of rational beings. As Kant put it, "But so far as animals are concerned, we have no direct duties. Animals are not self-conscious and are there merely as a means to an end. That end is man . . ." (Regan and Singer 1989, p. 23). This does not mean we should be cruel to animals, or treat them arbitrarily:

> Our duties towards animals are merely indirect duties towards humanity. Animal nature has analogies to human nature, and by doing our duties to animals in respect of manifestations of human nature, we indirectly do our duty towards humanity. Thus, if a dog has served his master long and faithfully, his service, on the analogy of human service, deserves reward, and when the dog is grown too old to serve, his master ought to keep him until he dies. Such action helps support us in our duties towards human beings. . . . If then any acts of animals are analogous to human acts and spring from the same principles, we have duties towards the animals because thus we cultivate the corresponding duties towards human beings. (p. 23)

In a way, then, the sentimental observations concerning nonhuman animals made by Montaigne, and echoed by Voltaire, are not morally irrelevant in the hands of Kant. They provide the analogical base on which to develop an account of *humane* behavior.

But in the hands of Kant, nonhuman animals lack intrinsic moral worth, and this implies that there are important limits to the way in which we should treat them. For as Kant himself observed, although tender feelings toward dumb animals develop humane feelings toward humankind,

> [v]ivisectionists, who use living animals for their experiments, certainly act cruelly, although their aim is praiseworthy, and they can justify their cruelty, since animals must be regarded as man's instruments; but any such cruelty for sport cannot be justified. A master who turns out his ass or his dog because the animal can no longer earn its keep manifests a small mind. (Regan and Singer 1989, p. 24)

If animal experimentation is in the service of the health and well-being of rational beings, as it may be in biomedical research, it can be justified. For in the end, nonhuman animals are not persons. There is no reason to believe that Kant held nonhuman animals to be unable to feel pain. In this sense he is no heir to Descartes or the Cartesian tradition in physiological research. Quite the reverse. That nonhuman animals behave the way we do when we are in pain was the basis for his indirect account of our limited obligations to them. But such pain, when inflicted in the service of science, for example, was not morally significant. Here the *end* can literally justify the *means.*

Bentham and the Moral Utility of Pain

Jeremy Bentham (1748–1832) developed a view of morality very different from that of Kant. For Kant, actions were not right or wrong because of their consequences—for example, that they promoted happiness. Rather, they were right or wrong according to whether they were done from the right motive dictated by practical reason accessible to all rational beings or were inconsistent with such dictates. For Bentham, by contrast, the consequences of action were all important. Many of his important ideas were laid out in *The Principles of Morals and Legislation* (1789).

But what sort of consequences are morally relevant? Bentham looked at the human condition:

> Nature has placed mankind under the governance of two sovereign masters, *pain* and *pleasure.* It is for them alone to point out what we ought to do, as well as to determine what we shall do. On the one hand the standard of right and wrong, on the other the chain of causes and effects, are fastened to their throne. (Lafleur 1948, p.1)

From this observation he arrived at the *principle of utility:*

> By the principle of utility is meant the principle which approves or disapproves every action whatsoever, according to the tendency which it appears to have to augment or diminish the happiness of the party whose interest is in question. . . . We say of every action whatsoever; and therefore not only of every action of a private individual, but of every measure of government. (p. 2)

Bentham continued:

> By utility is meant that property in any object, whereby it tends to produce benefit, advantage, pleasure, good, or happiness . . . or (what comes again to the same thing) to prevent the happening of mischief, pain, evil, or unhappiness to the party whose

> interest is considered: if that party be the community in general, then the happiness of the community: if a particular individual, then the happiness of that individual. (p. 2)

Bentham thus emerged as a champion of the moral relevance of pleasure and pain, as they arose in the *effects* of actions. For Bentham, in either public or private life, the value of action lay in the pleasures and pains produced. The resulting moral theory is known as *utilitarianism.*

Each individual is thus viewed as being concerned with the avoidance of pain and the achievement of pleasure. Bentham thought of pleasure and pain as quantifiable. For a given individual, pleasures and pains, by themselves, would be quantified by *intensity, duration,* degree of *certainty,* and *propinquity* or *remoteness.* When considering the consequences of actions, with respect to pleasure and pain, to this list of factors are added *fecundity* (the likelihood that an action producing a sensation is followed by sensations of the same kind) and *purity* (the likelihood that an action producing one kind of sensation has of *not* being followed by sensations of the opposite kind). When considering a number of persons, we also have to consider the *extent*—that is, the number of persons affected by an action (Lafleur 1948, pp. 29–30).

For an individual, Bentham suggested that we adopt the following procedure when considering the moral worth of an action:

> Sum up all the values of all the *pleasures* on the one side, and those of all the pains on the other. The balance, if it be on the side of pleasure, will give the *good* tendency of the act upon the whole, with respect to the interests of that individual person; if on the side of pain, the *bad* tendency of it upon the whole. (Lafleur 1948, p. 31)

For the community as a whole:

> Take an account of the *number* of persons whose interests appear to be concerned; and repeat the above process with respect to each. *Sum up* the numbers expressive of the degrees of *good* tendency, which the act has, with respect to each individual, in regard to whom the tendency of it is *good* upon the whole: do this again with respect to each individual, in regard to whom the tendency of it is *bad* upon the whole. Take the *balance;* which, if on the side of *pleasure,* will give the general *good* tendency of the act, with respect to the total number or community of individuals concerned; if on the side of pain, the general *evil tendency,* with respect to the same community. (p. 31)

These comments bring out at least two important points. First, moral consideration must be given to the pleasures and pains experienced by individuals. Second, pleasures and pains need to be assessed with respect to relevant communities of individuals.

Membership in a relevant community depends on whether one is likely to be an interested party—and this means liable to suffer painful or pleasurable consequences of a given action. It does not matter whether one is rational or can use language. The ability to experience painful and pleasurable sensations is enough. Bentham's moral theory was thus based on a much less ambitious estimate of cognitive sophistication needed for membership in a moral community than was Kant's.

Also relevant for present purposes are Bentham's views on punishment, since these views have more general applicability to cases where we deprive individuals of pleasure and deliberately inflict pain. Concerning the object of law, Bentham wrote:

> The general object which all laws have, or ought to have, in common, is to augment the total happiness of the community; and therefore, in the first place to exclude, as far as may be, everything that tends to subtract from that happiness: in other words to exclude mischief. (Lafleur 1948, p. 170)

But what of punishment itself? Bentham added:

> But all punishment is mischief: all punishment in itself is evil. Upon the principle of utility, if it ought at all to be admitted, it ought to be admitted in as far as it promises to exclude some greater evil. (p. 170)

So the community can be justified in inflicting pain on some of its members, if by these means greater evils can be prevented.

All of these issues bear on the moral status of animals. First, because animals feel pleasure and pain. This much was obvious by the end of the eighteenth century to all but the most obstinate Cartesian physiologists. Second, in agriculture, industry, and science, not to mention private households, animals were used in contexts that caused pain and suffering. Much cruelty was inflicted on animals and humans alike—slavery was not abolished in England until 1807 (1865 in the United States), and children under ten years of age could still be employed down English mines until the early 1840s. Though gradually being replaced by steam-powered technologies, and finally by the internal combustion engine, animal power was as important in Bentham's day as it had been in the medieval period. Membership in a moral community, moreover, was for Bentham determined by the capacity to experience pleasure and pain, not

Jeremy Bentham's (1748–1832) moral insight was relevant to the issue of the treatment of nonhuman animals. Their capacity to suffer, he believed, made them members of the moral community. (Hulton / Archive)

whether or not one was a member of the Kantian community of rational beings. Where did this leave animals?

Bentham thought animals were susceptible of happiness, but had nevertheless had their interests neglected because jurists had categorized them as *things.* Nevertheless, the principle of utility was to apply here as it did elsewhere, once, as Bentham said, allowance had been made for the "difference in point of sensibility" (Lafleur 1948, p. 311). Bentham was no vegetarian activist:

> If the being eaten were all, there is very good reason why we should be suffered to eat such of them as we like to eat: we are the better for it, and they are never the worse. They have none of those long-protracted anticipations of future misery which we have. The death they suffer in our hands commonly is, and always may be, a speedier, and by that means a less painful one, than that which would await them in the inevitable course of nature. If the being killed were all, there is very good reason why we should be suffered to kill such as molest us: we should be the worse for their living, and they are never the worse for being dead. (p. 311)

So Bentham thought that cognitive abilities were important when trying to understand how to apply the principle of utility to nonhuman species—unlike us, perhaps, they do not suffer the distress of anticipation of future pain. Our pleasures and pains must also be factored into the equation.

But what of wanton cruelty? What of the deliberate infliction of extreme pain and suffering? Here Bentham was ahead of his times:

> The day *may* come, when the rest of the animal creation may acquire those rights which never could have been withholden from them but by the hand of tyranny. The French have already discovered that the blackness of the skin is no reason why a human being should be abandoned without redress to the caprice of a tormentor. It may come one day to be recognized, that the number of legs, the villosity of the skin, or the termination of the *os sacrum,* are reasons equally insufficient for abandoning a sensitive being to the same fate. (Lafleur 1948 p. 311)

Importantly, Bentham concluded:

> What else is it that should trace the insuperable line? Is it the faculty of reason, or, perhaps, the faculty of discourse? But a full-grown horse or dog is beyond comparison a more rational, as well as a more conversable animal, than an infant of a day, or a week, or even a month, old. But suppose the case were otherwise, what would it avail? The question is not, Can they *reason?* Nor, Can they *talk?* But, Can they *suffer?* (p. 311)

We thus see in the work of Bentham that it is irrelevant whether animals can behave morally toward each other or toward us. The real issue is how should *we* behave toward creatures that can feel pain and that can suffer by our actions.

There are humans who are neither rational, nor who can behave morally, but who must nevertheless be treated with care by us, simply because they can suffer. Why not nonhuman animals too? As Bentham pointed out, if the *racist* cannot make a case for different treatment simply by pointing to the color of the skin, so the *speciesist* (one who discriminates on the basis of species membership) ought not to be able to make a case simply on the basis of membership of a nonhuman species. There is an enormous difference, then, between moral theories such as Kant's that single out species-specific traits, such as rationality or language use, to determine membership in the moral community and those such as Bentham's that are more inclusive by appealing to traits common to a great many animal species.

Who counts and who does not? Put differently, how is the moral community to be structured? There are two aspects of Bentham's thought that are worthy of mention in this regard. First, Bentham had a belief in equality: One person's happiness should count for no more than another person's happiness (Plamenatz, 1963b, p. 25). Second, when Bentham came to examine different forms of government, he noticed a curious discrepancy. The actual end of every government is the greatest happiness of the governors, whereas its proper end should be the *greatest happiness of the greatest number.* Bentham felt that representative democracy (the only kind possible in all but the simplest societies) was likely to make the actual end and the proper end of government coincide.

Bentham did not extend these ideas to nonhuman animals—and there is no reason to think that he thought that nonhuman species should somehow participate in democratic government! But there are two insights that are nevertheless important. First, no *individual* who suffers pain should count for more than any other such *individual.*

Perhaps the notion of *individual* should be taken to mean *individual regardless of species.* Many of us might balk at this kind of equality, but most people seem to think that nonhuman animals count for something, even if not as much as individual humans.

Second, Bentham noticed that individuals with power over the administration of policies leading to pleasure and pain often act so as to maximize their own pleasure and to minimize their own pain, often at the expense of other members of the community who may nevertheless experience pleasure and pain—and experience more of the latter and less of the former, as the result of these actions by the governors. Although Bentham sought forms of government in which the interests of the governors and the governed coincided, his moral insight is of relevance to the issue of the treatment of nonhuman animals. Their capacity to suffer makes them members of the moral community. Even though there is no form of government that can include them (as democracy can be extended from the limited franchise of late-eighteenth-century England to include ordinary working people, women, persons of color, and so on), it is a consequence of Bentham's moral theory that ethical legislation will not exclude nonhuman animals from equitable consideration of their pleasures and pains. Even today there are humans who have no political *voice*—children, the insane, and felons. They are not for this reason outside the moral community, even if they must be restrained for their safety and ours. Perhaps animals, who literally have no *voice,* also belong to the moral community.

Animal Protection Movements in the Nineteenth Century

It would not be correct to say that Bentham's views single-handedly led to a radical reappraisal of the place of nonhuman animals, especially with respect to the issue of pain and suffering. But the emergence of organized antivivisection movements occurred first in England, while Bentham was alive, and these movements reflected a growing awareness of the moral importance of pain and suffering. But as Rowan et al. noted:

> British utilitarian philosophy laid the groundwork for Victorian concerns about suffering and promoted admiration for those who showed they were "men of feeling." The women's movement was also politically important at this time and may have increased attention to questions of caring for the exploited, be they black, slaves, children or animals. (1995, p. 199)

Rowan also pointed to the role played by Protestant religions, such as Methodism, which held that animals had souls and would share in immortality (1984, pp. 48–49; see also Rowan et al. 1995, p. 199). By contrast, Roman Catholicism, steeped in the metaphysics and ethics of the medieval period, held that we had no particular moral obligations to nonhuman animals.

The first manifestation of an organization dedicated to the protection of animals was the Royal Society for the Prevention of Cruelty to Animals (RSPCA), which was founded in London, England, in 1824. The first piece of legislation to be passed in England regulating our interactions with nonhuman animals was the Cruelty to Animals Act of 1876. The act of 1876 regulated, but did not abolish, painful animal research practices (Rowan 1984, p. 49). Early legislation aimed at preventing the abuse of animals tended to concentrate on the treatment of animals used in spectacles and entertainment. (By contrast, the National Society for the Prevention of Cruelty to Children [NSPCC], also established in London, was not founded until 1889—the same year that saw the passage of the Prevention of Cruelty to Children Act—timing that reflected, perhaps, the deep and abiding concern for nonhuman animals for which the English are famous.)

In the United States, the American Society for the Prevention of Cruelty to Animals was founded in 1866, and the Massachusetts Society for the Prevention of Cruelty to Animals was founded in 1868. Rowan et al., noted:

> Their lead was followed by others and numerous humane groups were founded in the next twenty years. The American Humane Association was established in 1877 as a national organization representing the interests of the local societies at the national level. By 1900, animal protection and anti-vivisection enjoyed support from prominent leaders in America and were driven by similar concerns to those identified for Britain. (1995, p. 200)

So we see a growth in concern for the protection of animals on both sides of the Atlantic—a concern motivated by a recognition of animal suffering and animal pain.

But what role, if any, did the development of science play in these debates about the moral status and welfare of nonhuman animals? We will end this section by observing that the development of the biological and biomedical sciences in the nineteenth century contributed contradictory messages to the moral debates concerning the status of nonhuman animals. Although these contradictory messages will be the subject of the next two chapters, some comments are nevertheless in order here.

On the one hand, consider biomedical research. By the end of the nineteenth century, it was clear that biomedical research involving nonhuman animals had yielded some significant human benefits. Because of this, it is worth noting here that utilitarian moral philosophy is a sword that can be used to cut two ways. If there are incredible benefits to society from animal experimentation—and biomedical researchers claim just this—then perhaps these enormous benefits outweigh the pain and suffering of animal subjects. And this is especially true if one is inclined to believe that animal pain, though morally relevant, is not as relevant as human pain and that of nonhuman animals. though they have *some* moral standing, do not have the same standing as humans.

On the other hand, the publication of Darwin's *Origin of Species* in 1859 and *The Descent of Man* in 1871 served to erode the idea that humans had a special place in creation. Darwin argued that humans descended with modification from nonhuman ancestors. Humans and apes could trace evolutionary descent from common ancestors. We are phylogenetic kin, similar in some ways and different in others. Darwin, in particular, saw all manner of cognitive similarities between human and nonhuman animals. Where the vivisectionists wanted to emphasize cognitive differences between humans and nonhuman animals, Darwin emphasized similarities. Because of these perceived cognitive similarities, the very traits that make us find experimentation on humans morally repugnant should, even if found in lesser degree in nonhuman animals, make us uneasy about experimentation on them, too.

Thus nineteenth-century science contributed to later controversies surrounding the morality of animal experimentation and the moral status of nonhuman animals in at least two contradictory ways. Darwin's theory of evolution served to seriously erode claims that humans and nonhuman animals were radically different, especially in terms of cognitive abilities. Biomedical researchers, on the other hand, were finding animal experiments to be increasingly useful—in part because humans and nonhuman animals were perceived not to be radically different. The cognitive similarities between humans and nonhuman animals raises one set of moral questions, whereas the value of animal experimentation to humans raises another.

These contradictory messages from nineteenth-century science have been amplified in the course of the development of twentieth-century science, and they shape the contours of contemporary moral discussions. So how do the moral issues stand at the end of the twentieth century? It will be seen below that the contemporary participants in controversies concerning the moral status of animals are heirs to the scientific and philosophical legacies of the eighteenth and nineteenth centuries.

Some Contemporary Developments

In this section we will briefly examine the contemporary situation with respect to the moral status of nonhuman animals. We will focus on two basic issues. First, we will briefly look at the phenomenon of animal rights movements. Second, we will examine the contemporary philosophical and theoretical debate about the moral status of nonhuman animals.

My treatment of animal rights movements is necessarily brief, since they are not the focus of the arguments in this book. The contemporary phenomenon of animal rights movements is nevertheless worthy of scrutiny, however, because it reflects a growing social concern for the ways in which humans interact with nonhuman animals. These movements coevolved in the 1960s, 1970s, and 1980s with groups concerned about the environment (for example, Greenpeace), as well as vegetarian and feminist movements and other groups concerned with the promotion of civil rights (for example, groups promoting gay and lesbian rights). In this sense, the contemporary animal rights movements are part of a broader social phenomenon that calls into question the moral, social, and political status quo that exists in Western democracies.

The contemporary movements have been characterized by a willingness to engage in public protest and other forms of activism. The radical components of the animal rights movements have been accused of engaging in terrorist tactics when they raid animal research laboratories or free farm animals. Members of the same radical groups consider themselves to be freedom fighters acting to halt the oppression of other species.

Although the social phenomenon of animal rights movements is richly complex, it is nevertheless possible to produce a rough chronology of key events in its development (Finsen and Finsen 1998).

1963 Formation of the *British Hunt Saboteurs Association*. (Aim of the association was to disrupt fox hunting and confront hunters.)

1970 Richard Ryder coins the term *speciesism*. (Speciesism involves discrimination against a [sentient] organism in virtue of its [nonhuman] species membership, by analogy with sexism and racism.)

1972 Animal Liberation Front (ALF) begins operations in Britain. (ALF is an activist organization that has been involved in raids on laboratories to free nonhuman research subjects.)

1975 Philosopher Peter Singer publishes *Animal Liberation.* (In the decades since the 1970s, Singer and fellow philosopher Tom Regan publish articles and books that make the issue of animal rights, and more generally concerns for the moral status of nonhuman animals, of serious professional interest to academics. Their work has also been the source of much controversy.)

1977 ALF makes its appearance in the United States.

1980–1990 This decade sees the formation of national organizations such as People for the Ethical Treatment of Animals (PETA), and the Farm Animal Reform Movement (FARM).

1990 Seventy-five thousand people turn out for March of the Animals in Washington, D.C.

What sorts of impacts have the animal rights movements had? It is certainly true that government regulations governing the use and treatment of nonhuman animals, especially in scientific research, have become more stringent in recent decades. Laboratory animals today are generally treated better than they were three decades ago. In the eyes of activists, however, this may not be saying much. Animal testing in product development is down, and cosmetics, for example, that have not been tested on animals are generally more common than they used to be. In terms of public attitudes, reactions have been mixed. Thus Herzog has observed:

> Several surveys have reported that a majority of Americans have generally positive attitudes toward the animal rights movement. For example, a 1994 opinion poll reported that most respondents had either a very favorable (23%) or a mostly favorable (42%) view of the animal rights movement. On the other hand, only 7% of a 1990 survey said that they agreed with both the agenda of the animal rights movement and its strategies. Eighty-nine percent of the respondents felt that activists were well meaning, but either disagreed with the movement's positions on issues or on strategies for accomplishing specific goals. (1998, p. 54)

But what of the theoretical positions underpinning the animal rights movement? How have these developed in contemporary debates?

Members of the Farm Animal Welfare Society protesting against live animal exports outside the House of Commons, London, 1977. (Hulton/ Archive)

Contemporary philosophical debates about the moral status of nonhuman animals have been a source of much controversy. It is possible here only to isolate the central threads in these debates. Interested readers should consult the primary sources cited here for further elaboration of the basic positions. We will briefly characterize three identifiable positions concerning the moral status of nonhuman animals in contemporary thought: the *welfarists,* the *liberationsists,* and the *animal rightists.* Crucial for what follows in this book are the estimates of the cognitive capacities of members of nonhuman species that undergird patterns of moral reasoning in these complex debates.

Virtually everyone engaged in the moral debate concedes that animals feel pain. (There are exceptions, for example, Carruthers 1992; and there are some serious scientific issues here, as will be seen in Chapter 10). In research, for example, procedures are sometimes categorized in terms of how much pain they cause. The business of classifying the extent of pain and harm in experiments on nonhuman animals is of fairly recent origin. As Barbara Orlans (1998, p. 267) has recently pointed out, Sweden in 1979 was the first country to adopt a national policy on this matter. Since then Canada, the United Kingdom, Australia, Holland, Germany, Finland, and Switzerland, among others, have also introduced national policies. The United States is conspicuously absent from this list.

In Holland, for example, experimental procedures are classified on the basis of whether they have a minor, moderate, or severe effect on the animal subjects (Orlans, 1998, pp. 268–269). Procedures having only a minor effect might involve minor blood sampling, taking X-rays, or terminal experiments under anesthesia. Procedures with a moderate effect might include frequent blood sampling, immobilization or restraint, or the implantation of indwelling cannulae or catheters. Procedures involving severe effect on animals, by contrast, might include total bleeding without anesthesia; induction of genetic defects; deprivation of food, water, and sleep; tumor induction; LD50 tests; and the induction of pain in the context of *pain research.*

The welfarist position concedes that nonhuman animals (certainly mammals, birds, and arguably other vertebrates) have nonnegligible moral worth—the fact that they suffer pain means they should not be treated capriciously. Nevertheless, they do not have rights. Cognitively superior humans should promote animal welfare where possible, but this does not preclude using nonhuman animals in research. Promoting the welfare of nonhuman animals might involve the imposition of procedures governing animal experimentation, the alleviation of pain and distress during and after experiments, and broader rules concerning what is known as *animal care.*

Carl Cohen, having argued that animals do not have rights, represented a version of the welfarist position when he wrote:

> It does not follow . . . that we are morally free to do anything we please to animals. Certainly not. In our dealings with animals, as in our dealings with other human beings, we have obligations that do not arise from claims against us based on rights. . . . In our dealings with animals, few will deny that we are at least obliged to act humanely—that is to treat them with the decency and concern that we owe, as sensitive human beings, to other sentient creatures. (1986, p. 866)

There are echoes here of both Aquinas and Kant. Nevertheless, Cohen is an ardent defender of animal experimentation. The benefits for human health and well-being are so great that painful experiments, provided they are conducted humanely and with due concern for the welfare of the animal subjects, are morally permissible. In welfarist arguments, nonhuman animals are typically excluded from the moral community of equals—a community consisting of those who may claim rights and have duties and obligations to each other (see Fox 1986, p. 49).

The liberationist school of thought traces its roots back to the utilitarian philosophy of Jeremy Bentham. Perhaps the most notable proponent of the liberationist position is Peter Singer, who developed these themes in *Animal Liberation.* For Singer, membership in the moral community of equals is determined by an organism's capacity to feel pain and suffer:

> If a being suffers there can be no moral justification for refusing to take that suffering into consideration. No matter what the nature of the being, the principle of equality requires that its suffering be counted equally with like suffering—insofar as rough comparisons can be made—of any other being. If a being is not capable of suffering, or of experiencing enjoyment or happiness, there is nothing to be taken into account. So the limit of sentience . . . is the only defensible boundary of concern for the interests of others. To mark this boundary by some other characteristic like intelligence or rationality would be to mark it in an arbitrary manner. Why not choose some other characteristic, like skin color? (1990, pp. 8–9)

Thus, for Singer's argument to work, an organism in the community of moral equals must minimally have feeling-consciousness, even if it does not have the ability to use language or be counted as a rational agent.

And it is just here that the trouble arises. For defenders of animal experimentation have called into question just this very point: that animals suffer. Thus Carruthers observed:

> The fact that a creature has sense organs, and can be observed to display in its behavior sensitivity to the salient features of the environment, is insufficient to establish that it feels like anything to be that thing. It may be that the experiences of animals are wholly of the non-conscious variety. . . . If consciousness is like the turning on of a light, then it may be that their lives are nothing but darkness. (1992, p. 171)

Later in this book, we shall examine attempts to make this possibility —that nonhuman animals lack consciousness and thus cannot suffer— scientifically respectable. For this is one of the main controversies generated by the study of animals in the light of modern science.

It has been known for a long time that there is a basic problem with utilitarian arguments: That if harming an individual would nevertheless promote the greatest happiness of the greatest number, then it will be morally permissible to harm that individual on utilitarian grounds, even when the individual is a member of one's own species. If the individual in question is a member of a cognitively inferior (possibly even a cognitively deficient) species, the enormous benefits to humans may easily be seen to outweigh the pains suffered by these organisms. The animal rightist position is an attempt to evade this kind of reasoning. It was first worked out by Tom Regan (1987).

Earlier in this chapter it was seen that Kant thought that humans—as members of the moral community—must be thought of as ends in themselves and never as means to further ends. Thus for Kant, harming an individual that others may benefit is always wrong, no matter how great the benefits to other individuals. Kant excluded nonrational, nonhuman animals from the moral community of beings that count. But as Regan has recently noted, the animal rightist challenges this exclusion:

> The rights view takes Kant's position a step further than Kant himself. The rights view maintains that those animals raised to be eaten and used in laboratories, for example, should be treated as ends in themselves, never merely as means. Indeed, like humans, these animals have a basic moral right to be treated with respect, something we fail to do whenever we use our superior physical strength and general know-how to inflict harms on them in pursuit of benefits for humans. (1998, pp. 42–43)

But what, then, earns nonhuman animals membership in the moral community of beings that count?

Regan placed weight not on rationality and language use but on feeling-consciousness and self-consciousness. "These animals not only see and hear, not only feel pain and pleasure, they are also able to remember the past, anticipate the future, and act intentionally in order to secure what they want in the present. They have a biography, not merely a biology" (1998, p. 43). In later chapters, it will be seen that a large part of the contemporary debate about how to explain animal behavior hinges crucially on what can be inferred about the richness of their mental lives from their observable behavior.

Important participants in this debate deny, on the basis of an examination and analysis of behavioral evidence, that nonhuman animals have either feeling-consciousness or self-consciousness. It should be obvious from this chapter that these debates about what can and cannot be inferred from animal behavior are not merely of academic interest. The very claim that nonhuman animals have a moral status, be it from a welfarist position, a liberationist position, or an animal rightist position, is just what is at issue.

Further Reading

In addition to the primary sources cited in the text, readers wishing to know more about Kant's philosophical views might usefully consult J. G. Murphy, *Kant: The*

Philosophy of Right (London: Macmillan, 1970), or W. H. Werkmeister, *Kant: The Architectonic and Development of his Philosophy* (La Salle, IL: Open Court, 1980). A good account of Bentham's thought can be found in J. Plamenatz, *Man and Society,* volume 1 (London: Longman, 1963).

The classic texts of the contemporary animal rights movement are P. Singer, *Animal Liberation* (New York: Avon Books, 1990), and T. Regan, *The Case for Animal Rights* (Berkeley: The University of California Press, 1987). Many useful essays on issues raised by the animal rights controversies can be found in M. Bekoff, ed., *Encyclopedia of Animal Rights and Animal Welfare* (Westport, CT: Greenwood Press, 1998). Rowan et al., *The Animal Research Controversy: Protest, Process and Public Policy* (Medford, MA: Center for Animals and Public Policy, Tufts University School of Veterinary Medicine, 1995), also provide much valuable background material and historical information.

5

The Job of Physiology: Animals in Nineteenth-Century Medicine

There were three revolutions in nineteenth-century biology: the theory of evolution, genetics, and cell biology. Even though all these branches of biological inquiry were of immense scientific importance, Charles Darwin's theory of evolution was both scientifically and socially influential right from the moment it was published. Darwin was not the first biologist to speculate about the possibility of biological evolution, but his theory, published as *The Origin of Species* in 1859, was the first approach to biological evolution to be plausible from the standpoint of a driving mechanism. Darwin's theory of evolution will be the subject of the next chapter. Genetics, owing its origins to the work of Gregor Mendel (1822–1884), had to wait until the dawn of the twentieth century to have any significant biological impact and until the 1930s to begin to become deeply rooted in evolutionary biology. The impact of genetics on evolutionary biology will be discussed in more detail in Chapter 7.

Cell biology, by contrast, enabled investigators to follow the resolutio-compositive method one step further than it had gone in the Renaissance by analyzing tissues into their component cells and cell types. Cell biology really had to wait until the nineteenth century, for though Hooke had coined the term *cell* in his *Micrographia* (1665), it was the refinement of optical technologies in the nineteenth century that permitted the construction of microscopes with sufficient resolving power to discern the different cell types, and indeed intracellular structures, found in plants and animals. Thus the advent of cell biology permitted a continuation of biological inquiries into the nature of life as part of that older physiological tradition extending back to the Renaissance.

In this chapter we will be concerned to examine the role that human and nonhuman animals played in these nineteenth-century investigations into basic physiological processes—and why it was that experiments on animals were increasingly seen to be of great value to the promotion of human health and well-being. But before doing this, it will be helpful to pause and say something about the competing views of the nature of life that existed in the nineteenth century and that shaped the contours of scientific inquiry.

Vitalism and the Meaning of Life

From the seventeenth century to the end of the nineteenth century, there were at least two schools of thought about the nature of life: the *mechanists* (sometimes referred to as *physicalists* or *materialists*), who believed that living processes were to be explained in mechanical terms and ultimately reduced to physics and its account of matter in motion; and the *vitalists,* who believed that organisms had special biological features that could not be reduced to physics (or, indeed, chemistry). Both approaches drew support from scientific observations.

For the vitalists, there were essential differences between living things and nonliving things. Although the vitalists were a heterogeneous group—vitalism is more a cluster of related ideas and theories than a coherent and consistent body of doctrine—they can be thought of as a reactionary movement in science. They rejected the idea that organisms were *nothing but* machines, and they rejected the idea that life could be understood simply as *matter in motion*.

Some vitalists held that living organisms had within them a special *vital substance,* and it was possession of this vital substance that explained the difference between living things and nonliving things—by analogy with the way in which the Cartesians had postulated the possession of *thinking substance* (*res cogitans*) as a fundamental metaphysical characteristic of thinking beings, thereby differentiating them from nonthinking beings. Others believed in a *life force*—unseen, like gravity, yet animating living beings. Some focused on the holistic nature of organisms—watches can be taken to pieces and reassembled so as to be fully functional; not so organisms, which remain fully nonfunctional after disassembly (notwithstanding the literary speculation of Mary Shelley in *Frankenstein;* or the serious scientific work of Luigi Galvani, who reported in 1792 the successful animation of an amputated frog's leg with the aid of electrodes [Clendening 1960, pp. 593–594]).

Still others were impressed by the adaptive properties of organisms as well as the apparent *directedness* of biological processes observed in studies of organismal reproduction and development. And there was also a need to explain intelligence and memory—the phenomena are real, but scientists were increasingly wary of the Cartesian appeal to the existence of a nonphysical soul directing the human machine (Mayr 1997, pp. 8–12).

How did vitalism manage to flourish for so long? After all, it had prominent scientific representatives in all the major European nations in the eighteenth and nineteenth centuries. Mayr has made the following suggestions:

> Many of the arguments put forth by the vitalists were intended to explain specific characteristics of organisms which today are explained by the genetic program. They advanced a number of perfectly valid refutations of the machine theory, but, owing to the backward state of biological explanation available at the time, were unable to come up with the correct explanation of vital processes. . . . These vitalists not only knew that there was something missing in the mechanistic explanations but they also described in detail the nature of the phenomenon and processes the mechanists were unable to explain. (1997, p. 12)

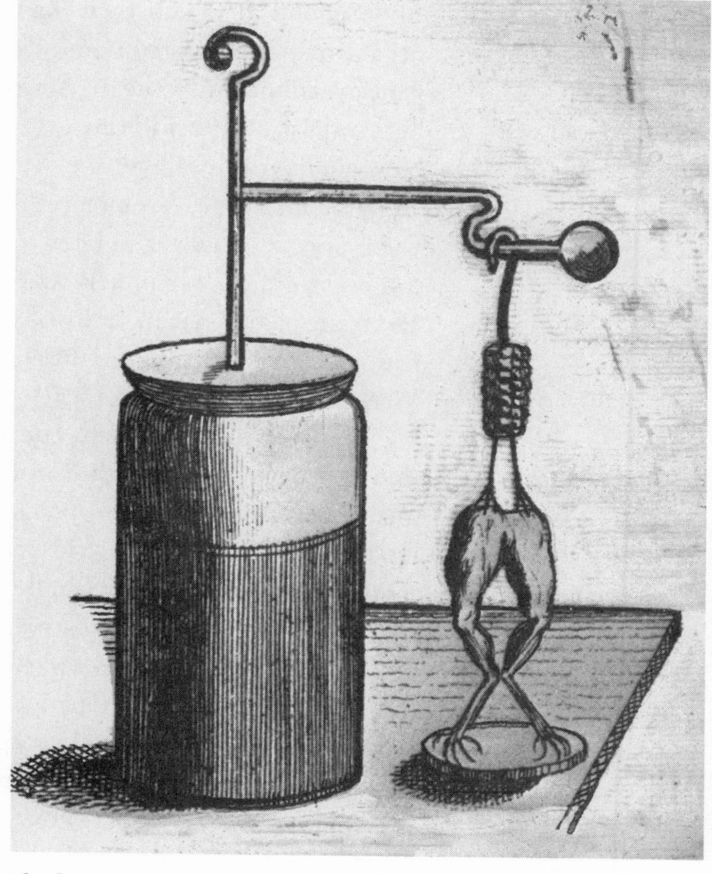

Galvani's discovery that an amputated frog's leg could be made to twitch when electrically stimulated suggested that there was more to the nervous system than a series of hydraulic tubes. (Bettmann / Corbis)

Vitalism survived because it was based on important, puzzling observations—observations seemingly beyond the reach of the mechanistic paradigm that had emerged in the Renaissance.

Yet vitalism, which survived into the early part of the twentieth century, was dealt a number of serious blows. First, the rise of modern chemistry in the eighteenth century showed that air, which had been viewed as an element in the seventeenth century, had parts, some of which supported life—then known as *respirable air* (now known as oxygen); some of which did not—then known as *mephitic air* (now known as nitrogen). The gradual emergence of the chemistry of oxidation was based in part on observed parallels between processes involving the oxidation of metals and chemical differences between inhaled and exhaled air—for example, in the work of Antoine Lavoisier (1743–1794) (Clendening 1960, pp. 590–593).

Second, the vitalists' belief that the chemical substances found in organisms—organic compounds—could only be made in living organisms animated by the mysterious *life force* and could not, therefore, be synthesized outside the organism, in the laboratory, was dealt a blow—albeit not a fatal one—by the laboratory synthesis of urea in 1828 by Friedrich Woehler (1800–1882). This achievement led first to organic chemistry and later to biochemistry. These studies opened the way for a proper understanding of the chemical basis of life and a rejection of the speculations of the vitalists—but, as noted above, it was a long battle, and one that ultimately required the rediscovery of Mendelian genetics after 1900.

Further blows were to come from the rise of thermodynamical thinking in the nineteenth century, which showed how to dispel the puzzles over observations of *animal heat*. Lavoisier had shown that the *ratio* of heat to carbon dioxide given out by animals was roughly the same as that produced by candle flames. But Lavoisier, though he did much to unravel the chemistry of oxidation, had no good theory of heat. He thought the phenomenon was brought about by the release of a special substance he termed *caloric*.

In the decades that followed Lavoisier's death, the idea gradually emerged that the heat energy observed in animals might result from the conversion of chemical energy in their food and not from the action of mysterious substances or forces. Woehler's friend Justus Liebig (1803–1873) advanced the idea that the mechanical energy of animals, as well as their heat, might result from the same process of energy conversion. Ultimately this research would culminate in the *kinetic theory of heat*—the idea that heat is to be explained as matter in motion (Mason 1970, ch. 39).

From these remarks, it will be evident that the scientific response to vitalism involved insights from the emerging chemical sciences as well as from physics. The mechanistic research program in the nineteenth century came to embody the idea that living organisms could be explained in terms of physics *and* chemistry, and not just physics. The chemistry of life, moreover, was increasingly seen as something that took place at the cellular level. Thus, in1881, some fifty years after the dawn of cell biology, Carl von Voit observed:

> The unknown causes of metabolism are found in the cells of the organism. The mass of these cells and their power to decompose materials determine the metabolism. . . . The metabolism continues in the cells until their power to metabolize is exhausted. All kinds of influences may act upon the cells to modify their ability to metabolize, some increasing it or others decreasing it.

> To the former category belong muscular work, cold of the environment (in warm-blooded animals), abundant food, and warming of the cells. To the latter, cooling the cells, certain poisons, etc. (Clendening 1960, pp. 598–599)

The gradual emergence and success of biochemistry, then, focused the attention of investigators on chemical processes at the cellular level, and away from mysterious substances and forces.

Another reason for the demise of vitalism was the enormous success—and promise of further success—of the expansion of the mechanistic program into the field of practical medicine. As evidence of this, consider that at the dawn of the nineteenth century, physicians could still diagnose disease in terms of imbalances in the four humors—a practice dating back to ancient Greece! By the 1840s, advancing chemical knowledge had led to the introduction of anesthesia. The 1870s saw the emergence of the modern *germ theory* of disease and the idea that specific diseases were caused by specific microorganisms. The germ theory of disease was due in no small measure to the work of investigators such as Louis Pasteur (1822–1895) and Robert Koch (1843–1910). The germ theory of disease led in turn to the realization of the importance of antiseptic measures—Joseph Lister showed that chemicals such as phenol could be used to eliminate disease-causing microorganisms.

Although the modern germ theory of disease finally nailed down the bacterial causes of diseases, it required extensive use of animal research subjects. This can be seen in Koch's postulates, which still guide bacteriological research. These are: (1) the hypothetical germ causing a disease must be discoverable in all cases of the disease and found in the body wherever the disease lies; (2) extracted from the body, the germ must grow in a pure culture, for several bacterial generations; and (3) this culture must give the disease to a susceptible animal, be recoverable from it in a pure culture, and transmit the disease to another such animal.

This new mechanistic program, with its emphasis on practical medicine, was grounded in a methodological philosophy that emphasized the centrality of experimentation on live animals. Many theorists contributed to the development of this new methodology in the nineteenth century. Some, such as Marshall Hall (1790–1857), were concerned for the welfare of animal subjects and the need to minimize pain. Others, such as Claude Bernard (1813–1878), were much less squeamish. And, since much work in the contemporary biomedical sciences descends (with due modification) from this nineteenth-century research program, it will be important for us to examine some aspects of its development in more detail. Along the way we will see how science became entangled with the emerging nineteenth-century concern for animal welfare.

Marshall Hall: The Compassionate Investigator

There is no escaping the fact that the biomedical sciences that emerged in the nineteenth century, based on physics and chemistry and deeply rooted in animal experimentation, revolutionized medicine. It was in the nineteenth century, then, that the claim gradually evolved that animal experiments are not performed simply for the sake of anatomical or physiological knowledge; they are performed for the sake of human clinical benefits. *Their pain is our gain!* The human benefits from animal experimentation must be placed in the utilitarian hopper, along with the pain of animal suffering. Perhaps the benefits of animal experimentation are so great that they outweigh the pain and suffering of the animal subjects. We will look at some examples to illustrate both the evolution of the mechanistic biomedical research program and the complexity of the issues surrounding animal experimentation.

In this regard, it will be useful to take a brief look at some events surrounding the development of the science of neurophysiology. For Descartes, the nerves had been viewed as hollow tubes in which a substance known as the *animal spirits* (supposedly a rarefied form of blood) was supposed to flow like hydraulic fluid. By the end of the eighteenth century, Galvani's discovery that an amputated frog's leg could be made to twitch when electrically stimulated suggested that there was more to the nervous system than a system of hydraulic tubes. Work in the 1830s by Marshall Hall led to the discovery of the reflex arc; finally, much of the mystery of the nervous system was dispelled by the work of Emil Dubois-Reymond (1818–1896), who offered a physical, electrical explanation of nerve activity.

Now we will take a closer look at Marshall Hall, whose pioneering work in the 1820s led to the discovery of the reflex arc. Hall's work took place at the dawn of the age of scientific medicine and will afford us a glimpse into the process by which the new mechanistic approach to biomedicine evolved. Hall was interested in reflexes. A reflex is an automatic response to a stimulus that is very rapid because it is mediated by a simple neural circuit called a reflex arc. The simplest involves a receptor linked to a sensory neuron, which synapses with a motor neuron in the spinal cord or brain. The physician who taps your knee with a hammer is testing one of your reflex arcs. How was this discovered? Here is Marshall Hall in his own words:

> The animals experimented on were salamanders, frogs and turtles. In the first of these, the tail, entirely separated from the body, moved as in the living animal, on being excited by the point of a needle passed lightly over its surface. The motion ceased on destroying the spinal marrow within the caudal *vertebrae.* The head of the frog having been removed . . . an eye of the separated head was touched: it was retracted and the eye lid closed—a similar movement being observed in the other eye. On removing the brain, these phenomena ceased. . . . The head of the turtle continues to move long after its separation from the body. . . . On pinching any part of the skin of the body, extremities or tail, the animal moves. The posterior extremities and tail being separated together, the former were immovable; the latter moved on the application of the flame of a lighted taper to the skin. All movements ceased in the tail also, on withdrawing the spinal marrow from its canal. (Clendening 1960, pp. 594–595)

Hall concluded that either the brain or the spinal marrow was essential for the link between sensory and motor nerves.

But Hall's pioneering work was being done around the time of the formation of the RSPCA, the world's first organization to promote animal welfare. The organization did not just pop up into existence out of nowhere, but reflected a growing concern and awareness in society of the pain and suffering that is endured by animals. Hall, unlike the Cartesian investigators of old, was fully aware of the issue of animal suffering. As he put it,

> Unhappily for the physiologist, the subjects of the principal department of his science, that of animal physiology, are sentient beings; and every experiment, every new or unusual situation of such being, is necessarily attended by pain or suffering of a bodily or mental kind. (Quoted in Rowan 1984, p. 46)

And right at the dawn of the new experimental approach to medicine, Hall believed that animal experimentation should be subject to regulation.

According to Rowan, Hall had several guidelines for the practice of animal experimentation. First, experiments should not be conducted if the results could be gained by observation alone. Second, experiments should always be performed with clearly defined and *attainable* objectives—"fishing expeditions" were to be discouraged. Third, unnecessary repetitions of experiments were to be avoided. Results, once reliably established, did not need to be endlessly and pointlessly replicated. Hall thought that experiments should be witnessed so as to reduce the need for replication. Fourth, experiments should be done

with a view to minimizing suffering. Writing before the advent of evolutionary biology, Hall also recommended the use of "lower" animals, believed to be less sentient, and hence less likely to suffer as extensively as the "higher" animals. As Rowan has observed, "In many ways, Hall displayed remarkable sensitivity to the issues of animal suffering considering the environment in which he worked. Certainly, many scientists later in the century could not lay claim to the same concern" (1984, p. 46).

One of these scientists was Claude Bernard, and since his work reflects the crystallization of the new methodology guiding animal research, it will be important for us to pay detailed attention to his work.

Marshall Hall's pioneering work in the 1820s on salamanders, frogs, and turtles led to the discovery of the reflex arc. Hall believed that animal experimentation should be subject to regulation. (National Library of Medicine)

Claude Bernard: The Methodology of Biomedical Research

Claude Bernard is without doubt one of the great figures in the history of science. Bernard made many contributions to the biomedical sciences in his own right, but for our purposes his *Introduction to the Study of Experimental Medicine* (1865) is an extremely important text since it laid out the philosophical and methodological rationale for the newly emerging biomedical sciences. More than a century after his death, his methods still guide experimental practices in the biomedical sciences. The American Medical Association, in its *Statement on the Use of Animals in Biomedical Research* on animal experimentation, praised Bernard for establishing the basic principles that guide the practice of experimental medicine (American Medical Association 1992). No lesser figure than the late immunologist and Nobel laureate Sir Peter Medawar commented that "[t]he wisest judgements on scientific method ever made by a working scientist were indeed those of a great biologist, Claude Bernard" (1984, p. 73).

Bernard denigrated clinical medicine—and at the time he was writing, it was still in a very backward state. But he believed that genuine biomedical science involved carefully controlled experiments on animals and that such experiments were directly relevant to human biology. In fact, Bernard wanted to do for medicine what other scientists had done for physics and chemistry:

> We cannot imagine a physicist or a chemist without his laboratory. But as for the physician, we are not yet in the habit of believing that he needs a laboratory; we think that hospitals and books should suffice. This is a mistake; clinical information no more suffices for physicians than knowledge of minerals suffices for chemists and physicists. ([1865] 1949, p. 148)

Claude Bernard, one of the great figures in the history of science. In its 1992 White Paper on animal experimentation, the American Medical Association praised Bernard for establishing the basic principles that guide the practice of experimental medicine. (National Library of Medicine)

Bernard did not deny that clinical observations had a place in the practice of medicine, but he did believe that science took place, not in the clinical context, but in the laboratory.

To understand the wedge that Bernard drove between clinical medicine and scientific medicine, we must examine his views about the methodology of science. In his opinion, the business of science was in the formulation and rigorous testing of hypotheses about phenomena of interest:

> The experimental hypothesis . . . must always be based on observation. Another essential of any hypothesis is that it must be as probable as may be and must be experimentally verifiable. Indeed, if we made an hypothesis which experiment could not verify, in that very act we should leave the experimental method and fall into the errors of the scholastics and makers of systems. ([1865] 1949, p. 33)

Clinical medicine could provide the observations that prompted the formulation of hypotheses, but it could not, in the nature of the case, be the context within which the hypotheses were tested.

The clinical setting does not permit the adequate control of experimental variables, and typically it is not a setting in which, for ethical reasons, those variables can be experimentally manipulated. Or as Bernard put it, "In a word, I consider hospitals only as the entrance to scientific medicine; they are the first field of observation which a physician enters; but the true sanctuary of medical science is a laboratory. . . . In leaving the hospital, a physician must . . . go into his laboratory" ([1865] 1949, pp. 146–147). But if the biomedical researcher is to conduct controlled experiments, carefully manipulating variables of interest, there must be appropriate subjects. Experiments on humans would be immoral. So Bernard became one of the most influential advocates (and practitioners) of animal experimentation.

But the focus on animals was not just motivated by a moral worry about human experimentation. Nonhuman animals were also scientifically appropriate objects of study:

> Experiments on animals, with deleterious substances or in harmful circumstances, are very useful and entirely conclusive for the toxicology and hygiene of man. Investigations of medicinal or of toxic substances are also wholly applicable to man from the therapeutic point of view; for as I have shown, the effects of these substances are the same on man as on animals, save for differences in degree. ([1865] 1949, p. 125)

To understand Bernard's views on the applicability of animal research, we must examine his commitment to *causal determinism.*

Causal determinism is a doctrine underlying a lot of experimental science, both within and outside of biology. Until the advent of indeterministic quantum mechanics, it lay at the heart of physics (and still does for many physical phenomena of interest). It is an idea based on two principles: *the principle of causality* according to which all events have causes and the *principle of uniformity* according to which, for qualitatively identical systems (systems identical with respect to scientifically relevant properties), same cause is followed by same effect.

The idea can be illustrated by a consideration of Newton's second law, according to which *force equals mass times acceleration.* The law tells us that we can evaluate forces by looking at the way masses accelerate. If you have two masses that accelerate in exactly the same way when the same force is identically applied, then the masses must be equal; if they accelerate differently under these conditions, then the masses must be different. Correspondingly, if the masses are known to be equal (these are the *initial conditions*), but accelerate differently when a force is applied in the same way, there must be a difference in the magnitude of the force applied. In accord with causal determinism, accelerations are caused by forces, and for identical masses (in this case, the *mass* of the object is scientifically relevant, but its *color,* for example, is not), same *cause* (force applied) is followed by same *effect* (acceleration of mass).

For Bernard, the application of this type of causal reasoning to the biomedical context was direct:

> If a phenomenon appears just once in a certain aspect, we are justified in holding that, in the same conditions, it must always

> appear in the same way. If, then, it differs in behavior, the conditions must be different. But indeterminism knows no laws; laws exist only in experimental determinism, and without laws there can be no science. ([1865] 1949, p. 139)

So, if seemingly identical systems behave differently, there must be a difference in initial conditions to account for the observed difference. A mature science should be able to account for such differences by analogy with the way we dealt with the differences in masses above. For Bernard, experimental medicine should yield deterministic laws akin to those of Newton.

Though living systems look different, and though their masses may vary, they obey the same physiological laws:

> Physiologists . . . deal with just one thing, the properties of living matter and the mechanism of life, in whatever form it shows itself. For them genus, species and class no longer exist. There are only living beings; and if they choose one of them for study, that is usually for convenience in experimentation. ([1965] 1949, p. 111)

For Bernard, differences between species were not physiologically relevant—all species obey the same laws:

> In living bodies, as in inorganic bodies, laws are immutable, and the phenomena governed by these laws are bound to the conditions on which they exist, by a necessary and absolute determinism. . . . Determinism in the conditions of vital phenomena should be one of the axioms of experimental physicians. ([1865] 1949, p. 69)

Of course, even if different species obey the same laws, we must be certain that there are no relevant differences between them that will undermine an extrapolation of results found in members of one species, for members of the other species.

Here we will discuss how Bernard conceptualized species differences, for his views have been enormously influential. But the issue has become the source of much controversy, especially in the light of modern evolutionary biology. So in Chapters 6 and 8 we will revisit these issues in the light of evolution—a theory whose ramifications for medicine were largely unknown in Bernard's day, but which is now starting to revolutionize not only the way in which investigators think about human health and disease but also the ways in which they think of the similarities and differences between humans and nonhuman animals.

Bernard knew that physiologists must confront problems different from those encountered by physicists and chemists when they investigate inorganic systems. The study of a living system requires the study not simply of the external factors and forces acting on the system but also of the *inner organic environment*—the organism viewed as a complex, dynamical system with many interacting parts:

> [*A*] *created organism is a machine* which necessarily works by virtue of the physico-chemical properties of its constituent elements. Today we differentiate three kinds of properties exhibited in the phenomena of living beings: physical properties, chemical properties and vital properties. But the term "vital properties" is itself only provisional; because we call properties vital which we have not yet been able to reduce to physico-chemical terms; but in that we shall doubtless succeed someday. ([1865] 1949, p. 93; my italics)

Once the inner environment is understood in appropriate physico-chemical terms, it will be possible to describe its behavior using physico-chemical laws—laws that are the same for all species. These laws of the inner environment will be the universal laws of the biological science.

Species differences—of which Bernard was well aware—were not impediments to biomedical research. Nevertheless, Bernard was aware that some investigators did think they were significant:

> Even today, many people choose dogs for experiments, not only because it is easier to procure this animal, but also because they think experiments performed on dogs can be more properly applied to man than those performed on frogs. ([1865] 1949, p. 123)

But this is a mistake. Bernard thought the mistake was generated because some experimenters mistook quantitative differences in initial conditions for fundamental qualitative differences between species. Bernard disagreed; he thought the fundamental properties of *vital units* (biological units) were the same for all species. Livers may come in different sizes and shapes, but they all respond to stimuli in basically the same way. In so far as there are differences, these seem to consist in slightly different arrangements of essentially similar building blocks:

> Now the vital units, being of like nature in all living beings, are subject to the same organic laws. They develop, live, become diseased and die under influences necessarily of like nature, though manifested by infinitely varying mechanisms. A poison or a mor-

> bid condition, acting on a definite histological unit, should attack it in like circumstances in all animals furnished with it; otherwise these units would cease to be of like nature; and if we went on considering as of like nature units reacting in different or opposite ways under the influence of normal or pathological vital reagents, we should not only deny science in general, but also bring into zoology confusion and darkness. ([1865] 1949, p. 124)

Of course there are idiosyncrasies and differences in the way members of different species behave when experimentally manipulated. Bernard thought these should be studied and brought under universal physiological laws.

Here is an analogy to explain his view. In the nineteenth century, idiosyncrasies had long been observed in the orbit of the planet Uranus (then believed to be the outermost planet). Did this mean that Newton's law of gravity was wrong? And hence that a large mass was not obeying the cosmic rules? No. In 1846, the French astronomer Urbain Leverrier realized that these orbital oddities could be explained, using Newton's laws, if Uranus were being acted on by the gravitational field of a hitherto unobserved planet. Calculations were done, observations were performed, and these led to the discovery of the planet Neptune. Idiosyncrasies do not necessarily show the laws are wrong, but they do signal the need for more science! If species behave differently when similarly stimulated, what we need to do is to seek the physiological "Neptunes" that explain the oddities in a lawlike manner.

In Bernard's approach to biomedical science, species differences were quantitative, and not qualitative, in nature. That is to say, once you made allowances for (quantitative) differences in weight of the animals, or in doses administered, same cause would be followed by same effect in humans and the test species. Bernard illustrated this idea in his discussion of toad venom. Doses of venom that stop the hearts of frogs do not stop the hearts of toads. Does this mean we have same cause followed by different effect? Bernard reasoned as follows:

> Now, in logic, we should necessarily have to admit that the muscular fibers of a toad's heart have a different nature from those of a frog's heart, since the poison which acts on the former does not act on the latter. That was impossible: for admitting that organic units identical in structure and in physiological characteristics are no longer identical in the presence of a toxic action identically the same would prove that phenomena have no necessary causation; and thus science would be denied. ([1865] 1949, p. 180)

So how do we accommodate both science and the puzzling observations? Bernard continued:

> Pursuant to these ideas, I rejected the above mentioned fact as irrational, and decided to repeat the experiments. . . . I then saw that toad's venom easily kills frogs with a dose that is wholly insufficient for a toad, but that the latter is nevertheless poisoned if we increase the dose enough. So that the difference described was reduced to a question of quantity and did not have the contradictory meaning that might be ascribed to it. (p. 180)

Bernard here laid down one of the basic principles of the science of toxicology: Once purely quantitative differences have been allowed for (differences in metabolic rate, body weight, surface area, and so on), same cause will be followed by same effect in members of a given species or in members of different species.

Bernard experimented on many species—domestic animals such as dogs, cats, rabbits, and pigs were explicitly mentioned, but, like Harvey and Hall before him, he experimented on frogs. And as John Parascandola has noted:

> Before the twentieth century, the frog played a major role in the history of research in the life sciences. The widespread use of frogs was due in part to their general hardiness for surviving severe operations and the excellent survival capacity for their isolated tissues. Frogs were experimented on so frequently that in 1845 Hermann Helmholtz referred to them as "the old martyrs of science." His fellow physiologist, Claude Bernard, called the frog "the Job of Physiology." (1995, p. 16)

In the twentieth and twenty-first centuries, murid rodents (rats and mice) have become the new Jobs of physiology, amounting to somewhere between 80 to 90 percent of the approximately 16 million research subjects used in biomedical research in the United States each year (American Medical Association, 1992, p. 15).

Bernard did not just think of organisms as matter in motion. He did not think physiology could be reduced to physics. But he did think that life could be understood in physico-chemical terms. He was an intellectual heir to the grand tradition of medicine that emerged from the Renaissance:

> It is doubtless correct to say that the constituent parts of an organism are physiologically inseparable from one another, and that they all contribute to a common vital result; but we may not conclude from this that the living machine must not be ana-

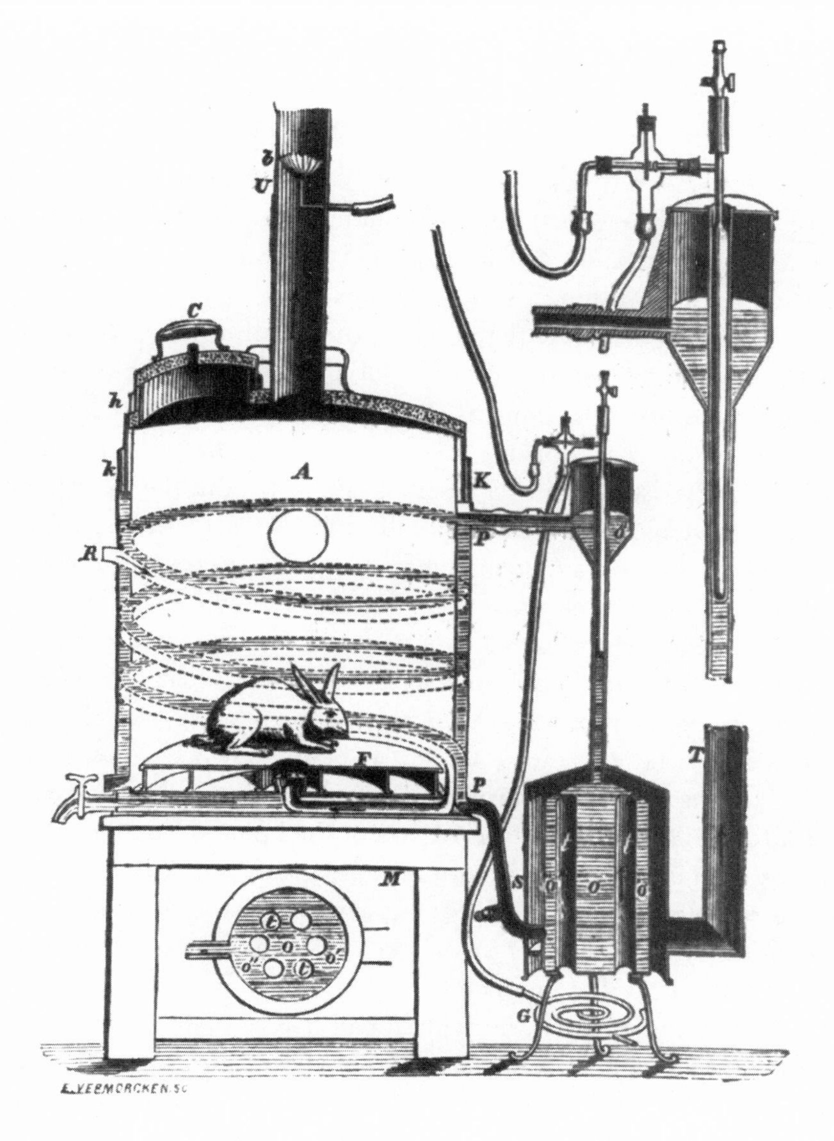

Claude Bernard believed that science took place in the laboratory, as illustrated here in one of his lab experiments with rabbits. Bernard experimented on many species, including dogs, cats, rabbits, pigs, and frogs. (National Library of Medicine)

> lyzed as we analyze a crude machine whose parts also have their role to play in a whole. ([1865] 1949, p. 89)

Though there is more to life than mere matter in motion, Bernard was aware that to understand the whole, you had to understand the properties of the parts and the mutual relationships between them. In this sense he was a true heir to the resolutio-compositive method.

The fundamental basis of Bernard's methodology was the tightly controlled laboratory experiment. The experiments themselves involved extensive vivisection.

> [As for vivisection on animals, no] hesitation is possible; the science of life can be established only through experiment, and we can save living beings from death only after sacrificing others.

> Experiments must be made on either man or on animals . . . the results obtained on animals may all be conclusive for man when we know how to experiment properly. ([1865] 1949, p. 102)

And as Bernard had pointed out, controlled experiments on humans were immoral, so animal experimentation was inevitable if medical science was to advance. The interspecific tradeoff is clear: If we are to save human lives, we must sacrifice animals.

Yet Bernard apparently did have some qualms about animal experiments. He refused to experiment on apes, for example, because of their resemblance to humans. And many in Bernard's day found his research highly objectionable. His wife and daughters even became militant antivivisectionists! (Rowan, 1984, p. 48) But the fact remains that animal experimentation, morally odious as it was increasingly perceived to be, was advancing medicine in the nineteenth century—and animals came to play many roles in the biomedical sciences.

In the long course of the twentieth century, animal research continued to be of great importance in the biomedical sciences. As in the nineteenth century, public concern about the practice was allayed by claims about the enormous medical benefits. Hugh LaFollette and I have examined these claims in our book, *Brute Science: The Dilemmas of Animal Experimentation* (1996). We are of the opinion that though animal experiments are not scientifically useless—certainly not as useless as some opponents of animal research suggest—claims about the enormous medical value of this kind of experimentation are complex claims that are hard to evaluate.

For example, given the widespread nature of the practice, you cannot simply point to a few textbook successes (or failures, for that matter) to establish the issue one way or another. What is needed—and what is virtually absent in most public discussion of this issue—is a careful examination of the practice of biomedical research, in proper historical context, and with due care to weigh up all the factors underlying the advance of the biomedical sciences. This is a complex issue indeed. We will not say more about it here, save to note that without such a careful, quantitative analysis, claims about the benefits of vivisection for human medicine will be hard to substantiate, one way or the other. For propagandists on either side of the fence, this does not matter. But for those concerned with the ethics of the practice, especially for those guided by utilitarian principles, this will be an important issue.

Here we will examine three cases where the new scientific approach to medicine yielded fruit. In one case, we will see medicine advance through the use of nonhuman animals; in another case we will see medicine advance through the use of human subjects. In the final case we will see humans and animals used together. This will help focus our attention on the dilemmas posed by the use of animal subjects in biomedical research.

In 1885 Louis Pasteur published a paper, *Prevention of Rabies,* which contained details of his groundbreaking research leading to the discovery of ways to prevent the onset of rabies after the patient has been bitten by a rabid animal. The virus causing rabies is neurotropic, which means that it has a particular affinity for nerve cells. The virus manipulates the behavior of the host animal, making it literally mad and furious; the virus also induces the characteristic "foaming at the mouth," so that when the host animal bites other animals, the virus is transferred in the saliva. The disease has a long incubation period in humans, averaging around sixty days, a span that provides a window of opportunity for medical intervention. After the first symptoms appear, treatment is generally pointless (Biddle 1995, p. 118).

Pasteur's experiments involved extensive use of rabbits and dogs. First he found he could induce rabies in rabbits by inoculating them with extracts from the spinal cords of infected dogs. The infected rabbit could then be used to infect a second rabbit, and this in turn could be used to infect a third rabbit, and so on. The first rabbit showed an incubation period of around fifteen says, but after the infection had been passed through a succession of rabbits, this incubation period diminished. Pasteur observed:

> After passing twenty or twenty five times from rabbit to rabbit, inoculation periods of eight days are met with, and continue for another interval, during which the virus is passed twenty or twenty five times from rabbit to rabbit. Then an incubation period of seven days is reached, which is encountered with striking regularity throughout a new series extending as far as the ninetieth animal. (Clendening 1960, p. 389)

These experiments enabled Pasteur to have a source of rabies with a constant degree of virulence in his laboratory.

He then noticed that if portions of the spinal cords of infected rabbits were removed and suspended in dry air, the virulence slowly diminished to the point of extinction. With this knowledge, Pasteur was able to render dogs refractory to rabies as follows:

Louis Pasteur in his laboratory, surrounded by cages of rabbits. Pasteur experimented on rabbits in an attempt to find a cure for rabies. (Hulton/Archive)

> Every day morsels of fresh infective spinal cord from a rabbit which has died of rabies developed after an incubation period of seven days, are suspended in a series of flasks, the air in which is kept dry. . . . Every day also a dog is inoculated under the skin with a Pravaz syringe full of sterilized broth, in which a small fragment of one of the spinal cords has been broken up, commencing with a spinal cord far enough removed in order of time from the day of the operation to render it certain that the cord was not virulent. . . . On the following days the same operation is performed with more recent cords, separated from each other by an interval of two days, until at last a very virulent cord, which has only been in the flask for two days, is used. (Clendening 1960, p. 390)

The end result is a dog that cannot get rabies, even if infected matter is placed directly on the animal's brain! Pasteur developed this technique on at least fifty dogs.

The human benefits were realized when a child, Joseph Meister, who had been severely bitten by a rabid dog, was presented to Pasteur, who noted that "[t]he death of this child appearing to be inevitable, I decided not without lively and sore anxiety, as may well be believed, to try upon Joseph Meister the method which I had found constantly successful with dogs . . ." (Clendening 1960, p. 391). The treatment was a great success. But equally importantly, it is an example of a case in which animal research, involving the deaths of numerous subjects, turned out to have benefits for humans. And this was a partial fulfillment of the promise of the new mechanistic program for biomedical research in the nineteenth century.

If you are going to experiment on live subjects, the only alternative to animal experimentation is human experimentation. Here is an example of human experimentation involving yellow fever. Yellow fever is caused by a virus transmitted by female *Aedes aegypti* mosquitoes. It kills about 30 percent of the people who become infected. A vaccine was not developed until 1937, but at the turn of the century there was uncertainty concerning the role played by the mosquitoes in the transmission of the disease. Walter Reed (1851–1902) performed a series of experiments on humans to establish the connection, and these results were reported in 1901 in his *The Propagation of Yellow Fever; Observations Based on Recent Researches.*

Reed conducted his research in Panama, where work on the Panama Canal was being disrupted by yellow fever outbreaks. Mosquitoes were contaminated by being allowed to feed on patients with yellow fever, who constituted a reservoir for the disease. Could such mosquitoes be responsible for the transmission of the disease?

> On the fifteenth day of our encampment . . . we concentrated our insects, so to speak, on one of these non-immunes—Kissinger by name—selecting five of our most promising mosquitoes for the purpose. . . . This inoculation was more successful, for at the expiration of three days and nine and one half hours the subject, who had been under strict quarantine during fifteen days, was suddenly seized with a chill about midnight, December 8th, which was the beginning of a well-marked attack of yellow fever. (Clendening 1960, p. 482)

This experiment was repeated several times with other humans. The results were the same. Further volunteers sat in rooms containing soiled clothing of fever patients—to make sure that it was mosquitoes and not *fomites* that were responsible for the spread of the disease.

Other controlled experiments were performed on human volunteers using a room divided by a wire screen that was impervious to mosquitoes but that would allow air to flow freely in the room. These experiments were designed to ascertain that it was the mosquitoes that caused the disease and not some extraneous factor:

> At noon on the same day, five minutes after the mosquitoes had been placed therein, a plucky Ohio boy, Moran by name, clad only in his night shirt, and fresh from a bath, entered the room containing the mosquitoes, where he lay down for a period of thirty minutes. On the opposite of the screen were two "controls" and one other non-immune. Within two minutes from Moran's entrance, he was being bitten about the face and hands

> by the insects that had promptly settled down upon him. . . . The building was then closed, except that the two non-immune "controls" continued to occupy the beds on the non-infected side of the screen. On Christmas morning at 11 A.M., this brave lad was stricken with yellow fever. . . . The two "controls" who had slept each night in this house, only protected by the wire screen, but breathing the common atmosphere of the building, had remained in good health. (Clendening 1960, p. 484)

With the role of the mosquito as a vector for the transmission of yellow fever now firmly established, eradication programs were undertaken—Col. William Gorgas used flame throwers, poisons, and fumigation methods and reduced substantially the incidence of yellow fever in the Panama Canal zone.

Elie Metchnikoff (1845–1916) used both animals and humans in his development of a cure for syphilis—this was the famous calomel ointment. Paul De Kruif described the Metchnikoff's experimental procedure as follows:

> He took two apes, inoculated them with the syphilitic virus fresh from a man, and then, one hour later, he rubbed the greyish ointment into that scratch spot on one of his apes. He watched the horrid signs of the disease appear on the unanointed beast, and saw all signs of the disease stay away from the one that had got the calomel. (Quoted in Copi 1982, p. 426)

But the effectiveness of the ointment still had to be demonstrated in humans. So, as De Kruif described it, Metchnikoff

> induced a young medical student, Maisonneuve, to volunteer to be scratched with syphilis from an infected man. . . . It was a more severe inoculation than any man would ever get in nature. The results of it might make him a thing of loathing, might send him, insane, to his death. . . . For one hour Maisonneuve waited, then Metchnikoff, full of confidence, rubbed the calomel ointment into the wounds—but not into those which had been made at the same time on a chimpanzee and a monkey. It was a superb success, for Maisonneuve showed never a sign of the ugly ulcer, while the simians, thirty days afterwards, developed the disease. . . . (Quoted in Copi 1982, p. 426)

In World War I, calomel ointment was used extensively to treat infected servicemen. It was estimated to be very effective. In this case, many people were spared the ravages of a serious sexually transmitted disease through experiments involving both humans and nonhuman animals.

The experiments described here are no doubt heroic examples from the golden age of scientific research, before animal and human experimentation became subject to rigorous regulation and bureaucratic oversight. But they nevertheless serve to bring out a basic point: If experimentation in biomedicine requires live subjects, what should we use? Humans or nonhuman animals? Since the end of World War II, and the adoption of principles limiting the use of humans as experimental subjects in the light of the Nuremberg medical trials, biomedical research has become deeply committed to the use of nonhuman animal subjects.

Further Reading

R. Gordon, *The Alarming History of Medicine: Amusing Anecdotes from Hippocrates to Heart Transplants* (New York: St. Martin's Press, 1993), provides an entertaining perspective on events and figures in nineteenth-century medicine. L. Clendening, ed., *Sourcebook of Medical History* (New York: Dover, 1960), provides excerpts from the writings of nineteenth-century medical investigators. Many valuable essays can also be found in N. Rupke, ed., *Vivisection in Historical Perspective* (New York: Croom Helm, 1987).

H. L. LaFollette and N. Shanks, *Brute Science: The Dilemmas of Animal Experimentation* (London: Routledge, 1996), discuss the role played by Claude Bernard in the development of biomedical theory. Ernst Mayr provides many interesting perspectives on the development of biological theory. See E. Mayr, *Toward a New Philosophy of Biology: Observations of an Evolutionist* (Cambridge: Harvard University Press, 1988); *One Long Argument: Charles Darwin and the Genesis of Modern Evolutionary Thought* (Cambridge: Harvard University Press, 1991); and *This Is Biology: The Science of the Living World* (Cambridge: Harvard University Press, 1997).

6

By Accident or Design: Darwin's Theory of Evolution

So far in this book we have seen how pervasive mechanical metaphors have been in the history of biology. But metaphors have limitations; they usually break down at some point or other. Thus the poet Robert Frost once remarked:

> Somebody said to me a little while ago, "It is easy enough for me to think of the universe as a machine as a mechanism."
> I said, "You mean the universe is like a machine?"
> He said, "No, I think it is one. . . . Well, it is like . . ."
> "I think you mean the universe is like a machine."
> "All right. Let it go at that."
> I asked him, "Did you ever see a machine without a pedal for the foot, or a lever for the hand, or a button for the finger?"
> He said, "No—no."
> I said, "All right. Is the universe like that?"
> And he said, "No. I mean it is a machine, only . . ."
> " . . . it is different from a machine," I said.
> He wanted to go just that far with that metaphor and no further. And so do we all. All metaphor breaks down somewhere. That is the beauty of it. It is touch and go with the metaphor, and until you have lived with it long enough you don't know when it is going. You don't know how much you can get out of it and when it will cease to yield. It is a very living thing. It is as life itself. (1972, pp. 40–41)

Darwin, who had studied Paley carefully, had lived with the metaphor of nature-as-a-machine long enough and had come to see where it had ceased to yield. In this chapter and the next we will see how evolution's implications call into question the mechanical metaphors that had been used to conceptualize organic nature since the dawn of modern science in the Renaissance.

In this chapter, we will examine the theory of evolution as it was introduced and understood by Charles Darwin. We will examine the implications of the theory of evolution for mechanistic physiology and will conclude with a consideration of the implications of evolution for the relative positions of humans and nonhuman animals in nature, especially with respect to mental lives—a matter that was of some importance to Darwin himself. The focus of this chapter is mainly on Darwin's achievement and its implications.

To understand Darwin's work, it is important to understand that he was the beneficiary of a new method in science—one that emerged during the late eighteenth and early nineteenth centuries. This new method shaped investigations in many fields, but it had special implications for geology and its estimate of the age of the earth and how the present state of the earth results from changes brought about by the operation of natural causes over long periods of time.

Geology and Method

In Chapter 3 we saw that theologians such as Ussher had argued that the earth was very young, created by God around 4004 B.C. (this view is known as *young earth creationism*). We also saw that theorists such as Burnet had argued that the present state of the earth reflected processes of ruination brought about as the result of the wrath of God. By the end of the eighteenth century, there was a growing realization, at least in educated circles, that the earth, though created, might be considerably older than Ussher's estimates implied (this view is known as *old earth creationism*).

In the late eighteenth and early nineteenth centuries, a new method found its way into science. Although this method provided a new way of thinking about some old scientific problems, it had important implications for the geological sciences. This method also shaped Darwin's thinking about biology, and for that reason, we must briefly examine it. William Whewell (1794–1866), in his *Novum Organon Renovatum* (1858), called it *the method of gradation* (see Butts 1989).

An important part of science—especially in the early stages of development—concerns the assignment of objects to various categories according as to whether they are similar or different with respect to some property of interest. But do these categorical differences between objects mean that objects in one category are different in kind from objects in another category, or do the categories represent extreme points on a continuous spectrum of cases? The method of gradation is used when there are questions about degrees of similarity and difference between objects of interest. As Whewell put it:

> The selection of instances which agree, and of instances which differ, in some prominent point or property, are important steps in the formation of science. But when classes of things and properties have been established in virtue of such comparisons, it may still be doubtful whether these classes are separated by distinctions of opposites, or by differences of degree. And to settle such questions, the *Method of Gradation* is employed; which consists in taking intermediate stages of the properties in question, so as to ascertain by experiment whether, in the transition from one class to another, we have to leap over a manifest gap, or to follow a continuous road. (Butts 1989, p. 240)

The method was used by Faraday, for example, to undermine an absolute distinction between electrical conductors and nonconductors (sulfur is a poor conductor, spermaceti is better than sulfur, water is better than spermaceti, and metals are better still).

But the method was to change thinking about events in the historical sciences, first in geology, and second in biology. Charles Lyell (1797–1875) published the *Principles of Geology* in three volumes between 1830 and 1833. Lyell's enterprise was none other than an attempt to explain how the earth had changed in the course of geological time by reference to causal processes that can be seen to be in operation today—for example, water erosion, wind erosion, vulcanism, earthquakes, and so on. The basic idea is that the small changes brought about by these causal processes can gradually accumulate over very long periods of time to result in substantial changes. Whewell, in a review written in 1832, called this idea *uniformitarianism*.

In the eighteenth century, geology explained the state of the earth today as the result of causes very different from those currently observed to be in operation. In particular, the *catastrophists* had tried to explain the current state of the earth as the result of a series of global catastrophes, Noah's flood being the most recent. For Lyell, the geological record resulted from the action, over long periods of time, of the kinds of natural cause that we see in operation today. The evidence of massive changes that we see today (mountains with fossils of sea creatures at their summits) thus resulted from the slow accumulation of small changes, occurring gradually over very long stretches of time, brought about by natural causes amenable to study today. The changes did not result from massive catastrophes brought about by supernatural agency in very short periods of time.

In his early work, Charles Lyell (1797–1875) believed that species became extinct as the conditions for which they were initially adapted changed. (National Library of Medicine)

But if this is true, then the physical environment on Earth is not static. If the physical environment is changing, then intelligently designed organismal machines, once fitted into their appointed place in nature, would, by staying the same over many successive generations, find themselves out of kilter with the natural environment in which they were embedded—yet this is not what we see. Instead, we see remarkable ranges of adaptation. In his early work, Lyell believed that species became extinct as the conditions for which they were initially adapted changed. The resulting gaps in nature were supposed to be refilled by the introduction of new species, presumably by supernatural means (Mayr 1991, p.16). Darwin, under the influence of Lyell's writings, saw the evidence of adaptation to a changing environment very differently.

Darwin and the Origin of Species

Charles Darwin (1809–1882) published *The Origin of Species* in 1859 (references here are to a 1970 edition). Darwin's work overturned the idea that organisms are machines deliberately designed to fit into a larger natural, ultimately cosmological, mechanism. Indeed, it was here that the metaphor of organism-as-machine found its first central challenge. For although grandfather clocks will not of themselves turn, as the result of the gradual accumulation of many small, unguided changes, into pocket watches, nevertheless, in the long course of evolutionary history, dinosaurs may well have evolved into the birds of the air that we see today!

Darwin was also an heir to the method of gradation—species

differences are not absolute discontinuities; they reflect varying degrees of similarity and difference. He was a beneficiary of Lyell's uniformitarianism—the differences we see between species today reflect the slow accumulation, over long periods of time, of small changes brought about by unguided, natural causes similar to those we see in operation today, or as Darwin himself put it:

Ethology, the study of the evolutionary biology of animal behavior, could be said to have begun with Charles Darwin (1809–1882) himself, since he saw behavioral similarities between humans and nonhuman animals as the result of evolutionary processes. (National Library of Medicine)

> But the chief cause of our natural unwillingness to admit that one species has given birth to clear and distinct species, is that we are always slow in admitting great changes of which we do not see the steps. The difficulty is the same as that felt by many geologists, when Lyell first insisted that long lines of inland cliffs had been formed, the great valleys excavated, by the agencies which we still see at work. The mind cannot possibly grasp the full meaning of the term of even a million years; it cannot add up and perceive the full effects of many slight variations accumulated during an almost infinite number of generations. (1970, pp. 116–117)

Evidence suggested that plants and animals had adapted to environmental changes, but prior to Darwin, there was no really good explanation for how these changes occurred. Here is the gist of Darwin's explanation.

A species is a cohesive population of individual organisms bound together by reproductive ties. Members of these populations show *variation* with respect to *heritable* traits, however. Since the dawn of agriculture, animal breeders have long exploited naturally occurring *intraspecific* variation to make new varieties: Only animals with desirable traits (wooliness of coat, milk yield, domesticity, and so on) have been allowed to reproduce and pass these traits into the next generation, where the process would be repeated. Natural varieties—Darwin called them *incipient species* (1970, p. 39)—pervade nature and not just the farmer's yard.

Without appealing to supernatural design, how could *nature* work with the variation in heritable traits found in natural populations to bring about adaptations and ultimately the origin of new species? Darwin's answer, based on ideas by Thomas Malthus (1766–1834) in his *An Essay on the Principle of Population* ([1798] 1966), was that a *struggle for existence* arises naturally from the fact that organisms tend to produce more offspring than can be supported by the environment:

> Every being, which during its natural lifetime produces several eggs or seeds, must suffer destruction during some period of its life, and during some season or occasional year, otherwise, on the principle of geometrical increase, its numbers would quickly become so inordinately great that no country could support the product. (Darwin 1970, p. 41)

It is here, in the context of the superabundance of organisms, that heritable variation plays its crucial role.

In this unrelenting struggle for existence, some variants will have characteristics that hamper their ability to survive and reproduce; other variants will have characteristics that enhance their ability to survive and reproduce. Such traits confer *fitness advantages:*

> Owing to this struggle, variations, however slight and from whatever cause proceeding, if they be in any degree profitable to the individuals of a species, in their infinitely complex relations to other organic beings and to the physical conditions of life, will tend to the preservation of such individuals, and will generally be inherited by the offspring. (Darwin 1970, p. 39)

For Darwin, this mechanism was the primary engine of evolution:

> This preservation of favorable individual differences and variations, and the destruction of those which are injurious, I have called Natural Selection, or the survival of the fittest. Variations neither useful nor injurious would not be affected by natural selection, and would be left either a fluctuating element, as perhaps we see in certain polymorphic species, or would ultimately become fixed, owing to the nature of the organism and the nature of the conditions. (p. 44)

Natural selection thus works on heritable variation found in *populations* of organisms. In the environment in which the struggle for existence takes place, the traits favored by selection increase in frequency over successive generations; they represent adaptations to the environment in which the struggle for existence occurs. Adaptations are those features of organisms that are the quintessential fruits of the operation of natural selection. In short, to use Henry Petroski's useful turn of phrase, "form follows failure" (1994, p. 22).

For Darwin, the same selective mechanisms that brought about adaptations within populations of organisms would also gradually bring about differences between populations great enough to constitute their designation as separate species. Over long periods, the result of this process would be increasing biodiversity:

> As each species tends by its geometrical rate of reproduction to increase inordinately in number; and as the modified descendents of each species will be enabled to increase by as much as they become more diversified in habits and structure, so as to be able to seize on many and widely different places in the economy of nature, there will be a constant tendency of natural selection to preserve the most divergent offspring of any one species. Hence, during a long continued course of modification, the slight differences characteristic of varieties of the same species, tend to be augmented into the greater differences characteristic of species of the same genus. (1970, p. 108)

This process of *adaptive radiation* explains what happens when animals from an ancestral species move into a multiplicity of *ecological niches,* each niche being characterized by a particular complex of features that affect an animal's way of making a living: nature and availability of food, nature and number of predators, pathogens and parasites, climates, and so on. Thus, as small differences between populations accumulate through adaptive specialization over many successive generations, the *invisible hand* of natural selection will accentuate differences between these populations until they are so distinct as to be recognized as different species.

The process of adaptive radiation described above will result in *evolutionary divergence*—the emergence of different characteristics in related organisms—for example, dogs and seals. But treatment of observed similarities and differences must proceed cautiously, and with all due regard for historical relationships between species, for the process of *evolutionary convergence* can result in unrelated organisms finding similar solutions to similar environmental problems—for example, the streamlined shapes of sharks and dolphins do not stand as evidence of close ancestry but rather of similar solutions to a common problem of negotiating an aquatic environment.

In 1871 Darwin published *The Descent of Man, and Selection in Relation to Sex* (the references here are to an 1896 edition). Here, in addition to the mechanism of natural selection, Darwin also considered the effects of *sexual selection:* "the advantage which certain individuals have over others of the same sex and species solely in respect of reproduction" (1896, p. 209). Darwin compared the two types of selection as follows:

> Sexual selection acts in a less rigorous manner than natural selection. The latter produces its effects by the life or death at all ages of the more or less successful individuals. . . . But generally the less successful male merely fails to obtain a female, or obtains a retarded and less vigorous female later in the season, or, if polygamous, obtains fewer females; so that they leave, fewer, less vigorous, or no offspring . . . but in regard to the structures adapted to make one male victorious over another, either in fighting or in charming the female, there is no definite limit to the amount of advantageous modification; so that as long as the proper variations arise the work of sexual selection will go on. (p. 226)

Sexual selection gives rise to secondary sexual characters, and these may be quite exaggerated, for example, the tail of the peacock or the mighty antlers of the Irish elk.

Evolution can be captured by the slogan, *descent with modification*. Descent means first that all species are derived from ancestral species, and ultimately from the common ancestor of all life of earth. Because evolution is a branching process, for any two species there must be a common ancestor. Humans, for example, did not descend from chimpanzees; rather humans and chimpanzees share a common ancestor from which both are descended—and the line that leads to modern chimpanzees and the line that leads to modern humans branched about 7 million years ago. This is the principle of *phylogenetic continuity*. Modification means that in the process of descent from common ancestors, species acquire differences. Evolution implies that all organisms will be similar in some respects and different in others. *Biodiversity* is the result of this branching process.

Evolution is thus not a steady march of progress to the attainment of preexisting biological goals, purposes, or functions. In the language of the ancient and medieval philosophers, there are no *final causes*. Evolution works purely in terms of the natural, *efficient* causes that result in differential reproductive success. Adaptations, the *functional* structures and processes in organisms, emerge as the result of what is essentially an unintentional, mindless process. In struggling to survive and reproduce, and in succeeding or failing in this (some more, some less), each member of a population of organisms unintentionally contributes to the emergence of adaptive changes in the population as a whole. It is precisely because organisms result from an unintentional, purposeless, mindless process that comparisons of organisms with (deliberately designed) machines are illegitimate.

As Gould has noted (1993, p. 146), Darwin's alternative to intelligent design of organisms builds adaptation *negatively* through the elimination of creatures lacking contingently fortuitous characteristics, permitting a much smaller number to pass their contingently fortuitous characteristics to future generations:

> [T]his . . . alternative is grossly inefficient and defies the logic of a clockwork universe, built by our standards and reasoning. No wonder it never entered Paley's head. The only thing going for this . . . view . . . is the curious fact that nature seems to operate this way. Nobody ever called this method elegant, but it gets the job done. (p. 146)

A gorilla wears a sandwich board that asks the question "Am I a Man and a Brother?" in answer to Charles Darwin's controversial theory on the evolution of the human race. (Hulton/Archive)

And in this observation is a glimpse of the significance of the Darwinian revolution: Evolution is a causal process, but not one that fits and coheres with a view of the universe as a designed machine. Darwin's theory thus represents a challenge not merely to a long theological tradition but also to a way of thinking about the objects of scientific inquiry: organisms as mechanical components of nature's grand machine.

Darwin and Mechanistic Physiology

Darwin's theory constituted a revolution in science precisely because it challenged an entire way of thinking that had guided the practice of science since the seventeenth century. Indeed, as Ernst Mayr has observed:

> Physiology lost its position as the exclusive paradigm of biology in 1859 when Darwin established evolutionary biology. When behavioral biology, ecology, population biology, and other

> branches of modern biology developed, it became even more evident how unsuitable mechanics was as the paradigm of biological science. . . . (1988, p. 12)

In other words, the appearance of evolutionary biology challenged a way of doing biology that had emerged from the rise of science itself during the Renaissance. And it did so by challenging the central metaphor of the biomedical approach to biology, that of the organism-as-machine. The question we must briefly address now is this: If Darwin challenged the metaphor of nature as a machine, then how does his theory stand with respect to the tradition of mechanistic physiology?

There are at least two things to examine here. First, how did investigators in the mechanistic tradition view Darwin's theory? Here we will focus on Claude Bernard's reaction to evolutionary theory. Second, what are the central implications of Darwinian evolution for important issues in physiology? Darwin himself did not deal with the implications of evolutionary biology for the picture of organisms presented by the mechanistic physiology of his day. But in the long course of the twentieth century, it has become clear to biologists that evolution does have important implications for physiology, and one way to bring these implications into focus is to reexamine the issue of the nature of species differences.

But first, how did Claude Bernard respond to Darwin's ideas? Bernard's concerns focused on issues of evidence and issues of method. Thus medical historian Paul Elliot has observed:

> Leading French biologists, such as Bernard himself . . . were resistant to the Darwinian theory of evolution. . . . [He] resisted these ideas because he saw them as the results of speculation unsupported by proper experimental evidence. The emergence of experimental physiology based on vivisection was therefore an integral part of a general trend in French science away from anything that could be interpreted as speculation towards a science based rigidly, too rigidly perhaps, on laboratory work and experiment. (1987, p. 73)

Part of the reason for this opposition to evolution was that Darwin was a naturalist, and his conclusions were generalizations based on the accumulation of vast amounts of circumstantial evidence. They did not grow directly out of controlled laboratory observation. This was a defect by Claude Bernard's lights.

For Bernard, who was trying to do to biology and physiology what Newton had done for physics, the dangers here were very real:

> [E]xcessive generalization creates an ideal science no longer connected with reality. This stumbling block, unimportant to a contemplative naturalist, is large for physicians who must first of all seek objective, practical truths. We must doubtless admire those great horizons dimly seen by the genius of a . . . Geoffroy Saint Hilaire, a Darwin. . . . Doubtless all these brilliant views do, after a fashion, guide and serve us. But if we gave ourselves up exclusively to hypothetical contemplation, we should soon turn our backs on reality; and in my opinion, we should misunderstand true scientific philosophy, by setting up a sort of opposition or exclusion between practice, which requires knowledge of particulars, and generalizations which tend to mingle all in all. ([1865] 1949, pp. 91–92)

Needless to say, for Claude Bernard, the knowledge of particulars that lay at the heart of true science was to come directly from tightly controlled laboratory experiments. It could not in the nature of the case be derived *indirectly* from circumstantial evidence, however great in quantity, about the hypothetical evolutionary relationships between organisms.

There are thus differences over what methods and what evidence are to count when doing biology. These differences persist in contemporary controversies in biology, where there is much heated debate over what can (and cannot) be learned from field observations and field studies and what can (and cannot) be learned from tightly controlled laboratory experiments. These controversies will loom large in Chapters 9, 10, and 11, where the focus is on what (if anything) can be learned about animal minds from the study of animal behavior.

The differences between Bernardian and Darwinian visions for biology are not just differences over what evidence and what methods are to count. In the preceding chapter it was seen that Bernard's vision for biology and physiology was shaped by his understanding of the methods and concepts used in mechanistic physics. In physics there was nothing analogous to a biological species, and hence to a difference between species. Moreover, Bernard, skeptical of evolution, had in the nature of the case no good account of the origin of species differences.

Being committed to causal determinism, Bernard downplayed the significance of species differences for the practice of physiology. He saw them as largely quantitative in nature, and this in turn implied that they were differences that could be compensated for to ensure that same cause was always followed by same effect. For example, a drug given to a big animal of one species should have the same effect as the same drug given to a smaller animal of another species, if it were given a bigger dose to compensate for quantitative weight and size differences. By contrast, two animals belonging to distinct species, but similar with respect to relevant quantitative factors, such as size and weight, should show the same effects after being exposed to the same causes. The analogy with physics is clear. If one mass is larger than another, then a larger force can be applied to the larger mass to make it accelerate in the same way as the smaller mass. By contrast, apply the same force to two identical masses, and they should accelerate in an identical manner. That one mass is red in color and the other is green, is neither here nor there. Similarly here. Two quantitatively similar organisms, belonging to different species, can be thought of as the same animal dressed up differently.

From the standpoint of evolutionary biology, however, species differences are biological effects that arise from the operation of various selective processes, and these are causal factors that have no analogs in the nonliving world of physics and chemistry. Natural selection, for example, is not a factor that a physicist or chemist has to worry about. There is no reason to suppose that evolutionary causes result in mere quantitative differences, as these might be understood by a physicist or a chemist.

That is, instead of being mere differences with respect to size, shape, mass, or surface area, evolved differences might involve differences with respect to complexity or with respect to complex organization. Thus Lewontin has observed that

> biological objects are internally heterogeneous in a way that is functionally relevant. While the earth has some variation in its internal composition, the behavior of the earth as a solar system object is unaffected by that heterogeneity. Organisms, on the other hand, are affected at every level of their functioning by their internal heterogeneity. (They can even change the way they fall "freely" through space as gravitational objects, as every sky diver knows). (1995, p. 8)

Evolved differences with respect to this internal heterogeneity can thus be expected to be important.

And it must never be forgotten that *variation* is ubiquitous in the biological world, both within and between species. And in complex, organized systems, even seemingly minor variation can lead to differences between organisms of great importance. For example, two members of different species that are nevertheless quantitatively similar—perhaps they are of the same size and weight—may react very differently when given the same drug.

Thus consider erucic acid, a component of canola (rapeseed) oil. Laboratory experiments revealed that erucic acid was highly toxic to rats and mice. As Lewontin has pointed out:

> Millions of dollars were spent breeding rapeseed to get rid of this compound, but it turns out that humans are not affected by erucic acid, nor indeed are laboratory animals like guinea pigs or rabbits. The sensitivity is a peculiarity of rat and mouse metabolism. Of course, it is better to err on the side of caution, but suppose it was humans who had the unique sensitivity. No animal model would have detected it. Until one has a detailed understanding of the physiology and metabolism of a species, it is not clear when one animal can be a model for another, even at the level of basic metabolism. (1995, p. 9)

Although the significance of evolved species differences is a matter to be examined in more detail in Chapters 7 and 8, it is worth observing here that Bernardian investigators doing physiology in the light of physics have a very different estimate of the nature and significance of species differences than do Darwinian investigators. It will be seen later that this is the source of some significant controversies in contemporary debates about the scientific value of animals as models for human biomedical phenomena.

This is all in the future, however, and for the present it is worth noting that Darwin's theory is of crucial importance for the concerns being discussed in this book because it has had a shaping influence on the broader implications of the similarities and differences between species. Thus Rowan et al. observed:

> Darwin's work also created a problem for those opposing animal experimentation because they sometimes argue (and still do) that animals were so unlike humans that they could not serve as experimental models of human physiology and biology. Of course, this leads to the discomforting corollary that, if animals are that dissimilar, then perhaps the moral questions are relatively minor. On the other side, scientists argued that the theory of evolution implied that animal biology would be sufficiently similar to human biology to render animals useful as research models, at least in some instances. Yet if animals are similar, then does it not raise moral questions about their use? This paradox seems to be especially severe for those interested in human psychology and cognition because of the moral weight usually accorded to cognitive abilities like rationality, speech, abstract thought and the like. (1995, p. 4)

But this only serves to focus our attention on what is meant by similarities and what is meant by differences. By now it should be obvious that the important similarities and differences in the moral debate about the use of animals are cognitive differences and similarities. Although Darwin did not say much about physiological similarities and differences between distinct species, he had much to say about cognitive similarities and differences.

Darwin on the Moral Status of Animals

Darwin's views about the mental capabilities of animals shaped, and in turn were shaped by, his views about morality. He saw morality as a device to overcome the natural limitations of human sympathies, and hence saw moral development as a process whereby these limited sympathies were gradually extended:

> As man advances in civilization, and small tribes are united into larger communities, the simplest reason would tell each individual that he ought to extend his social instincts and sympathies to all the members of the same nation, though personally unknown to him. This point being once reached, there is only an artificial barrier to prevent his sympathies extending to the men of all nations and races. If, indeed, such men are separated from him by great differences in appearance or habits, experience unfortunately shews us how long it is before we look at them as our fellow-creatures. (1896, pp. 122–123)

For Darwin, racism—the failure to extend one's sympathies to persons belonging to other races—resulted from a lack of moral development.

But for Darwin, what was true for racism was also true for what is known today as speciesism—the failure to extend one's sympathies to members of other species:

> Sympathy beyond the confines of man, that is, humanity to the lower animals, seems to be one of the latest moral acquisitions. It is unknown by savages, except towards their pets. How little the old Romans knew of it is shewn by their abhorrent gladiatorial exhibitions. The very idea of humanity, as far as I could observe, was new to most of the Gauchos of the pampas. This virtue, one of the noblest with which man is endowed, seems to arise incidentally from our sympathies becoming more tender and more widely diffused, until they are extended to all sentient beings. As soon as this virtue is honored and practiced by some few men, it spreads by instruction and example to the young,

> and eventually becomes incorporated into public opinion. (1896, p. 123)

From this it is clear that for Darwin, moral development was not the direct result of biological evolution but was in fact an artifact of cultural evolution, with moral sentiments spreading by instruction and example.

Darwin himself believed that our sympathies should be extended to members of nonhuman species. To this end, in 1863, he wrote an article in the popular monthly magazine *Gardner's Chronicle* on the evils of trapping animals:

> If we attempt to realize the sufferings of a cat, or other animal when caught, we must fancy what it would be to have a limb crushed during a whole long night, between the iron teeth of a trap, and with the agony increased by constant attempts to escape. Few men could endure to watch for five minutes an animal struggling in a trap with a crushed and torn limb; yet on the well-preserved estates throughout the kingdom, animals thus linger every night; and where game keepers are not humane, or have grown callous to the suffering constantly passing under their eyes, they have been known by an eyewitness to leave the traps unvisited for 24 or even 36 hours. (Quoted in Rachels 1991, p. 213)

Darwin's concern was for animal suffering; it was not dependent upon the species to which the suffering animal belonged. As was so with Bentham before him, it was the suffering that was morally relevant. But if concern for nonhuman, sentient animals represents the culmination of this developmental process, what of the moral status of vivisection in the context of physiological inquiry? Did Darwin think these were abhorrent surgical exhibitions?

It is worth pointing out that Darwin, who had a humane concern for nonhuman animals, was not blind to the needs of the biomedical sciences:

> I have all my life been a strong advocate for humanity to animals, and have done what I could in my writings to enforce this duty. . . . I know that physiology cannot progress except by means of experiments on living animals, and I feel the deepest conviction that he who retards the progress of physiology commits a crime against mankind. (Quoted in Rowan et al., 1994, p. 4)

And in a letter of 1871 he wrote:

> You ask about my opinion on vivisection. I quite agree that it is justifiable for real investigations on physiology; but not for mere

> damnable and detestable curiosity. It is a subject which makes me sick with horror, so I will not say another word about it, else I shall not sleep tonight. (Quoted in Rachels 1991, p. 214)

From this we see that the moral dilemma about the use of animals in science—a use that had provided Darwin himself with much valuable information of evolutionary significance—was keenly felt, but that when push came to shove, he was willing to bite the bullet and restrict the scope of his sympathies to his own species.

But there had to be a tradeoff: Experiments on animals had to yield information of *scientific* value—the crime against animals had to be offset by benefits for humans. The crime against animals in the light of physiological research had to be less of an evil than the crime against humanity resulting from the cessation of research.

Darwin was keenly aware not only of his great scientific debt to the science of experimental physiology but also of the value of genuine scientific knowledge:

> I have long thought physiology one of the greatest of sciences, sure sooner, or more probably later, greatly to benefit mankind; but, judging from all other sciences, the benefits will accrue only indirectly in the search for abstract truth. It is certain that physiology can progress only by experiments on living animals. Therefore the proposal to limit research to points of which we can now see the bearings in regard to health, etc., I look at as puerile. (Quoted in Rachels 1991, p. 216)

So for Darwin, the moral dilemma was not the hackneyed *your baby or your dog;* it was perhaps a more honest tradeoff involving *animal suffering for the sake of human knowledge of animals.* But in Darwin's case the dilemma was particularly severe because the resulting knowledge of animals had convinced him, through the development of the theory of evolution, that nonhuman animals, though very different in some respects, were very similar in other respects, especially cognitive respects. So much so that the pinnacle of moral development required their inclusion in the moral community.

Darwin stood, in fact, at the opposite end of the scientific spectrum from Descartes. For Descartes—and many of those who followed him—nonhuman animals were machines, phylogenetically and cognitively discontinuous from humans. For Darwin, evolution implied phylogenetic continuity, and this in turn implied a high degree of cognitive continuity. It is a good question—and one we shall return to in later chapters—as to whether phylogenetic continuity has implications for cognitive continuity, let alone judgments about relative extents of cognitive continuity. We will see later that no clear rules of proportionality exist. Exactly what evolutionary biology implies for differences and similarities is a complex matter.

In fact, as evolutionary studies of the mental lives of animals have progressed, one can find evidence of two opposing schools of thought: the *assimilators,* who see many analogies and similarities between human and animal behavior and who seek to ground these similarities in underlying cognitive similarities; and the *differentiators,* who seek to emphasize the differences between humans and nonhuman animals, especially cognitive differences. Darwin was definitely an assimilator.

Ethology, the study of the evolutionary biology of animal behavior, could be said to begin with Darwin himself, since he saw behavioral similarities between humans and nonhuman animals as the result of evolutionary processes. But Darwin did not pursue the issue, important to later ethologists, as to how animal behavior in its natural setting contributes to reproductive success (hence, how it might have evolved through natural selection). Had he given due consideration to evolved behavioral characteristics as solutions to species specific problems, it is possible that he might have been more cautious in his estimates of cognitive similarities.

Darwin on the Mental Status of Animals

Darwin employed the method of gradation to address the issue of the relative extents of cognitive development in animals:

> We have seen . . . that man bears in his bodily structure clear traces of his descent from some lower form; but it may be urged that, as man differs so greatly in his mental power from all other animals, there must be some error in this conclusion. . . . If no organic being excepting man had possessed any mental power, or if his powers had been of a wholly different nature from those of the lower animals, then we should never have been able to convince ourselves that our high faculties had been gradually developed. But it can be shewn that there is no fundamental difference of this kind. We must also admit that there is a much wider interval in mental power between one of the lowest fishes, as a lamprey or lancelet, and one of the higher apes, than between an ape and man; yet this interval is filled up by numerous gradations. (1896, p. 65)

In other words, nature affords numerous examples of cognitive gradation, and there is no reason to suppose that similar gradations do xist in the lineages leading to modern chimpanzees and modern nans respectively, after their divergence from a common ancestor (now estimated to be about 7 million years ago).

Today, with a better understanding of the fossil record than Darwin had, we have found more evidence of such cognitive gradation. This can be seen in the evidence of increasing cranial capacity in the various species found along the line that leads to modern humans. In the last 3.5 million years there has been a substantial increase in average cranial capacity. And there is further cultural evidence in terms of growing sophistication in tool-making and tool-using skills.

How did Darwin try to explain animal behavior? First of all, as a naturalist he was heavily dependent on his own field observations and on the anecdotes of other field observers. We will postpone until Chapters 10 and 11 a discussion of the controversies surrounding the use of field observations and anecdotal evidence, along with the issue as to whether Darwin's cognitively charitable conclusions have held up in the context of controlled laboratory experiments.

Also, and perhaps *because* he was an assimilator, Darwin had anthropomorphic tendencies. In this case it means that he tended to attribute human cognitive characteristics to nonhuman species. This can be seen most vividly in his comments on animal happiness, where he observes that "the lower animals, like man, manifestly feel pleasure and pain, happiness and misery. Happiness is never better exhibited than by young animals, such as puppies, kittens, lambs &c., when playing together, like our own children" (1896, p. 69). But more than this, emotions have similar physiological expressions in humans and nonhuman animals:

> Terror acts in the same manner on them as on us, causing the muscles to tremble, the heart to palpitate, the sphincters to be relaxed, and the hair to stand on end. Suspicion, the offspring of fear, is eminently characteristic of most wild animals . . . Courage and timidity are extremely variable qualities in the individuals of the same species, as is plainly seen in our dogs. (1896, p. 69)

And in this way, noticing behavioral analogies and similarities, Darwin gradually began to build up his case. Just as he had to amass a large quantity of circumstantial evidence to make an evidential case for the evolution of adaptations and the origin of species, so here he amassed a large quantity of circumstantial evidence to support his claims about cognitive continuity across species lines.

But Darwin's rationale for the search for evidence of cognitive similarity, regardless of whether his own methods were up to the task at hand, reflected the consequences of taking evolution seriously. In 1872 Darwin published *The Expression of the Emotions in Man and Animals* (the references here are to a 1965 edition). In this book he remarked:

> No doubt as long as man and all other animals are viewed as independent creations, an effectual stop is put to our natural desire to investigate as far as possible the causes of Expression. By this doctrine, anything and everything can be equally well explained; and it has proved as pernicious with respect to Expression as to every other branch of natural history. (1965, p. 12)

In these terms, the reason that Cartesian cognitive minimalism is wrong is not that animals have souls after all, but that we and they are phylogenetic kin—we have descended with modification from common ancestors.

But Darwin himself used the evidences concerning the expression of emotions in humans and nonhuman animals for a different purpose. First, Darwin noted:

> That the chief expressive actions, exhibited by man and by lower animals, are now innate or inherited,—that is, have not been learnt by the individual,—is admitted by everyone. So little has learning or imitation to do with several of them that they are from the earliest days and throughout life quite beyond our control; for instance, . . . the increased action of the heart in anger. (1965, p. 351)

It is not merely that the actions expressive of emotions are not cultural constructs that vary from context to context. They are inherited for a reason: They have selective value. Thus Darwin observed:

> The movements of expression in the face and body, whatever their origin may have been, are in themselves of much importance for our welfare. They serve as the first means of communication between the mother and her infant; she smiles approval, and thus encourages her child on the right path, or frowns disapproval. (p. 364)

Darwin saw in these observations further evidence for human evolution. Hence, they were of importance in the assessment of our relative place in nature.

Darwin began with some observations concerning the way in which the expression of emotions in humans transcends cultural boundaries:

> [A]ll the chief expressions exhibited by man are the same throughout the world. This fact is interesting, as it affords a new

> argument in favor of the several races being descended from a single parent-stock, which must have been almost completely human in structure, and to a large extent in mind, before the period at which the races diverged from each other. (1965, p. 360)

Darwin thus saw in cross-cultural similarities of expression a line of evidence supporting the conclusion that the distinct varieties of human being arose from a common ancestor. But his work also provided evidence of similarities of expression across species lines:

> We may confidently believe that laughter, as a sign of pleasure or enjoyment, was practiced by our progenitors long before they deserved to be called human; for very many kinds of monkeys, when pleased, utter a reiterated sound, clearly analogous to our laughter, often accompanied by vibratory movements of their jaws or lips, with the corners of the mouth drawn backwards and upwards, by the wrinkling of the cheeks, and even by the brightening of the eyes. (p. 360)

The many interspecific similarities with respect to the expression of emotions that Darwin cataloged were interpreted as evidence of descent from common ancestors. Any one of the similarities might have arisen by convergent evolution, but it is most improbable that all of them did.

Darwin thus saw the evidence concerning the expression of emotions as yet another argument supporting his thesis that humans have their origins in nonhuman ancestors. The point is not that humans are a cognitive gold standard, and we discover that nonhuman animals are similar in some respects to us. Rather, the similarities with respect to emotional expression are something we share with nonhuman animals, and the reason for the likeness has to do with descent from common ancestors—we are created from animals! As Darwin put it, "the study of the theory of expression confirms to a certain limited extent the conclusion that man is derived from some lower animal form, and supports the belief of the specific or subspecific unity of the several races" (1965, p. 365).

In addition to the similarities with respect to expression of the emotions, Darwin noted other similarities. For example, there are similarities with respect to *imitation*—and this means that nonhuman animals, like humans, are capable of learning, especially from parents. Darwin gives the example of a dog raised by a cat, imitating the action of the cat licking its paws. There are also similarities with respect to the faculty of attention, "as when a cat watches by a hole and prepares to spring on its prey" (1896, p. 73). For Darwin, this suggested that animals such as cats had a degree of consciousness of the objects in their environment and had expectations about the behavior of those objects. He also noted that nonhuman animals, like humans, often had excellent memories.

What, then, of the so-called higher faculties? For Darwin, the faculty of imagination enables us to bring together earlier images and ideas and to combine them in novel ways—sometimes, as in dreams, in ways independent of the will. The evidence of imagination in animals was seen by Darwin to lie in the observation that many mammals, and possibly birds, appear to dream. But what of the faculty of *reason,* that faculty that has traditionally been supposed to differentiate humans from nonhuman animals? Darwin observed:

> Only a few persons now dispute that animals possess some power of reasoning. Animals may constantly be seen to pause, deliberate and resolve. It is a significant fact, that the more the habits of any particular animal are studied by a naturalist, the more he attributes to reason and the less to unlearnt instincts. (1896, p. 75)

For Darwin, the similarities—at least in some nonhuman species—went right across all the faculties that we see in ourselves, though not necessarily developed to the same degree.

Nevertheless, there do appear to be fundamental differences between humans and nonhuman animals:

> It has been asserted that man alone is capable of progressive improvement; that he alone makes use of tools or fire, domesticates other animals, or possesses property; that no animal has the power of abstraction, or of forming general concepts, is self-conscious and comprehends itself; that no animal employs language; that man alone has a sense of beauty, is liable to caprice, has the feeling of gratitude, mystery &c.; believes in God, or is endowed with conscience. (1896, p. 79)

While noting that it might be hard to find behavioral evidence that would unambiguously settle these matters one way or another, Darwin contended that rudiments of these higher faculties and abilities could perhaps be observed—that it would be unwise, without further scrutiny, to dismiss the possibility:

> It may freely be admitted that no animal is self-conscious, if by this term it is implied, that he reflects on such points, as whence he comes or whither he will go, or what is life and death, and so forth. But how can we feel sure that an old dog with an excellent memory and some power of imagination, as shewn by his

> dreams, never reflects on his past pleasures or pains in the chase? And this would be a form of self-consciousness. (p. 83)

This is a matter we will return to later, since some prominent contemporary investigators have denied that animals can be self-conscious because they lack language. The spirit of Descartes is far from dead, notwithstanding the work of Darwin.

Where did Darwin stand on the issue of language? On this issue Darwin represented a return to the spirit of Montaigne. Animals are capable of minimal language use. Darwin recognized that many nonhuman animals have sophisticated communication systems but also that humans too use nonlinguistic communication systems—as in cries of pain, fear, anger, or surprise. Nevertheless there are differences:

> That which distinguishes man from lower animals is not the understanding of articulate sounds, for, as every one knows, dogs understand many words and sentences. In this respect they are at the same stage of development as infants, between the ages of ten and twelve months, who understand many words and short sentences, but cannot yet utter a single word. It is not the mere articulation which is our distinguishing character, for parrots and other birds possess this power. Nor is it the mere capacity of connecting definite sounds with definite ideas; for it is certain that some parrots, which have been taught to speak, connect unerringly words with things, and persons with events. The lower animals differ from man solely in his almost infinitely larger power of associating together the most diversified sounds and ideas; and this obviously depends on the high development of his mental powers. (1896, p. 86)

Once again, where others saw differences between humans and nonhuman animals, Darwin was an assimilator who saw similarities. The theory of evolution does not require that animals be rudimentary language users or that there be no characteristics that are unique to a species. The method of gradation might have inclined Darwin to see differences of degree where other saw differences of kind, but this does not settle the matter in his favor. And indeed, it is still a matter of considerable controversy.

So where did Darwin see the big differences between humans and nonhuman animals? He observed:

> I fully subscribe to the judgement of those writers who maintain that of all the differences between man and the lower animals, the moral sense or conscience is by far the most important. . . . It is the most noble of all the attributes of man, leading him

> without a moment's hesitation to risk his life for that of a fellow creature; or after due deliberation, impelled simply by the deep feeling of right or duty, to sacrifice it in some great cause. (1896, p. 97)

But Darwin could not resist the temptation of dealing with this issue from the standpoint of natural history. For if nonhuman animals are not moral agents, we can nevertheless see in some of them, at least, the evolution of characteristics that may serve as precursors for the development of a moral nature—a nature that allows us to ameliorate the natural limitations of our sympathies to others. Perhaps, then, our moral natures grow out of our animal natures!

So what might these evolutionary precursors of the moral sense actually look like. Upon what foundation could a moral sense evolve? Darwin suggested:

> The following proposition seems to me in a high degree probable—namely, that any animal whatever, endowed with well-marked social instincts, the parental and filial affections being here included, would inevitably acquire a moral sense or conscience, as soon as its intellectual powers had become as well, or nearly as well developed as in man. (1896, p. 98)

Darwin believed that several factors would be at work. First, that with sociality come feelings of sympathy for one's fellows. Second, with the development of mental faculties comes an appreciation of motives and consequences of past actions. Third, with language comes the ability to communicate the wishes of the community, and finally habits would be important since they bolster social instincts and feelings of sympathy.

Darwin was not here trying to say what is right and wrong—to derive what *ought* to be the case, from a moral standpoint, by examining what *is* the case from a biological point of view. Instead, he was trying to explain what it is about the biology of humans and nonhuman animals that enables humans to have a moral sense and enables some animals, at least, to be potentially capable of developing a moral sense in the fullness of evolutionary time. In considering this issue, Darwin warned that nonhuman social animals, with developed intellectual capabilities, might develop a moral sense different from ours (1896, p. 99). Perhaps this moral sense would differ by being tailored to the peculiarities of the social context confronting such a hypothetical creature, as well as the ecology in which it was embedded. Is there any evidence that Darwin might have been moving along the right lines in these speculations? We know of no nonhuman species whose

Dolphins are large-brained, intelligent creatures that live in societies where there is much mutual dependence. Connor and Norris (1982) identified three categories of altruistic care-giving behavior in dolphins. (Hulton/Archive)

members properly qualify as moral agents. But what about the existence of evolutionary precursors?

It is just possible that biologists after Darwin have indeed uncovered some relevant phenomena that may have bearing on the question of evolutionary precursors for the development of a moral sense. In particular, there have been observations, in mammals and other species, of altruism between unrelated individuals. Indeed there is now an enormous literature on altruistic behavior in nonhuman species and the conditions under which it can evolve. Altruistic behaviors are behaviors that benefit a (possibly unrelated) individual at some risk to the benefactor.

Dolphins afford a good example, for they are large-brained, intelligent creatures that live in societies where there is much mutual dependence. Connor and Norris (1982) identified three categories of altruistic care-giving behavior in dolphins. First, there are *standing-by behaviors,* where dolphins stay in the vicinity of individuals in need, thereby exposing themselves to the attention of predators. Second, there are *assistance-rendering behaviors,* whereby risks are taken in assisting a (possibly injured) fellow dolphin to evade a predator or escape capture, sometimes by attacking capture vessels. Third, there are *supporting behaviors,* in which, for example, a dolphin may help an injured individual get to the surface to breathe (see also Dugatkin 1997, p. 114).

Connor and Norris pointed out that there are some similarities between dolphin altruism and human altruism:

> In this case altruistic acts are dispensed freely and not necessarily to animals that can or will reciprocate. They need not necessarily be confined to the species of the altruistic individual. In human terms, a person can rescue a helpless fledgling that has fallen from a nest, and this does not imply any conscious intent

> toward the rest of society. The person is, instead, motivated first by a broad concept of distress and then by a complex of emotional responses, learning and social standards. To us the evidence from dolphins clearly fits this model. (1982, p. 370)

The dolphin case is worth mentioning for at least three reasons. First, because the line that leads to humans diverged from the line that leads to dolphins over 60 million years ago—unlike chimpanzees and other primates who show cooperative and altruistic behaviors, dolphins are not particularly close relatives. Second, the explanation of dolphin altruism offered by Connor and Norris attributes some degree of high-level cognitive processing to the dolphins (a matter to be reexamined in Chapter 11). Third, since part of the function of morality in human society is the amelioration of limited human sympathies for other humans (and at least some members of some nonhuman species), and since this seems to involve varying degrees of altruistic caregiving, perhaps dolphins do show the rudiments of a moral sense.

Further Reading

H. L. LaFollette and N. Shanks, *Brute Science: The Dilemmas of Animal Experimentation* (London: Routledge, 1996), discuss the role played by Charles Darwin in the development of biomedical theory. Ernst Mayr provides many interesting perspectives on Darwin's role in the development of biological theory. See E. Mayr, *Toward a New Philosophy of Biology: Observations of an Evolutionist* (Cambridge: Harvard University Press, 1988); *One Long Argument: Charles Darwin and the Genesis of Modern Evolutionary Thought* (Cambridge: Harvard University Press, 1991); and *This Is Biology: The Science of the Living World* (Cambridge: Harvard University Press, 1997).

The development of evolutionary thought, including work before and after Darwin, is given an excellent treatment by D. J. Depew and B. H. Weber, *Darwinism Evolving: Systems Dynamics and the Genealogy of Natural Selection* (Cambridge: MIT Press, 1995). Controversies surrounding Darwinism are discussed in M. Ruse, *The Evolution Wars: A Guide to the Debates* (Santa Barbara, CA: ABC-CLIO, 2000).

7

Darwinism Developed: The Ontogeny of an Idea

Darwin knew nothing about the mechanisms of inheritance. One of the most important developments in evolutionary biology in the twentieth century was the fusing together of evolutionary ideas about natural selection as a force driving change in populations over successive generations, with genetics, the science of heredity and variation in populations. An examination of these developments will enable us to see more closely how the evolutionary picture of humans and nonhuman animals differs from that in the tradition of mechanistic physiology.

This chapter will also be used to develop some ideas about Darwinian medicine, since evolutionary biology has a view of health and disease that is somewhat different from that found in the mechanistic physiological tradition. The detour into Darwinian medicine will pay dividends because in Darwinian medicine, human evolutionary history plays a crucial role in explaining what *we* are and how we differ from other organisms we interact with. We can learn much about what we are and where we stand in nature by looking at ourselves in sickness as well as in health.

The New Synthesis

The bringing together of Darwinian ideas about adaptive evolution by natural selection, and ideas from the science of genetics concerning variation and heritability in populations, results in something known as the *new synthesis* in evolutionary biology. These events took place over a thirty-year period beginning in the 1930s. Many theorists were involved in the formation of the synthesis, but of particular note are biologists such as R.A. Fisher, J. B .S. Haldane, Sewell Wright, Julian Huxley, Ernst Mayr, G. Ledyard Stebbins, George Simpson, and Theodosius Dobzhansky. Since the 1960s, thanks to the work of theorists such as George Williams and W. D. Hamilton, among many others, a thoroughly gene-centered view of evolution has emerged—and has been popularized by Richard Dawkins in *The Selfish Gene* (1989).

We saw in the preceding chapter that the grist for the evolutionary mill is provided by the existence of heritable variation in populations of interest. Cells carry genetic material. The particles of inheritance are called *genes,* and genes are made up of deoxyribonucleic acid (DNA). One of the most basic distinctions in genetics is that between *phenotype* and *genotype:*

> The "phenotype" of an organism is the class of which it is a member based upon the observable physical qualities of the organism, including its morphology, physiology, and behavior at all levels of description. The "genotype" of an organism is the class of which it is a member based upon the postulated state of its internal hereditary factors, the genes. (Lewontin 1992, p. 137)

Corresponding to this distinction is that between *genome* and *phenome:*

> The actual physical set of inherited genes, both in the nucleus and in various cytoplasmic particles such as mitochondria and chloroplasts, make up the *genome* of an individual, and it is the description of this genome that determines the genotype of which the individual is a token. In like manner there is a physical *phenome,* the actual manifestation of the organism, including its morphology, physiology and behavior. (p. 139)

The Genome Project has revealed that the human genome contains about 30,000 genes (compared to 13,600 for the fruit fly *Drosophila*) (Szathmáry, Jordan, and Pál 2001).

Genes are carried on *chromosomes,* which are threadlike structures in the nucleus of a cell consisting of DNA and associated proteins. DNA is composed of two chains of nucleotides. These are organic compounds consisting of a sugar, deoxyribose, linked to a nitrogen-containing *base.* The bases are adenine, cytosine, guanine, and thymine. The chains are wound round each other in the form of a spiral, ladder-shaped molecule—the famous *double helix.* The bases on each chain pair with a base on the other chain to form *base-pairs.* Adenine pairs with thymine, guanine with cytosine. Each base-pair can be thought of as a *bit,* or basic unit, of information. There are approximately 3.5×10^9 bits of information in the human genome.

In diploid organisms—organisms with two sets of chromosomes, one from each parent—the matched pairs of chromosomes are called *homologous chromosomes.* In humans, barring chromosomal abnormalities, each cell contains forty-six chromosomes (twenty-two matched pairs and one pair of sex chromosomes, with females having XX-pairs and males having XY-pairs). The number of matched pairs is called the *chromosome number, n,* the number of chromosomes. In humans, $n = 23$.

The *locus* of a gene is its position on a chromosome. For a given locus, a population of organisms may contain two or more variant forms of the gene associated with that locus. These variant forms of a gene are called *alleles.* In diploid organisms (for example, mammals), there are two alleles of any gene, one from each parent, which occupy the same relative position on homologous chromosomes. When one such allele is *dominant* and the other is said to be *recessive,* the dominant allele influences which particular characteristic will be found in the organism's phenome. (It is possible for both alleles to be fully expressed; this is called *codominance.* In humans, the AB blood group results from the expression of the A and B alleles. There are also cases where neither allele is fully expressed, and a characteristic results from the partial expression of each. This is called *incomplete dominance.*)

Important for evolution is the existence, in populations of organisms, of multiple alleles. A given allele may be found with a given statistical frequency in a population. Evolution occurs in a population when the relative frequency with which alleles are found in that population changes (for whatever reason). An important part of the new synthesis was the development of sophisticated techniques to analyze changing allele frequencies in populations of interest. The resulting theory is thoroughly *gene-centered.*

By this it is meant that what gets replicated are genes, and it is genes that travel down the generations—genes, barring mutations, that are *identical by descent.* Underlying the heritable variation in morphological, physiological, and behavioral characteristics observed in populations of interest is variation with respect to alleles. Parents pass on alleles to their offspring, which get 50 percent of their alleles from each parent. *Recombination,* the process whereby genes are shuffled during *meiosis*—the formation of either sperm or egg—results in offspring having a suite of characteristics different from those of their parents. *Germ-line mutation,* by contrast, results in heritable changes in an organism's genetic constitution. Both these processes add to variation in populations.

What parents pass on to their offspring are alleles. Alleles that contribute to reproductive success are more likely to find themselves in the next generation, in higher frequencies, than alleles that do not. Such alleles are said to confer *fitness* advantages. Members of a population of organisms typically differ from each other with respect to *relative fitness*. Differences with respect to relative fitness are defined in terms of differential reproductive success. Thus the effect of natural selection is to change the frequencies with which alleles are found in populations. As Ewald has noted, natural selection favors characteristics of organisms that increase the passing on of the genes (alleles) that code for those characteristics (1994, p. 4). Evolution works across generations in populations. It is populations that evolve, not the organisms that constitute them at any given time. The phenotypic characteristics favored by natural selection are called *adaptations*.

Consider *Staphylococcus aureus*, the microorganism responsible for much wound infection in hospitals. *S. aureus* can be treated with antibiotics. A given population of microorganisms colonizing a patient will typically vary with respect to susceptibility to a given dose of antibiotic. *S. aureus* reproduces asexually about every twenty or so minutes, giving rise to the next generation. The bacteria with alleles conferring tolerance to the clinical dose administered will reproduce and get those alleles into the next generation. The susceptible bacteria will be eliminated from the population. Over successive generations, alleles for antibiotic tolerance will increase in frequency. Antibiotic tolerance is thus a bacterial adaptation to hosts periodically flooded with antibiotics.

Taking a gene's-eye view of evolution also enables us to explain otherwise paradoxical features of organisms—for example, why individuals can behave altruistically toward one another. Altruism—behaving in ways that have the potential to reduce one's own fitness while increasing that of others (for example, a bird giving the alarm when it sights a predator potentially draws attention of the predator to itself while the others in the group escape)—looks like a paradox if evolution is construed as a theory about selfish *organisms*. How could genes disposing their bearers to altruism, once they had arisen, ever spread in a population? Would they not be penalized by natural selection?

The paradox is resolved through the realization that many animal populations consist of genetic relatives—kin. The fitness cost (offspring not produced), C, to an altruist in a group of organisms will be offset by a fitness benefit (offspring produced), B, to other members of the group, who are related in varying degrees r to the altruist ($r = 1$ for identical twins, 0.5 for full siblings, 0.125 for first cousins, and so on). In the simple case of one organism acting for the benefit of another, an altruism-disposing allele will increase in frequency in a population when $rB > C$. This is Hamilton's inequality (Hamilton 1964; see also Maynard Smith 1988). The basic idea is that one's relatives contain genes *identical by descent* to one's own to varying degrees. If being altruistic gets more copies of altruism-disposing alleles into the next generation, then altruism will be favored by natural selection, even though a given altruistic organism may perish. Genetic cost-benefit analysis shows under certain conditions that selfish genes do not necessarily dispose their bearers to be selfish. (There are many other puzzling phenomena that are explained by taking a gene's-eye view of evolution. See Maynard Smith 1988; Dawkins 1989.)

In emphasizing a gene-centered view of evolution, it should not be supposed that we are trying to *reduce* biology to genetics. Genes rarely *determine* phenotypic characteristics; rather the phenotype results from a complex, developmental interaction between the genome and the environment (including the maternal environment, and even the relative orientations of cells in the developing organism). It is far from being a simple matter of *all in our genes.* Moreover, the organism does not result from the independent action of individual genes. Genes typically work together in complexes and networks to make their contributions to the developing organism. Nor are we suggesting that genes are more important than the bodies that house them. In a real sense it is bodies that confront the selective environment, and genes are only as good as the bodies to which they contribute. Nevertheless, it is copies of genes, identical by descent or duly mutated, that travel down the generations and carry the heritable information needed for the operation of the evolutionary mechanisms that bring about changes within and between populations.

The new synthesis also resulted in an understanding of the importance of nonadaptive evolution. Allele frequencies can change for reasons unconnected with the operation of natural selection. Such changes can be effected by *gene flow*—the exchanges of alleles within and between populations—and the cessation of gene flow between populations can allow for the successive accumulation of significant genetic differences between those populations. Nonadaptive evolution can also result from *genetic drift*—changes in allele frequencies brought about by chance events. These changes can be very important, statistically speaking, in very small populations, where, just by chance, genes that are rare in large populations can become common. More will be said about this when we discuss *speciation*—the mechanisms by which new species are formed. There are other ways in which nonadaptive evolution can occur too, but the main point is that there are a multiplicity of ways in which allele frequencies can change, some of which involve selective mechanisms of various kinds, and some of which do not.

Evolutionary biologists believe that biodiversity results from speciation—the complex array of processes that give rise to new species. First, what is a species? As we saw in Chapter 1, the medieval view held that species were groups of organisms all of which had the same *essence* or *form*. For example, there was something that all dogs had in common, and in virtue of this they were dogs. On this view there are absolute discontinuities between species, and since the essences were generally viewed as unchanging, evolution of new species was a conceptual impossibility. The realization that the origin of new species was an observational fact, even if the precise mechanisms were obscure, led to the ultimate scientific demise of this view of species.

Other early species concepts, for example, the *morphological* species concept, maintained that species could be identified by their *shape* (especially the shapes of anatomical features) construed as a measure of form. These ideas fell into disrepute primarily through the observation of *polytypic* species, in which there is a great deal of variation with respect to characteristics, especially morphological characteristics, but also through the observation of *sibling species;* distinct species could be very nearly morphologically indistinguishable—implying that speciation can occur without change of form.

What is needed is a concept of species that reflects the genetic concerns of evolution. And at least for organisms that reproduce sexually, we have the *biological species concept* (BSC). Sibling species counted as distinct species because they were reproductively isolated from each other—and thus were incapable of interbreeding because of physiological or behavioral barriers. As Ernst Mayr has observed:

> A species . . . is a group of interbreeding natural populations that is reproductively (genetically) isolated from other such groups because of physiological or behavioral barriers. . . . Why are there species? Why do we not find in nature simply an unbroken continuum of similar or more widely diverging individuals, all in principle able to mate with one another? The study of hybrids provides the answer. If the parents are not in the same species (as in the case of horses and asses, for example), their offspring ("mules") will consist of hybrids that are usually more or less sterile and have reduced viability, at least in the second generation. Therefore there is a selective advantage to any mechanism that will favor the mating of individuals that are closely related (called conspecifics) and prevent mating among more distantly related individuals. This is achieved by the reproductive isolating mechanisms of species. A biological species is

> thus an institution for the protection of well-balanced, harmonious genotypes. (1997, p. 129)

Although this view of species is subject to various qualifications in the professional literature—for example, it could not apply to asexually reproducing organisms—it does provide us with enough of a grasp of what a species is to advance our discussion of the similarities and differences between humans and nonhuman animals.

In light of the BSC, we can begin to get a handle on what may be involved in speciation. Any mechanism that can bring about genetic divergence between populations to the extent that those populations are reproductively isolated (physiologically or behaviorally) will be a speciation mechanism. Many mechanisms can be devised and tested in the laboratory—typical experiments might involve short-lived organisms, such as fruit flies, which can be tracked in real time for fifty or more generations, to test the consequences of various hypotheses. Which of the possible mechanisms actually play a role in driving this process in nature is a current matter of scientific inquiry, one requiring careful field observations, for example, of Darwin's finches on the Galapagos islands (Weiner, 1994).

An important hypothesis in this regard is that of *allopatric speciation*—the view that the genetic divergence leading to speciation occurs in geographically isolated populations. Speciation resulting from geographic isolation could happen in at least two ways. Both ways result in a disruption of gene flow between previously interbreeding populations.

In what is known as *dichopatric* speciation, a previously continuous range of interbreeding populations is disrupted by a newly arisen barrier (it might be a new river), or there could be disruption through geological changes (separation of land masses, formation of mountains), or possibly disruption through the emergence of vegetational discontinuities. Mayr commented:

> Either strictly by chance, as in the case of chromosomal incompatibilities, or by a change of function in behavior as a consequence of sexual selection, or as an incidental byproduct of ecological shift, the two separated populations will become genetically more different in time and, as a correlate of this difference, will acquire isolating mechanisms that will cause them to behave as different species, when, later, they come again into contact with one another. (1997, pp. 182–183)

Such species may even be morphologically similar when they meet, but they qualify as species if there is reproductive isolation.

In what is known as *peripatric* speciation, cessation of gene flow and genetic drift play an important role. In this type of speciation, a very small founder population is established beyond the boundary of the species' previous range. For example, a bird is blown from the mainland to an island during a storm. Mayr commented:

> Such a population, founded by a single inseminated female, or by a few individuals, will contain only a small percentage of, and often an unusual combination of, the genes of the parent species. Simultaneously, it will be exposed to a new and frequently severe set of selection pressures owing to its changed physical and biotic environment. Such a founder population may undergo drastic genetic modification, and may speciate rapidly. (Mayr 1997, p. 183)

Again the result will be a population reproductively isolated from the ancestral population. There are other possible speciation mechanisms, but what they have in common is an exploration of the various ways in which populations can acquire enough genetic divergence to become reproductively isolated from each other.

Evolutionary biologists have come to realize that the discontinuities that constitute species differences are the results of complex dynamical processes. At the genetic level, these processes result in cessation of gene flow between populations. When genetic differences between populations accumulate to the point of reproductive isolation—for whatever reason, and it may be a combination of adaptive and nonadaptive evolution—speciation is the result. But if, from the standpoint of evolution, the differences and similarities between species are genetic differences and similarities, how much genetic difference is enough?

Consider humans and chimpanzees. It is sometimes said that there is a 99 percent genetic (base-pair) similarity between humans and chimpanzees. Does this not make them fundamentally similar to us—humans in ape suits, perhaps? The issue is rather more complex than it might at first appear. First of all, a lot of our DNA is not expressed and has no functional significance—the so-called *junk DNA*. Such DNA diverges between species at a constant rate, and differences and similarities with respect to the degree of this divergence may record little about differences and similarities between species but may merely convey information about the time since divergence. In the present case all it may mean is that the line that leads to modern humans diverged from the line leading to modern chimpanzees about 7 to 10 million years ago (Lewontin 1995, pp. 15–16).

A medical researcher holding a chimpanzee. There is said to be a 99 percent genetic (base-pair) similarity between humans and chimpanzees. However, this doesn't make chimpanzees fundamentally similar to us because a lot of our DNA is not expressed and has no functional significance. (National Library of Medicine)

Nevertheless, if we are so similar to chimps at the genetic level, we are also clearly different both morphologically and behaviorally. How could this be explained? At the genetic level, a distinction has recently emerged between *structural genes* (that make protein when turned on) and *regulator genes* (that turn the structural genes on and off). In tandem with this distinction, the idea has also arisen that genes do not work in isolation but work together in complex, interconnected networks—in fact, the study of this phenomenon belongs to a new branch of biology known as *genomics* (Carroll *et al.* 2001; Davidson 2001).

In this context, what matters is not enormous similarity with respect to structural genes but rather a few changes in regulator genes affecting the patterns of activity of those genes. As King and Wilson put it:

> Small differences in the timing of activation or in the level of activity of a single gene could in principle influence considerably the systems controlling embryonic development. The organismal differences between chimpanzees and humans would then result chiefly from genetic changes in a few regulatory systems. . . . (1975, p. 114)

This is another way of making the point that organisms exhibit a high degree of causally significant organized complexity. In this case, the complexity is the interactive complexity of genetic switching networks. Important parts of the biological significance of species differences between organisms arise because of differences with respect to this particular kind of organized complexity.

In such interconnected networks, a single mutation in a regulator gene could have very large effects, bringing about changes in large patterns of gene expression (Kauffman 1993, p. 412). Two organisms may be said to be similar not simply because they have roughly similar numbers of similar genes, but because in addition to these similarities, they have the same types and numbers of interactions between them. But now, when comparing such interactive systems across species lines, there will be no simple, linear quantitative rules relating degree of genetic (base-pair) similarity in organisms of distinct species to degree of phenotypic similarity.

Another way to make this point is to consider not humans and chimpanzees, but humans and insects. Over the past ten years many genes (including the so-called *Hox* genes) have been found to regulate similar developmental roles in animals as distantly related as mammals and insects. And developmental biologists have been confronted with a puzzle known as the Hox paradox:

> How can bodies as different as those of an insect and a mammal be patterned by the same developmental regulatory genes? Very few anatomical structures in arthropods and chordates can be traced back to a common ancestor with any confidence. Yet to a rough approximation, we humans share most of our developmental regulatory genes not only with flies, but also with such humble creatures as nematodes and such decidedly peculiar ones as sea urchins. (Wray 2001, p. 2256)

One approach to this paradox was to simply deny that distantly related animals were that different after all. It has since become clear, however, that developmental regulatory genes have acquired new roles in both insect and mammal lineages since divergence from a common ancestor. This has led to a new approach to the Hox paradox in which it is recognized that though developmental regulatory genes have been conserved—so that similar genes are found in distantly related organisms—their interactions have not been. Theorists now contend that many of the changes we see in animal evolution are the result of rewiring developmental gene networks (p. 2256). Thus Carroll et al. observed:

> The recurring theme among the diverse examples of evolutionary novelties . . . is the creative role played by evolutionary changes in gene regulation. The evolution of new regulatory linkages—between signaling pathways and target genes, transcriptional regulators and structural genes, and so on—has created new regulatory circuits that have shaped the development of myriad functionally important structures. These regulatory circuits also serve as the foundation of further diversification. (2001, pp. 167–168)

There is a good sense, then, in which developmental biology is showing that diversity is in the details of the genetic interactions!

The perils of being misled by degrees of genetic similarity will come back to haunt us in the next chapter, where the focus will be on nonhuman animals as models of human biomedical phenomena, and in Chapter 11 in the context of a discussion of similarities and differences with respect to cognition, between humans and nonhuman animals. For the present, it is important to note that our current understanding of evolutionary biology has added more weight to the idea, emerging from Darwin's own work, that organisms are very different from intelligently designed machines. One way to approach this issue is to look at the issue of biological development and to reconsider the question as to whether humans and nonhuman animals are machines.

Genes, Machines, and Development

Darwin's original theory of evolution laid down a powerful challenge to the claim that organisms were machines. Adaptations—the very features of organisms that seemed to cry out for an account in terms of deliberate intelligent design—could be accounted for in terms of the operation of natural processes, and especially natural selection. Biology after Darwin has continued to challenge the viability of mechanical conceptions of organisms—this time from the standpoint of reproduction and development.

Mechanistically minded biologists of the seventeenth and eighteenth centuries had to explain the apparent generation and development of new organisms. How could one machine, the mother, give rise to other machines, the offspring? Mechanistic biologists formulated the theory of *preformationism* as an answer to this question. According to this theory, organisms are fully formed and differentiated in the seeds from which they are derived, with the developmental process being viewed as a process by which the preformed, miniature organism simply increases in size. In effect the little person expands into a big person.

There were two schools of preformationism. One, led by Jan Swammerdam (1637–1680), held that individuals were preformed in the egg. He argued that an egg contained all future generations as preformed miniatures—a bit like Russian dolls, with one doll inside another, and so on. Another school, based on the work of Antonie van Leeuwenhoek (1632–1723) and Nicolas Hartsoecker (1656–1725), saw the preformed humans (or *Homunculi*) as residing in sperm.

The mechanists saw organisms as machines but could not see how mechanical principles involving matter in motion could explain reproduction. Preformationism sidesteps the issue by seeing organisms as fully formed in their seeds, with all future generations of each species being preformed in miniature, one within another, at the time of initial design and creation by God. As Albrecht von Haller (1707–1777) put it:

> The ovary of an ancestress will contain not only her daughter, but also her granddaughter, her great granddaughter, and her great-great granddaughter, and if it is once proved that an ovary can contain many generations, there is no absurdity in saying that it contains them all. (Quoted in Mason 1962, p. 367)

The preformationist school in effect solves the problems of reproduction and development by denying that reproduction occurs (future generations are already there in miniature) and conceiving of development as an expansion of a preformed individual. Or as Swammerdam put it:

> In nature there is no generation but only propagation, the growth of parts. Thus original sin is explained, for all men were contained in the organs of Adam and Eve. When their stock of eggs is finished the human race will cease to be. (Quoted in Mason 1962, p. 364)

Needless to say, modern biology has found no evidence of preformed individuals in either sperm or egg. Nevertheless, other options are possible for those who wish to see organisms as machines.

To understand what is going on, it is useful to go back to the writings of William Paley. Paley, it will be recalled from Chapter 3, thought of organisms as intelligently designed systems to be understood through an analogy with machines such as pocket watches. But watches, unlike organisms, do not reproduce and develop. Paley anticipated this objection as follows:

> Suppose, in the next place, that the person who found the watch should after some time discover that, in addition to the properties which he had hitherto observed in it, it possessed the unexpected property of producing in the course of its movement another watch like itself—the thing is conceivable; that it contained within it a mechanism, a system of parts—a mould, for instance, or a complex adjustment of lathes, files and other tools—evidently and separately calculated for this purpose. . . . ([1801] 1850, p. 14)

This is the *machina ex machina hypothesis.* In Paley's self-replicating machine, it is imagined that the machine has a mechanical program and equipment to first manufacture the components of a watch and then to assemble these parts into a new, functioning, offspring watch that inherits the ability to replicate itself from the parent watch. Paley's theory has the defect that although it offers an explanation of reproduction—it does not sidestep the issue as the preformationists did—it makes development mysterious. Mammalian parents do not make fully grown copies of themselves. It is as though big clocks make little pocket watches that somehow turn into big clocks.

Nevertheless, Paley's *machina ex machina* hypothesis is not entirely without merit. Something like it was developed independently by John von Neumann (1903–1957), one of the architects of modern computer theory. Von Neumann offered a theory of self-reproducing automata—hypothetical devices now known as *von Neumann machines.* It will be useful to briefly examine this modern version of Paley's hypothesis. Like many seventeenth-century theorists (for example, Hobbes and Leibniz), von Neumann made a distinction between *natural automata* (organisms) and *artificial automata* (sophisticated machines, possibly of our own manufacture). Von Neumann's development of a *machina ex machina* hypothesis did not have a theological motivation. Instead, it grew out of a puzzle in computer theory.

In the theory of *machine computability,* an ideal computer—a hypothetical computer capable of performing all possible computations—is known to mathematicians as a *Turing machine* (named after the mathematician Alan Turing, 1912–1954). As J. N. Crossley et al. noted, Turing

> delineated the concept of a machine, a computing machine. He gave arguments to suggest that any computation a person could do could also be done by one of these machines. He analyzed all the steps that you do when you make a computation: writing things down, scanning blocks of symbols, making notes, going back to things you had done before, making lists, and all this sort of thing, and he devised a kind of machine that could do all these operations in a very simple way. The machine only needs to be able to look at squares on a tape and to identify symbols in the squares. It looks at one square at a time and it can move one square to the left and one square to the right and it can change the symbol. It has a program telling it what to do if it sees a certain symbol and it has a finite program; that is all there is to it. (1979, p. 8)

A computer, then, has inputs and outputs in the form of a tape (or some other medium, for example, electrical signals). And it has internal states that enable it to respond to the tape. As von Neumann pointed out, such a machine merely modifies something—the tape—which is completely different from it.

Von Neumann saw the above conception of automata as incomplete. He wanted to include in an account of machine computability, machines whose operations are directed toward themselves:

> A complete discussion of automata can be obtained only by taking a broader view of these things and considering automata which can have outputs something like themselves. Now one has to be careful what one means by this. There is no question of producing matter out of nothing. Rather one imagines automata which can modify objects similar to themselves, or effect syntheses by picking up parts and putting them together, or take synthesized entities apart. (1966, p. 75)

The von Neumann machine, then, finds or synthesizes parts similar to those it is made of. And it then assembles those parts into a copy of itself. In effect, the machine is programmed to interact with its environment to make more machines just like itself, which inherit the program to interact with the environment and make copies of themselves, and so on. The von Neumann machine can be thought of as an abstract mathematical model for something like Paley's self-replicating watch.

Although von Neumann was clearly interested in the implications of his research in the field of machine computability for biological questions (1966, p. 77), he nevertheless saw differences between his hypothetical self-replicating machines and real self-replicating organisms. For example, there is a fundamental difference between organisms and computers with respect to error. Von Neumann points out that a single error will often stop a computer (users of personal computers know this all too well). By contrast:

Jon von Neumann saw differences between his hypothetical self-replicating machines and real self-replicating organisms. He hoped to shed light on processes in organisms by considering processes in his abstract, hypothetical machines. (Bettmann / Corbis)

> The fact that natural organisms have such a radically different attitude about errors and behave so differently when an error occurs is probably connected with some other traits of natural organisms, which are entirely absent from our automata. The ability of a natural organism to survive in spite of a high incidence of error (which our automata are incapable of) probably requires a very high flexibility and ability of the [natural] automaton to watch itself and reorganize itself. And this probably requires a very considerable autonomy of parts. (1966, p. 73)

Von Neumann nevertheless hoped to shed light on processes in organisms by considering processes in his abstract, hypothetical machines.

In fact, it turns out that animal development is not very much like machine assembly at all. Development does not proceed through the initial fashioning of parts and subsequent assembly of those parts by the craftsman or even, in the Paley–von Neumann case, by the parent machine. It is actually a process far more intriguing than a machine assembly process.

Animal development proceeds from the fusion of sperm and egg to form a zygote. This process typically requires an appropriate maternal environment, but the mother does not deliberately bring about this fusion as a watchmaker (or self-replicating watch) might join two components together. The zygote undergoes mitosis, giving rise to two daughter cells, each having a nucleus containing the same number and kind of chromosomes as the cell from which they are derived. These cells in turn divide and form a *blastula*—a hollow ball with the *blastomeres* (dividing cells of the embryo) forming a *blastderm* (layer) around a *blastocoel* (or central cavity).

The *gastrula* stage of development succeeds the blastula stage and is characterized by the production of *germ layers*—layers of cells from which the animal's organs will be derived in the course of developmental time. At this stage of development the embryo gets converted into a cup-shaped structure with a hollow cavity. The important point to be borne in mind is that all the cells in the developing embryo are genetically identical, and the question naturally arises as to how cells become specialized into liver cells, brain cells, kidney cells, and so on. It does not appear that the parent deliberately fashions differently specialized cells and then assembles them into an organism.

We now believe that the process of cell differentiation depends on different genes being active in different cells. *Structural genes* (genes that make the proteins constitutive of the developing body's infrastructure) get turned on or off by proteins made by *regulator genes* (and there can be complex cascades of switching activity). Regulator genes are turned on and off in complex ways by chemicals in their environments. Different cell types result from different patterns of switching activity. As Maynard Smith and Szathmáry have recently noted:

> In the cells of multicellular animals and plants, genes tend to have many different regulatory sequences, and are affected by many regulatory genes. Hence the activity of a particular gene, in a particular cell, can be under both positive and negative control from different sources, and can depend on the stage of development and of the cell cycle, on the cell's tissue type, on its immediate neighbors, and so on. (1999, p. 113)

The developing embryo thus makes cells that, with appropriate environmental cues, *self-organize* into specialized cells and tissues. They do not require an external guiding hand to account for their origin.

What we also need is some account of the mechanisms by which the initial developmental symmetry (a cluster of identical cells) becomes an asymmetric, spatially patterned structure containing specialized cells with distinct morphologies. Although we still have much to learn about the mechanisms underlying developmental processes, some things have been discovered through studies on mice and the fruit fly *Drosophila*. One mechanism by which specialization can occur is called *embryonic induction,* which Maynard Smith and Szathmáry explained through the following example:

> [T]he lens of the vertebrate eye is formed by the differentiation of typical epithelial cells. What makes these cells different from other epithelial cells is that they come into contact with the eye cup, an outgrowth of the developing brain that will become the

> retina and the optic nerve. Thus a group of cells that would otherwise have become a normal component of the skin are induced to form a lens by contact with the eye cup. This has the desirable consequence that the lens forms exactly in front of the retina. (1999, pp. 117–118)

But embryonic induction implies at least the prior development of minimal structure. What other mechanisms could there be?

We now know that one mechanism involves the developmental roles played by chemical gradients, a concept initially proposed by Lewis Wolpert. Thus, if a chemical substance (referred to as a *morphogen*) is made at a given point in an embryo, chemical diffusion will carry the substance away from the point of origin and will establish a chemical concentration gradient. The basic idea is that cells can respond to the local concentration level, with different genes getting switched on or off in different regions of chemical concentration. As noted by Maynard Smith and Szathmáry:

> Typically, gradients arise from within the embryo itself. In principle, a single gradient could specify many embryonic regions, with different concentrations switching on different genes. In practice, however, it seems that not more than two, or at most three, regions are specified by a single gradient. . . . Thus only a small amount of spatial complexity is generated in a single step. Embryonic development depends on a series of steps, with genes that are switched on in one step being the source of signals in the next. (1999, p. 119)

Thus as a morphogen diffuses through a sheet of cells, the chemical gradient (ranging say, from high to medium to low concentration) excites or induces cells in different regions to respond differently, resulting, for example, in differentiation and the emergence of spatial asymmetries.

At this point we are a long way from parental watches assembling offspring watches. The embryo develops as the result of its genes, its complex interactions with its environment, and its subsequent modifications of its local environment, including itself. In other words, there are complex processes of self-organization occurring in a developing system that has complex exchanges with its surroundings. The modern understanding of replication and development is thus very different from the picture derived from the *machina ex machina* hypothesis.

Indeed, though organismal development has turned out to be very different from computerized machine assembly, one of von Neumann's hypothetical devices—the *cellular automaton*—has come to play an important role in the computer modeling of the behavior of real biological systems. A cellular automaton is basically a unit system that interacts with its environment, including its immediate neighbors, in accord with simple rules involving switches (1966, pp. 93–110). If you followed the discussion of development above, you will have noticed that real cells in embryos also interact with their neighbors in ways that have consequences for switching, this time *genetic switching*.

Von Neumann was particularly interested with the behavior of lattices or sheets of cellular automata. Such a sheet forms what is now known as an *excitable medium*. A cell (or region) in an excitable medium might be in one of several possible states: It might be unexcited (but excitable upon appropriate stimulation), excited (owing to appropriate stimulation), or temporarily refractory to excitation (owing to prior excitation). Sheets of cellular automata are often employed in computer simulations of complex biological processes, including the origins of life and evolution, as well as development (see Goodwin 1996; Kauffman 1993; 1995). These sheets are of interest to biologists precisely because they can exhibit the phenomenon of self-organization: the formation of complex spatial and temporal patterns and structures as a result of initial stimulation and dynamical interactions internal to the system—patterns and structures that do not require a guiding intelligence for their formation.

Much recent interest has focused on sheets of cells that are excitable and that, once excited, fall into a refractory state before becoming excitable again. In such a sheet, if a cell is stimulated and excited (switches are turned on in accord with rules), it then interacts with its immediate neighbors, and they respond in turn to this excitation, and interact with their immediate neighbors, and so on. Waves of interaction effects can then propagate out from initial points of excitation in the sheet because cells just excited fall into a temporary, unexcitable state. Since real chemical and biological systems behave like excitable media (Shanks 2001), computer-generated cellular automata can be used to model them. Contemporary interest in modeling self-organization and evolution using cellular automata is referred to under the heading of *artificial life* (Goodwin 1996, p. 184).

We now have good reason to believe that self-organization is a phenomenon crucial to the origin of life on earth, as well as the development of organisms, and that evolution is a natural, unguided process, exploiting the fruits of self-organization, by which we get the diversity of life. Yet the natural processes that give rise to these complex organismal systems are now beginning to be unraveled and understood with the aid of machines—some employing the idea of cellular automata. As in modern physics, so too in modern biology: Machines (for example, computers) are tools to understand nature, but the resulting understanding of nature is no longer as a grand machine, with humans and animals as intelligently designed parts.

Contemporary developments in evolutionary biology thus add further weight to the conclusion that humans and nonhuman animals are not literally machines. It was seen in Chapters 5 and 6 that the mechanistic tradition in physiology and medicine grew out of a mechanical view of nature. Since evolutionary biology has done much to undermine the view of organism-as-machine, it is now important to consider again the implications of evolution for human physiology and medicine. For we can learn much about ourselves and our place in nature by considering how evolution, as a historical process, is currently shaping our understanding of ourselves in sickness and in health. This discussion will also set the scene for the next chapter, in which controversies concerning the roles of nonhuman animals in contemporary biomedical research will be examined.

Evolutionary Medicine at a Glance

First, let us take a brief look at what evolutionary medicine actually involves. Evolutionary medicine is based on the idea that current biomedical effects and their proximate biological causes reflect the operation of causal processes operating over long periods of historical time. The effects of history show up in at least two distinct kinds of ways: first, in terms of the specific adaptations organisms have acquired and the conditions under which they were acquired; second, in terms of adaptations possessed by organisms they interact with, especially predators, prey, pathogens, and parasites.

For any population of organisms, evolution is a dynamical process reflecting not just adaptation to the changing physical environment but also coevolution with the (evolving) biotic environment. The historical consequences of this dynamical interaction do much to shape the characteristics of the extant organisms we see today. When evolutionary trajectories differ, as they do for species that have diverged from common ancestors, the differences between the members of those species we see today will reflect not just quantitative differences with respect to mass and surface area but also the effects of these more distant, dynamical, interactive processes that have forced successive generations of the respective populations to solve (or perish in the face of) unique environmental problems.

Humans are the product of the action of evolutionary processes. It was seen in the preceding chapter that the similarities and differences between species reflect the consequences of these processes. Trevathan et al. bring out the significance of this by observing:

> Our biology and many of our behaviors, are the result of millions of years of evolutionary history, beginning with the earliest forms of life. Characteristics that identify us as mammals (e.g., viviparity, mammary glands, homoiothermy, hair) may have appeared with the earliest mammals 225 million years ago. Characteristics that place us in the taxonomic order Primates (e.g., prehensile digits, structure of the middle ear, arboreal adaptations) trace their origins to the earliest primates 65 million years ago. Selection for visual acuity, grasping hands and feet, greater encephalization (larger brains relative to body size), and long periods of maturation may have begun 40 million years ago or more with the origin of the "higher primates" (monkeys, apes, and humans). Characteristics of our taxonomic family, Hominidae, began to be selected in our ancestors approximately 5 million years ago. (1999, p. 4)

The medical consequences of evolution are not hard to uncover. For example, bipedalism—a characteristic identifying us as members of the family Hominidae—began to be selected for about 5 million years ago (mya). It is a characteristic responsible for a whole host of medical problems, ranging from back ache to hemorrhoids.

Australopithecus is the earliest known genus in the family Hominidae, and fossils of several species have now been identified, dating from 4.4 to 1 mya. *Australopithecus* had a pelvis that allowed upright posture. Between 2.5 and 2 mya the genus *Homo* emerged, and around 0.2 mya there is clear evidence of our own species, *Homo sapiens* (Trevathan et al. 1999, p. 4). Our Australopithecine ancestors had a brain volume (average) of 450 cubic centimeters (cc), not much bigger than that of a modern chimpanzee. In the genus *Homo,* our ancestor *H. habilis* (approx. 2 mya) had an average brain volume of about 660 cc. *H. erectus* (approx. 1.6 mya) had a brain volume of 950 cc. Modern humans (approx. 0.5 mya) have an average brain volume of 1,400 cc. Brain size has more than tripled in 3 million years. Large brains require large brain cases, and this makes for difficult human births—a good example, in fact, of the medical consequences of evolution.

There is evidence of the use of crude stone tools by *Australopithecus,* and sophistication in tool manufacture and use has accompanied the increases in brain size. As Price has observed:

> Brain size is obviously associated with the coordination of some very complex processes. One is the sensitivity of the human hand. Another is the need for learning to remember design technique in tool making and hunting, and the rudiments of culture and social norms. As skills and culture increased, so the need for communication, with the ultimate development of language, involving complex integration of lips, tongue, larynx and a large commitment in the brain. (1996, pp. 250–251)

This rapid evolution of increasing brain size made it possible for us to become language users and to become capable of forming and living in sophisticated cultural settings. The transition from life in hunter-gatherer communities to life in agricultural communities—the agricultural revolution—seems to have occurred about 10,000 years ago. Although no one knows exactly when language evolved, it almost certainly preceded the agricultural revolution, with the evolution of the anatomical structures that would eventually support the evolution of a capacity for language themselves beginning with the appearance of the genus *Homo* itself.

Whether you can have any thought without language is a debatable issue, but one thing is almost certain, and that is that sophisticated thought and language use go together. But the ability to think—the possession of problem-solving intelligence—is biologically important for any creature. For humans:

> Thinking has conferred on us a priceless adaptive advantage. Evolutionarily speaking, we are successful because our ability to think has enabled us to remain physically unspecialized. We are the supreme generalists. We prove it by our ability to live anywhere and make our living in a hundred different ways. We don't grow thick coats; we get them from other animals. We don't grow long necks; we invent ladders. We don't have teeth as big as apes do, nor are we as strong, pound for pound. We don't see as well as hawks. We don't run as fast as any large quadruped. But by our wits, and more recently by the devices we make, we can outperform all of them in every way. (Edey and Johanson 1989, pp. 383–384)

Clearly a discussion of our large brains cannot be separated from a discussion of how we are cognitively different from nonhuman animals and also of how we are physiologically different. The evolutionary trajectory we have taken means that we and our nonhuman relatives are not simply the same animal dressed up differently, either psychologically or physiologically.

Our large brains have recently enabled us to live much longer.

Model of an Australopithecus skull. Our Australopithecine ancestors had a brain volume (average) of 450 cc, not much bigger than that of a modern chimpanzee. Brain size has more than tripled in 3 million years. (Karen Huntt H. Mason / Corbis)

We are not the only species to benefit from cultural evolution (Bonner 1980), but we are beneficiaries of a peculiarly sophisticated form of cultural evolution that has given us modern science in the last 400 years and effective, life-saving medicine in the last 150 years. But the same cultural evolution that has given us these benefits has also transformed the conditions of life, especially since the agricultural revolution 10,000 years ago.

By this, it is meant that the environment in which we are embedded has a physical component, which, barring natural disasters, changes relatively slowly. It has a biotic component that is coevolving with us—today it is made up primarily of prey (our food), parasites, and pathogens (in earlier times, it also included predators). There is also a cultural component that changes very rapidly. There has probably not been much evolutionary change in the human brain in the last 2,000 years, but we can see cultural conditions today that are very different from those experienced by our ancestors two millennia in the past. In these cultural transformations lie the source of many peculiarly human medical problems. For as Trevathan et al. have observed:

> Taking a conservative estimate of the origin of humanness as 200,000 years ago, we can safely say that more than 95% of our [uniquely human] biology, and presumably many of our behaviors, were shaped during the time period in which our ancestors lived as gatherers of wild food resources. This lifestyle included low fat diets, low technology, adequate but not overly abundant food resources, low birth rate, near-exclusive breastfeeding for the first two years, long (4 year) birth intervals, low population,

> high mobility, and living in small (25–50 people), kin-based social groups. Natural selection thus favored characteristics and behaviors that were advantageous and suitable for that kind of lifestyle. (1999, p. 4)

They go on to point out that this evolutionary fact, coupled with our capacity for fast cultural evolution, gives rise to important features of our current medical predicament:

> The human environment changes with the origin of agriculture, however, and many of those behaviors are no longer advantageous. These changes include sedentism, a narrowing of the food base, increase in fat intake, larger communities, infectious diseases, decreased birth interval, increased birth rate, increased total population, and changes in interpersonal relations (e.g., male-female and intercommunity dynamics). (p. 5)

In other words, the environment of adaptation—the environment for which most of our uniquely human characteristics were selected—was a hunter-gatherer environment in which very few of us live today.

One consequence of this is that some types of disease result from the fact that we are maladapted for the environment in which we are currently embedded (for example, heart diseases, diabetes, many cancers). Moreover, in a hunter-gatherer culture, overabundance of food was rare. A selective advantage in this environment lies in the ability to quickly store calories as fat, so that there are reserves to be drawn on in times of scarcity. Yet in these physically active hunter-gatherer groups, obesity was rarely a problem, because reserves were frequently consumed. Not so in modern, fat-saturated, sedentary cultures. What was at one time an adaptation conferring a fitness advantage is now a liability—and not one that selection can readily weed out, since its bad effects rarely show up in time to disrupt reproductive success (except, perhaps, in very extreme cases).

Notwithstanding the remarks above, it is nevertheless important to note that modern evolutionary biology has had a relatively limited impact on biomedical research. Moreover, where it has had an impact, it is often in a form that demonstrates misunderstandings of evolution as a process and evolutionary biology as a scientific theory about the nature and implications of that process. As Day has recently observed:

> It is one of the minor mysteries of medicine that the publication by Darwin (1859) that identified evolution as a continuing biological process and that accounted for the diversity and successive change in living organisms was not immediately taken up by

> the physicians of the day. Their daily work then, as now, related to the whole animal biology of one species. This failure to recognize the significance of the idea of evolution and the explanatory power of its mechanism, natural selection, may relate to the opposition of the established church, which focused all its disbelief upon the perceived contradiction the theory posed to the biblical explanation of the universe as set out in the Book of Genesis. Be that as it may, it is still a matter of wonder that it has, until recently, been rare to hear an evolutionary reason put forward as an aetiological factor in any but a few disorders, and when they have been cited, it has often disclosed an ignorance of adaptation as a process—for example, backache is caused by *incomplete* evolution to upright posture. (1999, p. vii)

Nevertheless, an effort is currently underway to reshape the biomedical sciences and the practice of medicine to reflect the facts and theories of evolutionary biology.

The result of this effort is a comparatively new discipline called *evolutionary medicine.* But as Trevathan et al. observed:

> Evolutionary medicine takes the view that many contemporary social, psychological, and physical ills are related to incompatibility between the lifestyles and environments in which humans currently live and the conditions under which they evolved. Unfortunately much of modern medical practice demonstrates a misunderstanding of the evolution of physical responses to stresses that were faced by our ancestors. (1999, p. 3)

In a similar vein Ewald has commented:

> Evolutionary biology is so firmly integrated with the rest of biology that it is not possible to mark a boundary between them. But modern medicine has been a peninsula. It is broadly and firmly connected with most regions of biology such as anatomy, physiology, biochemistry, molecular biology and genetics, but has just a few thin bridges traversing the gulf to evolutionary biology. Knowledge about the evolution of antibiotic resistance is perhaps the best developed bridge between the disciplines. The discovery of the evolutionary basis for sickle cell anemia—protection against malaria—is another. (1994, p. 7)

There are several things worth noting here. First, though the perceived connections between evolutionary biology and biomedicine exist, they are few and far between. Second, until comparatively recently, where evolution has been invoked in the context of medicine, it was often with misunderstandings of what evolution was or implied. Third, where the implications of evolution have been properly thought through—as in the case of antibiotic resistance—it has yielded much valuable fruit.

Before proceeding further, it is instructive to see how the medical implications of evolution have been misunderstood by the medical community. The issues go well beyond errors about incomplete evolution to upright posture (as though evolution were a steady march toward a preestablished goal of postural perfection, as opposed to a tinkering, trial-and-error process, with no end in sight, just differential reproductive success). They include conceptual issues about the very way to understand evolution.

A good example concerns the phenomenon of *virulence*—the degree of harm caused to a host by a parasitic or pathogenic organism. In evolutionary biology, this phenomenon is discussed under the heading of host-parasite coevolution. Ewald (1994) provided a good, detailed treatment of this topic. But as he was at pains to point out, the medical literature on the evolution of virulence sees it as a process involving a steady march to a state of benign coexistence. Thus René Dubos could write: "Given enough time a state of peaceful coexistence eventually becomes established between any host and parasite" (quoted in Ewald 1994, p. 3), and Lewis Thomas could say, "Disease usually represents the inconclusive negotiations for symbiosis . . . a biological misinterpretation of borders" (quoted in Ewald 1994, p. 3).

The rationale offered for this obligate evolution to benign coexistence is that natural selection favors what is best for a parasite species or for most individuals within a parasite species. The argument runs like this: If a parasite species exhibits a high degree of virulence and incapacitates and kills its host organism speedily, this will be bad for the large number of individuals in the host—and bad for the species as a whole that parasitizes such hosts. Lower degrees of virulence will allow more parasites to live longer in their chosen hosts. On this view, selection operates for the benefit of the *species* (for example, Burnet and White 1972)—or the greatest number of individuals over the longest period of time (for example, Simon 1960).

But evolutionary biologists do not think this way. As Ewald pointed out in his discussion of virulence, evolution does not operate with goals such as benign coexistence in mind:

> Natural selection favors characteristics that increase the passing on of the genes that code for the characteristics. If more rapid replication of a virus inside a person leads to a greater passing on of the genes that code for that rapid replication, then replication rate will increase even if the more rapid growth of the virus

> population within a person causes the person to be severely ill, or leads to an overall decrease in the numbers of the virus among people, or hastens the eventual extinction of the virus. (1994, p. 4)

It is wrong to think that virulent viruses are necessarily inefficient, leaving behind lots of progeny in dead hosts. Ewald offered the following analogy:

> We might as well say that maple trees are poorly adapted because 990 out of every 1000 helicopter-like seeds are doomed to an early death. The number relevant to natural selection is the number of genes passed into the succeeding generation. Would the genes coding for production of 1000 seeds be left in greater numbers than genes coding for production of 100 seeds? The 1000 seed strategy may be vastly more wasteful in terms of seed death and tissue destruction, but if it ultimately yields more trees in succeeding generations, it will be more efficient in terms of evolutionary success. (p. 5)

This is but one important, medically relevant consequence of seeing evolution from a gene's-eye point of view—a matter discussed earlier in this chapter. What is at work here is the grim calculus of the heritability of traits leading to reproductive success.

This discussion of host-parasite coevolution not only serves to show how evolution's implications can be misunderstood, even by investigators friendly to evolution; it also serves as an introduction to an important source of species differences that are biomedically significant. For host-parasite coevolution, even though it has resulted in the evolution of virulence in some populations of microorganisms, has also led to the formation of mutually beneficial relationships between animals and microorganisms. It will be seen in the next chapter that differences in these evolved relationships have implications for patterns of metabolism across species lines. The resulting differences between species will not reflect quantitative differences to be simply scaled away.

The human medical predicament, like the biological and cognitive predicaments, is in fact deeply intertwined with the causes and effects of human evolution. And this is a changing medical predicament, for fast cultural evolution has changed the selection pressures on our ancient pathogenic foes, as when, with the advent of antibiotic technologies, populations of microorganisms responded with the evolution of drug resistance. But none of this changes the fact that what we are biologically, cognitively, or medically reflects the long course of evolutionary causation.

Differences between ourselves and nonhuman animals with respect to cognition, biology, or medicine also reflect these same causal factors. So one way to look at what we are and how we are different from nonhuman animals is to look at us in evolutionary perspective. In the next chapter, it will be seen how taking an evolutionary perspective about the nature of species differences has had some radical implications for estimates of the usefulness on nonhuman animals as models for human biomedical phenomena. This is one of the key places where the contemporary conflict between evolutionary biology and the mechanistic tradition in physiology is currently been played out.

Further Reading

D. J. Depew and B. H. Weber, *Darwinism Evolving: Systems Dynamics and the Genealogy of Natural Selection* (Cambridge: MIT Press, 1995), provide a helpful review of developments in evolutionary theory since Darwin's day. The new gene-centered view of evolution has been popularized by R. Dawkins in *The Selfish Gene* (Oxford: Oxford University Press, 1989). Perhaps the very best textbook concerning contemporary thinking in evolutionary biology is found in D. Futuyma, *Evolutionary Biology* (Sunderland, MA: Sinauer Associates, 1998).

Theoretical extensions of Darwinian principles are discussed in an accessible manner in G. Cziko, *Without Miracles: Universal Selection Theory and the Second Darwinian Revolution* (Cambridge: MIT Press, 1995). Contemporary thinking about the interface between developmental and evolutionary biology is reviewed in S. B. Carroll et al., *From DNA to Diversity: Molecular Genetics and the Evolution of Animal Design* (Malden, MA: Blackwell Science, 2001). The role of self-organization in biology is discussed in S. A. Kauffman, *The Origins of Order: Self-Organization and Selection in Evolution* (Oxford: Oxford University Press, 1993), and *At Home in the Universe: The Search for the Laws of Self-Organization and Complexity* (Oxford: Oxford University Press, 1995). In this regard I also recommend B. Goodwin, *How the Leopard Changed Its Spots: The Evolution of Complexity* (New York: Harper Torchbooks, 1996). The basic ideas behind the new Darwinian approach to medicine are reviewed in R. M. Nesse and G. C. Williams, *Why We Get Sick: The New Science of Darwinian Medicine* (New York: Vintage Books, 1995).

8

The Mouse as Man Writ Small: Animals in Modern Medicine

Contemporary biomedical science in the United States and Europe, though subject to laws and regulations governing the treatment of animal subjects, is deeply committed to the use of such animals as research subjects. The American Medical Association (AMA) sees animal research as being essential to progress in human medicine (1992, p. 11). Sigma Xi, the scientific research society, has defended the use of animals in biomedical research by citing the enormous benefits:

> Results from work with animals have led to understanding mechanisms of bodily function in humans, with substantial and tangible applications to medicine and surgery (e.g., antibiotics, imaging technologies, coronary bypass surgery, anti-cancer therapies), public health (e.g., nutrition, agriculture, immunization, toxicology, and product safety). . . . As the Surgeon General has stated, research with animals has made possible most of the advances in medicine that we today take for granted. An end to animal research would mean an end to our best hope for finding treatments that still elude us. (1992, p. 74)

For many purposes, biomedical researchers argue forcefully that they need to be able to experiment on live animal systems.

As we saw in Chapter 4, opponents of animal experimentation find the practice to be morally objectionable. They wonder whether human health and well-being could be promoted equally well by adopting policies and regulations aimed at promoting public health. They emphasize measures aimed at the prevention of disease—measures that would reduce the need for animal experiments to find cures for disease. Nevertheless, though public health measures have been useful, the American Medical Association has concluded:

> [F]or most infectious diseases, improved public health and nutrition have played only a minor role. This is clear when one

> considers the marked reduction in the incidence of infectious diseases such as whooping cough, rubella, measles and poliomyelitis. Despite advances in public health and nutrition, eradication or control of these and most other infectious diseases was not achieved until the development of vaccines and drugs through research using animals. (1992, pp. 11–12)

From the standpoint of the AMA, the history of biomedicine in the twentieth century was a vindication of the methodological precepts laid down around the middle of the nineteenth century by Claude Bernard.

Animal Use in Modern Science

The AMA estimated that somewhere between 16 and 22 million animals were used annually for biomedical research and testing in the United States (1992, p. 15). Rowan et al. suggested that a conservative estimate would range between 14 and 21 million (1995, p. 14). Of these animals, 85 to 90 percent were probably rats and mice. The ratio of commercial to noncommercial to governmental use of animals in the United States is 45:40:15 (Rowan et al. 1994, p. 19). Fewer than 1 percent of animal subjects are used in testing of household and personal care products. In the commercial context, most are used for discovering, developing, and testing new medicines and therapies. Overall use of animals in the lab can be broken down into the following categories: education, drug discovery and toxicity testing, development and toxicity testing of other products, testing of biological agents, diagnosis, and research in the biomedical sciences. Of these categories, Rowan et al. commented:

> [D]iagnosis and education probably account for less than 5% of the total each. Toxicity testing of other products will account for around 10% of the total (with more such testing involved in drug discovery and biological production). Drug discovery and biological production may account for between 30–40% of all animal use with other research accounting for the remainder. (p. 20)

Though the use of animals in various research contexts is subject to governmental regulation, defenders of the practice tout its direct benefits to humans; opponents cite the pain and suffering endured by nonhuman animals.

The mechanistic physiological tradition has tended to simultaneously emphasize biological similarities between mice and humans on the one hand, and great differences with respect to cognition, on the

Of the 16 and 22 million animals used annually for biomedical research and testing in the U.S., 85 percent to 90 percent are probably rats and mice. Most animals are used for discovering, developing, and testing new medicines and therapies. (National Library of Medicine)

other. In contemporary biomedical research one can still hear echoes of Claude Bernard, but it is a Bernard who has been forced to acknowledge, if not to fully appreciate, the theory of evolution. The remainder of this chapter is concerned with controversies surrounding evolution's implications for the relevance of experiments on nonhuman animals for human health and well-being. Issues surrounding evolution's implications for morally relevant cognitive similarities and differences between humans and nonhuman animals will be examined in Chapters 10 and 11.

Man and Beast: Similar yet Different

It is instructive to examine the scientific rationale for using nonhuman animals, typically rats and mice, as experimental models of humans. Most researchers will probably acknowledge that there are indeed biological differences between humans and the animals used to model them. As Sir William Patton has pointed out, however:

> [O]ne should not over-estimate these differences. We have only to look at an account of evolution or at textbooks in comparative

> anatomy to see how much we have in common: hearts, lungs, kidneys, brains, endocrine glands, nerves, muscles, digestive systems, all built on the same plan. This homology goes back still further as one moves down to the biological elements like the nucleus, the mitochondria, or the cell membrane, out of which the higher organisms are built. The only differences [in drug actions] that appear are in dose required, duration of action, sometimes in the way action manifests itself, and sometimes in side-effects. (1993, p. 166)

The similarities between organisms to which Patton referred are explained as the result of biological evolution and the fact that humans and the animals used to model them in experiments share common evolutionary ancestry.

The modern mechanistic physiologist does not deny the facts of evolution but sees in its implications two methodological rules. Geneticist Richard Lewontin characterized these as follows:

> First, it is assumed that the detailed similarity between organisms increases as one goes to more and more basic cellular processes within organisms. So, a rat and a human may not look alike, but they are assumed to have similar general structures for their nervous systems, extremely similar chemistry for the actual firing of individual nerve cells, and identical chemistry for the copying of genes that code for the production of these chemicals. So the model maker feels comfortable carrying over the results from rats to people if it is a question of basic cellular processes. That is why rats are used for drug trials, at least to screen for harmful effects of new drugs. Second, it is assumed that whatever differences exist at any level, these get smaller and smaller as one compares organisms that are more closely related in evolution. Thus the model maker feels confident about carrying over nearly anything seen in chimpanzees to people. (1995, p. 9).

To borrow words from biochemist A.G. Cairns-Smith (1985, p. 50), the first assumption implies that at the biochemical level humans and their nonhuman animal models are "the same animal dressed up differently." The second assumption brings mechanistic physiology's linear, quantitative estimate of the significance of species differences to a discussion of the significance of evolved differences. These two assumptions represent the way in which mechanistic physiologists have tried to accommodate evolution's implications for their experimental practices.

It is a good question as to whether either of these assumptions is correct. On the one hand, Richard Lewontin pointed out that "the

Experimental obesity studies are performed on rats. According to geneticist Richard Lewontin, "a rat and a human may not look alike, but they are assumed to have similar general structures for their nervous systems, extremely similar chemistry for the actual firing of individual nerve cells, and identical chemistry for the copying of genes that code for the production of these chemicals. . . ." (National Library of Medicine)

problem is that neither of these rules turns out to be reliable. There is an extraordinary diversity among organisms and between different aspects of the same organism even for basic cellular processes. . . . At the level of cell metabolism, there is even less carryover from species to species" (1995, p. 9). On the other hand, we have Sir William Patton, the medical establishment, and vivisection's utilitarian apologists all speaking of the great utility and direct benefit of experiments involving nonhuman animals for human health and well-being.

If ever there was a controversy in science, this is it. All sides to this debate accept the theory of evolution. Nevertheless there are profound differences concerning its implications for the biological relevance of research conducted on nonhuman animal subjects for human biomedical phenomena. Who is right? To discuss this issue, it will be important to look a bit more closely at the nature of the similarities and differences between humans and nonhuman animals.

Organisms are not just complex systems; they are systems that exhibit hierarchical complexity. By this it is meant that complexity can be found at various levels in an organism. At the bottom of the organizational hierarchy, there are the macromolecular components of organisms. They are individually complex, and there is much complexity in their mutual chemical relationships. Macromolecules are organized in turn into intracellular structures, which in turn are organized into cells of various types. Cells in turn are organized into tissues, tissues into organs, and organs into the organism. Complexity is found at each level of the hierarchy, and the levels affect each other in complex ways.

We have just seen that the first assumption undergirding the use of animals as models for human biomedical phenomena hinges crucially on some types of evolution having been relatively conservative, especially at the lower levels of the biological hierarchy. Thus it is argued that, at the biochemical level, there are enormous similarities between organisms. Organisms are made up of the same types of chemical building blocks, and the central metabolic pathways are very similar in all known forms of life. Other types of evolution have been less conservative; for example, there are numerous and striking evolved differences in morphology and behavior. So to what extent are humans and the animals used to model them in biomedical research simply "the same animal dressed up differently"?

In biomedical research, an ideal animal model should be the same as the human with respect to (1) symptoms, (2) postulated etiology, (3) neurobiological mechanism, and (4) treatment response (for example, Crawley et al. 1985, p. 300). Only humans could closely approximate these *desiderata*. In practice we have to work with animal models that are different, to varying degrees, in one or more of these four respects. This means that researchers must find ways to accommodate species differences.

But how, exactly, are species differences to be accommodated? As we have seen in earlier chapters, this depends in large measure on how you understand the species differences themselves. And as should be evident by now, this is a matter concerning which there are serious differences of opinion. The mechanistic physiological tradition extending back to Bernard, but permeating much modern biomedical research, tends to see species differences as being largely quantitative, reflecting differences in such factors as size, mass, and surface area.

Consequently, if this is the nature of the difference between mice and humans, we should be able to accommodate the differences through consideration of *scaling factors:* allow for differences in size, mass, and/or dose, and same cause in human and model will be followed by same effect. In fact, scaling has become an integral component of research methods in toxicology. In carcinogenicity studies, surface area (calculated as body weight$^{2/3}$) is the scaling factor of choice (Klaassen and Eaton 1993, p. 43). When due allowance has been made for relevant differences in scale, it is a basic principle of modern toxicology—as it was for Bernard—that effects produced by compounds in laboratory animals are applicable to humans (p. 31).

The research tradition in mechanistic physiology helps shape its conception of species differences. That tradition emphasizes the ahistorical, physico-chemical description of organisms, thus focusing almost exclusively on the immediate, *proximate* causes of biomedical phenomena. This view of species differences is different from that found in evolutionary biology, where the effects of history in the form of *distal* causes give rise to the very organisms whose proximate causation is at issue. It is to the effects of history that we must look if we are to see species differences in the correct evolutionary light.

Thus, in evolutionary biology the adaptive features of organisms that were once thought to be the results of intelligent design have an explanation in terms of natural, mindless causes operating over vast stretches of time. And if organisms are not deliberately designed, then it is no longer appropriate to think of them as machines, in particular as machines differing primarily with respect to quantitative factors. In the debates over evolutionary biology, much more is at stake than the issue of intelligent design and creation. Rather, the very metaphor that has undergirded the rise of the modern biomedical sciences—that of the *organism-as-machine*—is at issue.

Evolutionary biology is a research specialty in its own right. Yet only recently has the biomedical community even begun to rigorously explore the practical implications of evolutionary biology for medicine, which extend well beyond the business of cataloging the genetic causes of disease. In what follows we will see that there exists some degree of tension between estimates of the nature and significance of species differences from the standpoints of mechanistic physiology and evolutionary biology.

For the evolutionary biologist, species differences arise from the operation of causal mechanisms, such as natural selection, that have no role in the physics and chemistry of the inanimate world, and they give rise to effects in organisms that reflect long historical consequences of interactions between populations of organisms and the environments (biotic and abiotic) in which they are embedded. Adaptation as well as nonadaptive evolution can yield complex differences in structure and organization at all levels of the biological hierarchy in an organism, from molecules, intracellular structures, cells, tissues, and organs, ultimately to the organism itself. On this view, organisms will exhibit differences not just with respect to quantitative factors such as mass and surface area but also with respect to internal complex organization.

The importance of this last remark can be brought out by considering a couple of examples from the field of comparative endocrinology. If the modelers are right, the chemical action of hormones ought to be broadly similar across species lines. And anatomical considerations seem to undergird this claim. However, as Gorbman et al. have observed:

> The comparative anatomy of endocrine glands also shares a fairly conservative evolution, so that it can be said that generally the same or very similar hormones are produced by the corresponding glands of different vertebrates. Despite the general similarities, hormones do many different things in different vertebrates. The diverse functions of the pituitary hormone prolactin is perhaps one of the most extreme examples. (1983, p. 33)

In other words, similarities in glandular anatomy do not imply underlying functional similarities of the hormones they produce.

For a specific example, consider the thyroid gland. As Gorbman et al. have pointed out, anatomical similarities are pervasive:

> The thyroid in all adult vertebrates is follicular. With few exceptions it would be difficult to differentiate among species on the basis of thyroid histology. Even on the level of electron microscopy, parallelism of thyroid structure has been found between lower vertebrates and mammals. (p. 43)

Nevertheless, despite these similarities, there are significant differences with respect to physiology and pathology. Thus, McClain observed:

> The marked species differences between rodents and primates in thyroid gland physiology, the spontaneous incidence of thyroid gland neoplasia and the apparent susceptibility to neoplasia secondary to simple hyperthyroidism support the conclusion that thyroid gland neoplasia secondary to hormone imbalance is species specific. (1992, p. 401)

Indeed, observations of species differences in the context of comparative endocrinology have led at least some observers to give serious consideration to evolution's consequences. Thus Hart has commented:

> It has proved heuristically useful in studies on estrogens . . . to adopt the unifying concept that species differences in estrogen toxicity mirror species differences . . . in estrogen endocrinol-

> ogy. The poor predictiveness of animal studies for humans thus becomes comprehensible in terms of interspecies variations in endocrinology. (1990, p. 213)

In view of observations such as these, one might have expected the biomedical community to have paid more attention to the implications of evolution. After all, understanding the medical implications of how we evolved may enable us to see a bit more clearly how we may not simply be mice writ large.

Species Differences Again

Notwithstanding the earlier remarks about the mechanistic physiological tradition extending back to the work of Claude Bernard, one might have thought that comparative physiology would be that branch of physiological inquiry most sensitive to the consequences of evolution. Not so, according to evolutionary biologists Burggren and Bemis:

> Unfortunately, comparative physiology traditionally has been, and continues to be, outside the framework of contemporary evolutionary biology, often embracing theories, positions or approaches that contemporary morphologists, evolutionary biologists, and geneticists have abandoned. (1990, p. 193)

The problem identified here manifests itself in several ways, all relevant to the conduct of research and to its biological implications.

One problem is that the relationships between species tend to be of secondary importance in the context of comparative physiology owing to its concern with the mechanical structure of organisms:

> [A]nimals are usually chosen for comparative physiological experimentation either on the basis of extreme physiological characters or because the animal is conducive to a certain physiological technique (i.e., squid axons 1mm in diameter can be punctured relatively easily by microelectrodes). This manner of choosing animals . . . is not concerned with whether a species occupies a key position (or any position!) within a putative evolutionary sequence. (Burggren and Bemis 1990, p. 205)

Yet, for the evolutionary biologist, phylogenetic relationships are of crucial importance. It is against the background of such relationships, once they are uncovered, that the biologist structures experiments and explains the significance of observations. It is against this background that species differences can be properly understood.

Where the evolutionary biologist is extremely sensitive to *inter-specific* differences—differences between species—and to diversity within major biological groups, the comparative physiologist seeks model species—*types*—to represent major groups notwithstanding that diversity:

> Yet the use of "the cockroach as insect," "the frog as amphibian," or "the turtle as reptile" persists, in spite of clear evidence of the dangers of this approach. Not surprisingly, this type of comparative physiology has neither contributed much to evolutionary theories nor drawn upon them to formulate and test hypotheses in evolutionary physiology. (Burggren and Bemis 1990, p. 206)

Seeing species differences as largely quantitative leads the mechanistic physiologist to downplay the significance of diversity.

For the evolutionary biologist, the study of intraspecific variation—the way characteristics vary within populations of a given species—is of paramount importance. But as Burggren and Bemis pointed out, "While comparative physiologists have made an art of avoiding the study of variation, such heritable variation nonetheless is the source for evolutionary changes in physiology as well as for all other types of characters" (1990, p. 201). We will shortly see just how important intraspecific variation is, and the reason is simple: The existence of such variation *undermines* the view that species differences (which in evolutionary biology are viewed as the result of amplification, in historical time, of variation initially found in populations of a given species) can be understood using quantitative, mechanistic scaling formulae.

Evolutionary Toxicology

An example will help. Let us look at variation in the context of the metabolism of xenobiotics (pharmacologically or toxicologically active substances foreign to the body). Different animal species often (though not always) look very different, yet if they really are "the same animal dressed up differently" we would expect similarities at the level of biochemistry and metabolism. And broad similarities have indeed been observed:

> New biochemical pathways seldom come into existence. As every biochemist knows, the biochemical characteristics of organisms are far less diverse than morphological features. Many physiological adaptations entail not biochemical but behavioral or structural changes. . . . The basic biochemical pathways and

> even the kinds of cells that make up an animal are almost invariant throughout the Metazoa; and the evolution of changes in morphology involves changes in the developmental patterning of cellular mechanisms, not of the cellular mechanisms themselves. (Futuyma 1986, pp. 409–410)

The point here is that there are extensive similarities between species. But similarities are not identities. And there are indeed differences. But here is the rub, for in complex, interactive systems such as organisms, small differences can have significant nonlinear effects so that two organisms exhibiting a high degree of quantitative similarity can nevertheless show very different effects when identically stimulated. As always, the devil is in the details of the differences.

The metabolism of xenobiotics can occur in one or both of two phases. In phase I metabolism, a substance (the substrate) is metabolized through chemical reaction involving reduction, oxidation, or hydrolysis. The resulting substance (product) is a phase I *metabolite.* In phase II metabolism, the original substance or a phase I metabolite is joined (conjugated) with an endogenous molecule (one produced by the organism). A substance may be excreted in unchanged form, as a phase I or a phase II metabolite, or as a metabolite arising from phase I and phase II metabolism (Sipes and Gandolfi 1993, pp. 88–89). A *metabolic pathway* is the chemical route—a series of reactions—by means of which a substance is metabolized.

Organisms are very similar with respect to basic metabolic pathways, and this has led to the hope that they could be used to predict the threat posed to humans of chemical substances to which we are exposed. But things are not quite so simple:

> It is a matter of common experience that the actions of the major classes of drugs, which in the main work by interfering with the normal function of physiological systems, are the same throughout mammals and most other living organisms. . . . Despite this commonality of fundamental mechanisms of drug action, it is now appreciated that there are numerous situations where the effects of a drug or a chemical on the body depend on the animal species in question. (Caldwell, 1992, p. 652)

Species differences are ubiquitous and biomedically important. How are we to think of such differences?

Toxicologists differentiate between *quantitative* differences and *qualitative* differences. Qualitative differences are generated by metabolic reactions that are unique to a given species. Perhaps members of one species are capable of achieving metabolic reactions not found in members of other species, or perhaps members of one species cannot achieve a particular reaction common to members of other species. For example, phenol (carbolic acid) can be excreted either through conjugation with glucuronic acid or conjugation with sulfate. Cats are incapable of the former reaction, and pigs of the latter.

Quantitative differences occur when members of different species use the same reactions to metabolize a given product but differ with respect to the relative extents of these reactions. If we look at humans and rats, both species can metabolize phenol by conjugation with both glucuronic acid and sulfate. But if we look at the ratio of conjugation with sulfate to conjugation with glucuronic acid in terms of percent excreted in twenty-four hours, we see that it is 80:12 percent in humans whereas it is 45:40 percent in rats (Caldwell 1980). These sorts of differences are quite common in mammals.

It is instructive to look at some examples. The first is drawn from the use of nonhuman animal models in *teratology,* the study of birth defects, especially as they are induced by developmental toxins. After the thalidomide disaster in the late 1950s, when thousands of children were born with horrendous birth defects as the result of maternal ingestion of the anti–morning sickness drug, thalidomide, scientists started to test a wide range of substances in nonhuman animals to evaluate their teratogenic potential.

Human studies of teratogenesis would be deeply immoral, but they would also be difficult to perform. For as teratologist J. L. Schardein has noted:

> It is unlikely that testing of chemicals (even drugs) in the pregnant woman will ever be acceptable. Testing agents for teratogenicity only in impending therapeutic abortions may be more desirable, but since vast numbers of subjects would need to be treated, even this method would not necessarily determine a chemical's teratogenic potential. To establish at the 95% confidence level that a given agent changes by 1% the naturally occurring frequency of congenital deformity would require a sequential trial involving an estimated 35,000 patients. . . . (1985, pp. 27–28)

But if using humans is impractical, what are we to do? Schardein suggested that we need to seek nonhuman animal species that satisfy the following *desiderata* (p. 19):

- Absorbs, metabolizes, and eliminates test substances like man.
- Transmits test substances and their metabolites across the placenta like man.

- Has embryos and fetuses with developmental and metabolic patterns similar to those of man.
- Breeds easily and has large litters and short gestation.
- Is inexpensively maintained under laboratory conditions.
- Does not bite, scratch, kick, howl or scream.

In other words we need nonhuman animal species that are similar in some respects to humans, but different in other respects.

That this may be a tall order has been noted by other teratologists. For example, discussing the question of choice of test species, Palmer has noted:

> A great deal of time and effort has been expended discussing the most suitable species for teratology studies, and it is time that a few fallacies were laid to rest. First there is no such thing as an ideal test species, particularly if the intent is to extrapolate the results to man. The ideal is approximated only when testing veterinary products or new food materials in the domestic species for which it is intended. . . . Even then, the value of using the ultimate recipient of the test material may be severely limited by practical difficulties of obtaining sufficient numbers of animals and sufficient background information to interpret the results. (1978, p. 219)

This is a theme that has recurred in the literature on teratology. There are similarities between humans and nonhuman animals, but there are relevant differences also.

That humans and nonhuman test species may not be simply "the same animal dressed up differently" is something that emerges from the teratology literature. For as Schardein himself has observed, when thalidomide was tested in animals:

> In approximately 10 strains of rats, 15 strains of mice, eleven breeds of rabbits, two breeds of dogs, three strains of hamsters, eight species of primates and in other varied species as cats, armadillos, guinea pigs, swine and ferrets in which thalidomide has been tested teratogenic effects have been induced only occasionally. (1976, p. 5)

None of this is to imply we should not conduct scientific experiments on animals. But it is to imply that there are biomedically significant differences between humans and the animals used to model us. Mice are not little humans, nor are humans large mice.

This observation is important because it is the direct biomedical relevance of results in the mouse for medicine in the human that is cited in typical moral justifications of animal experimentation. The examples discussed here should serve as a warning that the real biological situation may well be rather more complicated. The line that leads to modern humans diverged from the line leading to mice more than 60 million years ago, for a total of at least 120 million years of independent evolution. Evolutionary biology sees the process of evolution as one involving descent from common ancestors with subsequent evolutionary modification. There has been much time for the successive accumulation of modifications along the divergent evolutionary trajectories leading to mice and humans respectively.

So the first point that needs to be made is that although biochemical evolution has been conservative, so that mammals, for example, share a common biochemical framework, adaptive evolution has nevertheless occurred. As Lehninger et al. observed:

> Although there is a fundamental unity to life, it is important to recognize at the outset that very few generalizations about living organisms are absolutely correct for every organism under every condition. The range of habitats in which organisms live . . . is matched by a correspondingly wide range of specific biochemical adaptations. These adaptations are integrated within the fundamental chemical frameworks shared by all organisms. (1993, p. 5)

Biochemical adaptations come into existence when natural selection acts on heritable variation in biochemistry that occurs in a population of organisms interacting with an environment.

When some organisms survive and reproduce better than others because of possession of useful biochemical characteristics, then over many generations genes coding for those biochemical features will tend to increase in frequency. In other populations, solving different environmental problems, selection may act to preserve and propagate different variants. Biochemical evolution is constrained by the need to accommodate any changes that occur within the basic metabolic framework, which is the same for all organisms. But it is not thereby rendered impossible.

Species differences with respect to the metabolism of drugs and other novel compounds do not directly reflect the effects of evolutionary mechanisms. For example, if mice are not sensitive to some drug that is highly toxic to humans, it does not follow that mice have evolved insensitivity to the drug through the operation of natural selection over many generations (or even artificial selection in the lab). Such differences between mice and men will reflect the *indirect effects* of different biochemical and metabolic adaptations. Differences between species brought about by the operation of evolutionary causes over long periods of time may then have unexpected effects in the presence of novel substances. And to some extent, this is what toxicologists have seen:

> It has been obvious for some time that there is generally no evolutionary basis behind the particular drug metabolizing ability of a particular species. Indeed among rodents and primates, zoologically closely related species exhibit markedly different patterns of metabolism. (Caldwell 1980, p. 106)

The traits giving rise to the observed differences typically will not have evolved to solve the particular problem posed by the introduction of the novel drug into the organisms in question. But differences in these evolved traits may have profound and unpredictable effects when organisms are exposed to novel substances.

For an evolutionary biologist, the fact that species differences in biochemistry occur in organisms of different sizes, masses, surface areas, and so on does not imply (1) that the biochemical differences simply reflect these quantitative differences or (2) that the biochemical differences can be compensated for simply by making allowance for the quantitative differences. This last point is particularly important, because mice, facing and solving different biological problems than humans, have many *different* adaptations, not just smaller versions of human adaptations.

In addition to the observation that species may differ with respect to biochemical adaptations, there is a second point to be made. Organisms have not merely evolved in the context of a physical environment; they have evolved also in a biotic environment and have thus evolved complex relationships with other species. Some of these relationships are metabolically important. Many organisms are colonized by other organisms. Sometimes these relationships are parasitic and associated with disease states; sometimes they are benign—and relationships of codependence can evolve.

Especially important for our purposes are relationships with intestinal flora, for these reflect the effects of organismal coevolution and will not be mere quantitative differences to be scaled away. As Sipes and Gandolfi noted, these evolutionary phenomena are metabolically important:

> An aspect of *in vivo* extrahepatic biotransformation of xenobiotics frequently overlooked is modification by intestinal microbes. It has been estimated that the gut microbes have the potential for biotransformation of xenobiotics equivalent to or

> greater than the liver. With over 400 bacterial species known to exist in the intestinal tract, differences in gut flora content as a result of species variation, age, diet, and disease states would be expected to influence xenobiotic modification. (1993, p. 109)

These effects are known to be significant. For even when one is working with primates, because of their close evolutionary relationships to humans, care has to be taken in using vegetarian species, precisely because of differences in intestinal flora (Mitruka et al. 1976, p. 342).

This latter observation is important because it has a bearing on Lewontin's second assumption underlying the use of nonhuman animal models that was identified earlier in this chapter. That is, that the more closely organisms are related from a phylogenetic standpoint, the more likely they are to show the same effect when subjected to a similar cause. Hugh LaFollette and I have referred to this assumption as the *modeler's phylogenetic fallacy* (1996, pp. 136–137). The consequences of the fallacy are brought out by Mitruka et al., who have observed:

> Nonhuman primates offer the closest approximation to human teratological conditions because of phylogenetic similarities. . . . However, a review of the literature indicates that except for a few teratogens (sex hormones, thalidomide, radiation, etc.) the results in nonhuman primates are not comparable to those in humans. (1976, pp. 467–468)

Neither phylogenetic continuity nor phylogenetic *closeness* guarantees that the results in the animal model will be relevant for human subjects.

But there is more going on here from an evolutionary point of view. We all too readily speak of mice and humans as if all mice and all humans were the same. This too is an error from an evolutionary perspective, for there are differences in the form of heritable variation within populations. There are both individual differences and strain differences. These will not be explicable, in general, by the usual *quantitative* suspects invoked in the context of scaling—differences in mass and so on. For example, Sipes and Gandolfi pointed out:

> In humans, large individual differences exist in the acetylation of the antituberculosis drug isoniazid. The population is bimodally distributed into rapid and slow acetylators. . . . The incidence of slow and rapid activators is not the same in all racial groups. Among Caucasians, a slightly higher percentage of slow acetylators predominate. . . . In contrast, in Orientals, rapid acetylators predominate. (1993, p. 116)

Molnar has provided an extensive discussion, well informed by evolutionary biology, of the nature and extent of variation in human populations of interest (1998). And with reference to rodents, rather than humans, one of the observations to emerge from animal studies of the effects of thalidomide was not merely that different species reacted differently but that different strains of the same species showed highly variable sensitivity (Manson and Wise 1993, p. 228).

There is also metabolic variation reflecting gender differences, though this is not as marked in humans as in rodents and other mammals. For example:

> Chloroform is converted to a reactive intermediate (phosgene) ten times faster by microsomes obtained from the kidneys of male mice than those of female mice. Male mice are susceptible to chloroform-induced nephrotoxicity, whereas female mice are resistant. (Manson and Wise 1993, p. 228)

These variable differences in response reflect underlying biochemical differences. And in fact some of the differences have been uncovered.

The biochemical system that is most important in phase I metabolism is the cytochrome P-450 enzyme system. There are several forms of cytochrome P-450, and types and amounts vary within a species. In humans these genetic polymorphisms can manifest themselves in the form of intraspecific differences in drug metabolism. Sipes and Gandolfi observed that with respect to the antihypertensive agent debrisoquine, some 3 to 10 percent of Caucasians were poor metabolizers. Over half the poor metabolizers appeared to have a specific genetic defect in the P450IID1 gene. In the case of the antiepileptic drug mephenytoin, more than 20 percent of the Japanese population (compared to about 3 percent of the Caucasian population) were poor metabolizers (1993, pp. 95–96).

There is also variation across species lines. And so, even though antibody probes have been used to identify human counterparts to rat liver cytochromes P-450, Sipes and Gandolfi cautioned:

> It should be emphasized, however, that the function of structurally related cytochromes P-450 is not always conserved across species lines and that the same function may be served by structurally unrelated cytochromes P-450. Therefore, an antibody to a rat liver cytochrome P-450 responsible for a particular reaction in that species may not recognize the cytochrome P-450 responsible for the same reaction in humans. (1993, p. 95)

This type of variation should caution us in at least two ways. First, it cannot simply be assumed that structurally similar molecules will necessarily achieve the same functions across species lines—at the molecular level function does not necessarily follow form. Nor can it be assumed that where the same function is achieved, it is achieved in the same way. To make either of these assumptions without independent evidence is to commit what Hugh LaFollette and I have termed the *modeler's functional fallacy* (1996, pp. 134–136). And since it is evident that you have to do more than merely point to structural similarities to prove the existence of functional similarities, this greatly complicates the assessment of the extent of metabolic similarity between species. Second, the existence of variation alerts us to the effects of the long, historical processes that result in evolutionary differences between species.

Another way to see the relevance of evolved metabolic differences between species is to consider the fate of xenobiotics *in vivo,* in an intact animal, and *in vitro,* in a test tube containing ground-up animal cells. In the situation *in vitro,* the grinding up of the cells destroys some of the evolved, organizational differences between species. As Caldwell has pointed out:

> [I]n order to be metabolized *in vivo,* compounds must pass several membranes to reach the metabolizing enzymes, and in only a few cases are only one set of enzymes involved. Although the oxidation of foreign compounds occurs principally (but not exclusively) in the micosomes, as does the hydration of epoxides and glucuronic acid conjugation, many reductases are present in the cytosol, as are the sulfate conjugating enzymes. . . . Additionally in the whole animal many organs other than the liver can contribute to metabolism, sometimes catalyzing reactions that cannot occur in the liver, e.g., dog kidney can conjugate benzoic acid with glycine, while the liver cannot and gut flora can perform several reactions not carried out by the tissues. (1980, p. 109)

The context in which enzymatic metabolism takes place is an evolved, complex system: the organism.

In the situation *in vitro,* where reactions might take place in a test tube containing an aqueous solution of ground-up liver cells, or indeed, in the context of purified enzymes acting on model substrates, the organismal, and indeed cellular, context is destroyed. Although such studies are important for biochemists,

> it must always be remembered that the inside of a cell is quite different from the inside of a test tube. The "interfering" components eliminated by purification may be critical to the biological

> function or regulation of the molecule purified. *In vitro* studies of pure enzymes are commonly done at low concentrations in thoroughly stirred aqueous solutions. In the cell, an enzyme is dissolved or suspended in a gel-like cytosol with thousands of other proteins, some of which bind to the enzyme and influence its activity. Within cells some enzymes are parts of multienzyme complexes in which the reactants are channeled from one enzyme to another without ever entering the bulk solvent. . . . In short, a given molecule may function somewhat differently within the cell than it does *in vitro.* (Lehninger et al. 1993, p. 48)

These differences blur, in the context of *in vitro* studies, evolved differences between species. For example, in the context of *in vitro* studies of microsomal metabolism in rhesus monkeys, squirrel monkeys, tree shrews, pigs, and rats, only minor differences were found when model substrates such as phenol or amphetamine were examined. In *in vivo* studies of the same species with respect to the same substrates the situation is very different: "Since it is clear from data quoted previously that enormous differences exist in the fate of foreign compounds in these species *in vivo,* it is important to consider the difficulties in extrapolation from *in vitro* to *in vivo* studies" (Caldwell 1980, p. 109).

But it is not just an extrapolation problem from *in vitro* to *in vivo.* For what this highlights is the importance of the evolved organismal context in which metabolism takes place. Organisms are not passive vessels in which reactions occur, with differences between species to be explained away by scaling for differences in mass, surface area, and so on. In a test animal, the environmental circumstances in which it finds itself, along with its long and complex evolutionary history (implying broad similarities with members of other species, but also the cumulative effects of differences between this animal and members of other species), constrain the metabolism of xenobiotics in ways that are biomedically significant.

This chapter is intended to serve several purposes. First, since examples of significant species differences concern metabolic and biochemical phenomena, we should be cautious of the assumption that at the molecular and cellular levels of the biological hierarchy humans and nonhuman animals are simply the same animal dressed up in different morphological clothes. Second, the degree of biomedical relevance of results in animal subjects for human biomedicine is not correlated in any simple way with degree of phylogenetic closeness. Evolution's consequences go all the way down the hierarchy of organization. Moreover, because evolutionary processes have yielded organisms that are internally complex, organized interactive systems,

Goat subjected to electric shocks. Defenders of animal research tout its direct benefits to humans, while opponents cite the pain and suffering endured by nonhuman animals. (Hulton-Deutsch Collection / Corbis)

even small organizational differences between such systems can have effects, even systemic effects, that are not simply proportional to the degree of difference. The resulting differences in effect, consequent upon the application of similar causes, can exhibit a high degree of nonlinearity.

In turn, this should make us cautious of claims about the direct and immediate relevance and benefits of animal experiments for human health and well-being, especially if they concern molecular matters. If experiments on nonhuman animals play a prominent role in human biomedical inquiry, it is more reasonable to assume that the benefits, whatever they are, are indirect—a point made by Darwin long ago. This may be of only minor concern to a scientist. But it should have a sobering effect on those who think that a utilitarian justification of animal experimentation is a simple matter of toting up the benefits to humans and weighing them against the costs to the nonhuman animal subjects.

The fact that evolutionary biology has revealed a rather more subtle and sophisticated estimate of the nature of differences and similarities between species than is found in mechanistic physiology should not blind us to the complications that will be faced when assessing similarities and differences between humans and nonhuman animals with respect to cognitive abilities. As in the case of biomedicine, so too here. Evolution's implications have turned out to be complex and far from easy to fathom. Indeed, they have been a source of much vexatious controversy. To examine these cognitive matters in more detail, it will be important to first discuss how biologists and psychologists view the task of explaining animal behavior. This is the subject of the next chapter.

Further Reading

Biomedical research practices using animals have been defended in a number of places. Among relevant public policy documents are American Medical Association, *Statement on the Use of Animals in Biomedical Research: The Challenge and Response* (Chicago: American Medical Association, 1992), and Sigma Xi, "Statement of the Use of Animals in Research" (*American Scientist* 80: 73–76, 1992). A more technical discussion is provided by W. Patton, *Mouse and Man* (Oxford: Oxford University Press, 1993).

The complexity of these issues is brought out by A. N. Rowan, *Of Mice, Models, and Men: A Critical Examination of Animal Research* (Albany: SUNY Press, 1984). H. L. LaFollette and N. Shanks, in *Brute Science: The Dilemmas of Animal Experimentation* (London: Routledge, 1996), have raised serious questions about the standard defenses of animal research practices from an evolutionary point of view.

9

Mice, Mazes, and Minds: Explaining Animal Behavior

How are we to explain our observations of animal behavior from the standpoint of modern science? There is no simple answer to this question, though a variety of approaches were developed in the twentieth century from the standpoint of both psychology and biology. Much of this work was motivated by considerations rooted, with varying degrees of subtlety, in evolutionary theory. Shaping much of this research is something I will term *evidential behaviorism*. According to evidential behaviorism, the *evidence* for or against a given explanation of animal behavior must be derived from unambiguous behavioral assays. Evidential behaviorism must not be confused with either *classical behaviorism* or *radical behaviorism*. These are psychological theories and will be discussed extensively in the second, third, and fourth sections of this chapter. By contrast, evidential behaviorism is a claim about the nature of relevant evidence for or against given explanations of behavior—it does not prescribe theoretical limits on the content of such explanations.

Theoretical Perspectives

As we have seen, there are two sides to the evolutionary coin: phylogenetic continuity between species and evolved difference and diversity between species. In our discussion of Darwin's own views, it was seen that he was an assimilator who believed that phylogenetic continuity between species implied psychological continuity:

> It has, I think, now been shewn that man and the higher animals, especially the primates, have some few instincts in common. All have the same senses, intuitions, and sensations,—similar passions, affections, and emotions, even the more complex ones, such as jealously, suspicion, emulation, gratitude, and magnanimity; they practise deceit and are revengeful; they are sometimes

> susceptible to ridicule, and even have a sense of humor; they feel wonder and curiosity; they possess the same faculties of imitation, attention, deliberation, choice, memory, imagination, the association of ideas, and reason, though in very different degrees. ([1871] 1896, p. 79)

Darwin's approach was thoroughly *anthropomorphic* in that he had a strong tendency to see animal behavior in human terms and categories. As we will see, there is much contemporary debate about whether this is both conceptually feasible and evidentially plausible. To provide a background for this discussion, it will be useful to review some of the major theoretical perspectives that have been used to explain animal behavior.

Some of Darwin's insights about the relationships between human and nonhuman animal behavior were incorporated into some of the early-twentieth-century work in *comparative psychology,* which grew out of the idea that there were psychological similarities between the species. Much of this work was *anthropocentric,* however, in that animals were used as experimental subjects to model humans and test hypotheses about human behavior. It is a good question as to whether rats are humans writ small.

By contrast, *ethologists* who study the biology of animal behavior see behavioral traits, like other evolved organismal traits, as the fruits of evolution by natural selection. Ethologists are interested in animal behaviors as adaptations to features of the environments in which organisms are embedded. This insight was to apply to humans as well as other animals, for as Konrad Lorentz remarked:

> Our categories and forms of perception, fixed prior to experience, are adapted to the external world for exactly the same reason as the hoof of the horse is already adapted to the ground of the steppe before the horse is born and the fin of the fish is adapted to the water before the fish hatches. (Quoted in Cziko 1995, p. 73)

Classical ethology grew out of the work of Lorentz (1903–1989) and Niko Tinbergen (1907–1988). Classical ethology, as developed by Tinbergen, for example, confronts four basic issues in the explanation of animal behavior. These are:

1. *Mechanistic.* Understanding the mechanisms (for example, neural, physiological, psychological) underlying the expression of a trait.
2. *Ontogenetic.* Determining the genetic and environmental fac-

tors that guide the development of a trait.

3. *Functional.* Looking at a trait in terms of its effects on survival and reproduction (that is, its fitness consequences).
4. *Phylogenetic.* Unraveling the evolutionary history of the species so that the structure of the trait can be evaluated in the light of ancestral features (Hauser, 1997, p. 2).

Dr. Jane Goodall is a British ethologist and chimpanzee specialist. Ethologists are interested in animal behaviors as adaptations to features of the environments in which organisms are embedded. (Bettmann / Corbis)

The classical ethologists showed their commitment to evidential behaviorism through a demand for objective descriptions of animal behavior in response to environmental stimuli. But as the four points above show, an objective description of behavior in response to stimuli was only the beginning of a far more complex and wide-ranging explanatory enterprise.

Classical ethology thus has its focus on patterns of behavior displayed by organisms under natural conditions and on how these patterns of behavior adapt organisms to particular environmental challenges. But as noted by E. O. Wilson,

> Increasingly, modern ethology is being linked to studies of the nervous system and the effects of hormones on behavior. Its investigators have become deeply involved with developmental processes and even learning, formerly the nearly exclusive domain of psychology, and they have begun to include man among the species most closely scrutinized. The emphasis of ethology remains on the individual organism and the physiology of organisms. (1978, p. 16)

In the last forty years new disciplines have emerged that have shifted this focus characteristic of classical ethology.

As Shettleworth has recently pointed out (1998, pp. 18–20), in the 1960s and 1970s the study of the evolution and adaptive value of animal behavior developed into *behavioral ecology* and *sociobiology.* The focus of early behavioral ecology was on the evolution of adaptive behavioral traits. The central question concerned the prediction of behavior from evolutionary principles using models of the consequences of behavior for fitness (p. 20).

For example, Alcock—discussing John Krebs's work with great tits (European songbirds)—commented:

> In the spring, males of this species attempt to establish territories in woodland habitat. This is where singing males are first heard in each new year and where fights, chases and threat displays can first be observed. . . . Krebs' study area became completely filled at a time when there were still other birds searching for breeding sites. These individuals settled in hedgerows adjacent to the woodland. They too established territories and attempted to breed. Given the bird's apparent preference for woodland, Krebs predicted that great tits holding territories in this habitat would outreproduce birds with hedgerow territories. This proved to be the case, as only 2 of 9 nests (22 percent) succeeded in producing fledglings in the hedgerow location whereas 54 of 59 woodland nests (92 percent) yielded fledglings. (1984, p. 253)

It is clear that a study of this sort, by simply describing natural behavior in ecological context, along with its reproductive consequences, does not focus on the cognitive capacities of the animals under study.

More recently, behavioral ecologists have started to examine causal hypotheses concerning cognitive capacities. There are well-documented animal behaviors whose explanation seems to call out for references to a range of cognitive abilities. For example, consider allogrooming among impala as an example of *reciprocal altruism.* In allogrooming, one impala removes ticks from another, with the recipient of this healthful benefit returning the favor to its benefactor. It has been observed that individuals receive back from a partner as many grooming bouts as they give out, and bouts of allogrooming end when one partner ceases to reciprocate—so something like tit-for-tat reciprocation appears to be involved (Dugatkin 1997, p. 92). Allogrooming impala are typically not related, so kinship is probably not the explanation.

The impala seem to have the cognitive ability to *identify* individuals and *remember* who did what. As Dugatkin has observed:

> In particular, if mammals are in some sense more behaviorally complex . . . [d]oes this not at least suggest that they are cogni-

> tively more complex, and doesn't that suggest they are better able to remember events and individuals? . . . I believe the answer to this question may be "yes," in that reciprocity is the category of cooperation that is most closely associated with cognitive functions such as memory and individual recognition. (1997, p. 90)

So although behavioral ecology did not originally begin with a consideration of causal roles for cognitive capacities, it is at least now giving the matter some consideration. The study of the evolutionary role of cognition in the explanation of animal behavior has come to be known as *cognitive ecology*.

By contrast, sociobiology is the study of the evolution of social adaptations (social behaviors, social groups, and societies). As described by E. O. Wilson:

> Sociobiology . . . is a more hybrid discipline that incorporates knowledge from ethology (the naturalistic study of whole patterns of behavior), ecology (the study of the relationships of organisms to their environment), and genetics in order to derive general principles concerning the biological properties of entire societies. What is truly new about sociobiology is the way it has extracted the most important facts about social organization from their traditional matrix of ethology and psychology and reassembled them on a foundation of ecology and genetics studied at the population level in order to show how social groups adapt to the environment by evolution. (1978, pp. 16–17)

Sociobiology matured after the demise of the radical behaviorist school of thought and has a comparative interest in sociality in human and nonhuman animals. So although some of its conclusions are highly controversial, it did not have a history of excluding references to cognitive capacities as causal factors in the explanation of social behavior.

The growing realization of the importance of cognitive abilities as components of causal explanations of animal behavior has recently led to the development of an ecological approach to the study of comparative cognition. The older tradition in comparative psychology was anthropocentric in that it tended to see animals as humans writ small (and simple).

Against this, as Shettleworth has observed, the ecological approach to comparative cognition focuses on how species use cognition in the wild:

> The ecological approach includes explicitly comparative studies designed to analyze the evolution and adaptive value of particular

> cognitive abilities. The species compared may be close relatives that face different cognitive demands in the wild and therefore are expected to have *diverged* in cognitive ability. . . . Alternatively, species may be compared that are not very close relatives but face similar cognitive demands in the wild. Such species are expected to have converged in the ability of interest. (1998, pp. 20–22)

The cognitive approach to comparative psychology thus removes an implicit focus on human ability in the study of the evolution of cognitive capacities in nonhuman animals.

And the particular evolutionary study of cognition in humans has come to be known as *evolutionary psychology.* Rejecting the idea that the human mind is a sort of general purpose problem-solving device, the basic idea behind evolutionary psychology is that the human mind consists of (heritable) cognitive modules that evolved as solutions to adaptive problems confronting our Pleistocene ancestors. To see how these two views differ, consider the mathematical operations involved in differential calculus (or any other branch of higher mathematics that many people find difficult). On the *general-purpose problem-solver* view of the mind, there is variation in the human population with respect to general problem-solving ability—as there is variation in height—and this is what renders calculus beyond the reach of many people. By contrast, evolutionary psychology sees calculus as a recent cultural artifact. There has been no evolution for abilities to solve differential equations. If the equations can be solved at all by a given human, it is as an accidental by-product of the possession of cognitive modules that evolved to solve other tasks.

These hypothetical cognitive modules are believed to be specialized for different cognitive tasks as our organs are specialized for different physiological tasks (Futuyma 1998, p. 744). Evolutionary psychology carries with it the implication that cognitive differences between humans and nonhuman animals arise in part because they have confronted different adaptive problems in unique ways. This is of importance because if the mind is indeed modular—if it is not a general purpose problem solver—then there is no longer any reason to expect that animal minds are simply scaled-down general-purpose problem-solving devices either. There may be qualitative cognitive differences between species, not simply quantitative differences with regard to computing power.

Moreover, if the hypothetical cognitive modules did indeed emerge as adaptive solutions to problems confronting our Pleistocene ancestors, then it is possible that some types of mental illness might result from discord with the very different social conditions that prevail in modern societies, leading to the idea of *evolutionary psychiatry.* Thus Randolph Nesse has recently observed:

> Psychiatry is now where ethology was in the 1960s: trying to compensate for the lack of theoretical foundation by striving for scientific rigor in detailed observations and measurements about carefully defined categories. As soon as ethology found its bedrock in the principles of evolutionary biology, it took off in a still-growing burst of scientific advances. The hope that similar progress may be possible for understanding human behavior explains the early interest of many psychiatrists in an evolutionary approach. (1999, p. 352)

Evolutionary psychiatry is a discipline still in early infancy, and although its tentative conclusions have generated much interest, it is only slowly emerging from the realm of the speculative.

Nevertheless, two things emerge from study of human and nonhuman animal behavior, from the standpoint of evolutionary psychology and ecological approaches to the study of comparative cognition. First, there is a growing interest in evolutionary accounts of animal behavior. Second, there is a growing interest in the role of cognitive abilities as factors in causal explanations of animal behavior. Why did it take so long for the focus on cognitive abilities to emerge? To answer this, it will be instructive to examine the influence of various behaviorist schools of psychological thought.

Behaviorism and Evidence

Why do we say that humans have minds? We do not experience directly what other people feel: their joy, pain, or suffering; what they are aware of; their sense of identity; what they remember about the night before; and what they are thinking as they interact with us. Yet we readily formulate hypotheses about all these matters. We freely indulge in analogical reasoning—"I know just how you feel," "I'm glad I'm not in her shoes," "If I were you I'd watch out for Smith"—and we justify ourselves in all of this by attention to the behavior of those we interact with. Behavior, linguistic or otherwise, is our only bed of evidence, and we are remarkably adept at tracking it. If we are to justify claims about cognitive similarities or differences between ourselves and nonhuman animals, we will have to ground our account of these similarities or differences in an account of animal behavior. That will constitute our evidential base.

Evidential behaviorism is the view that the only evidence that exists to support or falsify hypotheses about animal minds (human or otherwise), ranging from Cartesian minimalism to the full-blown cognitive generosity and expansiveness of Darwin or Montaigne, is behavioral evidence. Much attention has been devoted in the recent literature on animal behavior to the search for unambiguous behavioral evidence to support generous estimates of the cognitive abilities of nonhuman animals. One of the issues we will confront is that much of the evidence typically cited in this regard can often be explained in other, cognitively less generous ways.

In the debate about animal cognition, much will hinge on how evidential behaviorism is to be understood. Here I will differentiate between two possible interpretations of this evidential prescription:

1. *Strong evidential behaviorism:* All conclusions about animal minds must be based on behavioral evidence drawn from unambiguous, tightly controlled laboratory experiments on animals.
2. *Weak evidential behaviorism:* In addition to data derived from unambiguous, tightly controlled behavioral experiments, some conclusions about animal minds can be reliably based on reports derived from field observations of animals in the wild or in captivity.

In what follows, especially in the next two chapters, it will be seen that a portion of the controversy concerning the cognitive capacities of nonhuman animals is generated by disagreements over which version of evidential behaviorism is appropriate for the study of animal behavior.

By contrast, *classical behaviorism* and *radical behaviorism* are psychological theories about the interpretation and significance of animal behavior, and just what can, and mostly cannot, be inferred from it. In the first half of the twentieth century, behaviorism, of one stripe or another, resulted in the perception that animal bodies (human or otherwise) were black boxes that either did not contain minds, or if they did, were minds beyond the scope of legitimate scientific scrutiny. B. F. Skinner, for example, although he did not deny the existence of internal states, or even their importance, maintained until his death that if psychology were to be a science, it must study only variables that could be directly observed, and that direct experimental manipulation of these variables, particularly rate of emission of behavior, was the only scientific way to proceed.

The problems with Darwin's approach to nonhuman animal psychology were, in part, its reliance on anecdotal evidence and also its tendency to be highly anthropomorphic. In the context of human behavior, in everyday life we form expectations about the behavior of others and explain that behavior quite intuitively by attributing beliefs, hopes, fears, and desires to them. *They* appear to do what *we* would do under similar circumstances, therefore they must be doing it for much the same reason. This has come to be known as *folk psychology*—it is something we all engage in as we negotiate a complex and messy social world. Darwin's approach to the study of animal behavior is suggestive of *folk ethology*—nonhuman animal behavior resembles human behavior in certain respects, and this resemblance implies that it is to be predicted and explained in much the same kinds of way, once due allowance has been made for difference of degree. The evolutionary descendant of this view (duly modified)—known as *cognitive ethology*—will be examined later in the light of the present discussion of animal behavior.

The behaviorist school of thought arose partly as a response to the idea that psychology was the *science of consciousness* (as physics was the study of matter in motion), but it was also a response to the methodological failings of anecdotal, anthropomorphic studies of animal behavior. Behaviorism has a long and interesting history, and I will restrict my discussion here to some ideas that are relevant for the issue of differences and similarities between humans and nonhuman animals. Behaviorism can be thought of as a theory (or collection of theories) about the ways in which humans and nonhuman animals respond (behave) in the light of stimuli in the environment. But it is more than this, for it also embodies ideas about the significance and interpretation of these observations.

Lying at the heart of early behaviorist thought was the concept of *classical conditioning*. This is what underlies *Pavlovian conditioning*, named after Ivan Pavlov (1849–1936). Dogs salivate at the taste and smell of food. This is not something that is learned; it appears to be a hardwired reflex arc connecting perception of stimulus (food) to physiological response (salivation). Since the taste of food almost always elicits the salivatory response, both stimulus and response are said to be *unconditional*—that is, not dependent upon prior learning.

Pavlov discovered that a *neutral stimulus*—for example, the ringing of a bell or the flashing of a light—that initially elicited no particular response from an animal could be paired with an uncondi-

In his work with dogs, Ivan Pavlov (1849–1936) discovered that a neutral stimulus—the ringing of a bell—could be paired with an unconditioned stimulus such as the taste and smell of food. After repeated pairing, the neutral stimulus could induce salivation even in the absence of food. (National Library of Medicine)

tioned stimulus (for example, the taste, and to a lesser degree smell, of food). After repeated pairing, the neutral stimulus alone could induce salivation in the absence of food. Pavlov believed that by the dog's learning to salivate at the sound of a bell, the original reflex arc had been modified.

Unlike Pavlov, Edward Thorndike (1874–1949) was interested in how animals learned new behaviors. In what is known as *instrumental conditioning,* an animal is rewarded or punished for making a particular response. The speed of the response is observed to increase or decrease accordingly. The response is instrumental in getting the reward (or avoiding punishment). Some of Thorndike's research centered on puzzle boxes from which an animal could escape by performing a simple response.

The reward for escape might be food in the animal's view, but outside the box. At first the animal blunders around, seemingly at random, but then, by chance it stumbles on the means of escape and gets rewarded. When placed in the same box, it takes less and less time to escape. Thorndike commented:

> Of the several responses to the same situation, those which are accompanied or closely followed by satisfaction to the animal will, other things being equal, be more firmly connected with the situation, so that, when it recurs, they will be more likely to recur; those which are accompanied by or closely followed by discomfort to the animal will, other things being equal, have their connections with that situation weakened, so that, when it recurs, they will be less likely to recur. (Quoted in Cziko 1995, p. 91)

Thorndike thus explained the phenomenon of instrumental conditioning in terms of the animal's forming *associations* between stimulus (*S*) and response (*R*). The stronger the association between stimulus and response, the more rapid will be the response to stimuli. This is known as *S-R learning.*

Skinner and Radical Behaviorism

Perhaps the name most commonly associated with behaviorism itself is that of B. F. Skinner (1904–1990). Skinner was an advocate of radical behaviorism. His goal was to restrict the domain of psychology to that which was open to direct observation: stimuli, responses, and consequences of responses. Skinner is known for the theory of *operant conditioning*.

Skinner hoped to do for psychology what Darwin had done for biology. It will not go amiss to follow his chain of reasoning. He described Darwin's achievement as follows:

> Darwin simply discovered the role of selection, a kind of causality very different from the push-pull mechanisms of science up to that time. The origin of the fantastic variety of things could be explained by the contribution which novel features, possibly of random provenance, made to survival. (1976, p. 41)

And survival, Skinner told us, is contingent upon certain types of behavior (involving, for example, courtship, reproduction, rearing of young, coping with predators, and so on). Natural selection, operating across the generations, could result in behavioral adaptations—dispositions to behave in certain kinds of ways, given certain stimuli in the environment. But in the nature of the case, such responses would be relatively inflexible. Organisms sometimes face rapidly changing environments and novel situations that cannot be anticipated in the genome. It was Skinner's contention that organisms capable of learning had thus acquired adaptations that enabled them to *adapt* to changing environments and *cope* with novel circumstances.

The ability to cope with novelty was where Skinner felt operant conditioning came into its own:

> Many things in the environment, such as food and water, sexual contact, and escape from harm, are crucial for the survival of the individual and the species, and any behavior which produces them therefore has survival value. Through processes of operant conditioning, behavior having this kind of consequences becomes more likely to occur. The behavior is said to be *strengthened* by its consequences, and for that reason the consequences themselves are called "reinforcers." Thus, when a hungry organism exhibits behavior that *produces* food, the behavior is reinforced by that consequence and is more likely to recur. (1976, p. 44)

Operant conditioning can thus lead to the development of behavioral adaptations to changing environments and novel circumstances

Pigeons in an operant conditioning test. Two pigeons in a box developed by psychologist B. F. Skinner are studied as part of research into operant conditioning. (Bettmann/Corbis)

encountered by an organism—environmental contingencies that cannot be directly anticipated by the genome. Genes permitting organisms this kind of behavioral plasticity appear to have been subject to selection.

Humans and nonhuman animals are subject to operant conditioning. Does this mean that Skinner thought there were no differences between humans and nonhuman animals, except, perhaps, with respect to what has been reinforced and punished? Skinner did not go as far as behaviorist John B. Watson in denying that there was such a thing as consciousness. Skinner thought there were such things as private stimuli—stimuli internal to an individual. But although not directly accessible to the external community, such stimuli could often be inferred from observation of an individual's behavior. Macphail commented:

> Skinner's next step is to claim that human consciousness consists simply in the use of verbal responses to private stimuli. So, nonhuman (non-verbal) animals may be conscious in the sense that they are awake and use their senses, but they are not conscious in the sense of being *aware* that they are seeing anything, or doing anything. (1998, p. 89)

If this is right, then even for a behaviorist, there is an important difference between the psychology of humans and the psychology of nonhuman animals, and it lies—as it has since the seventeenth century—in the issue of language use.

There is a particular irony here, since the demise of radical behaviorism, at least as it applies to humans, was brought about by considerations of just what was involved in language use. For Skinner, a child's acquisition of language, like its acquisition of many other behavioral traits, was the result of operant conditioning—a matter of stimulus, response, and reinforcement, albeit in more subtle, less direct kinds of way—and it was here that Skinner got into trouble. As Gary Cziko has recently observed (1995, p. 192), to explain language learning, Skinner had to stretch the notion of *reinforcement* to amazing lengths—verbal behaviors of the child were sometimes said to be "automatically self-reinforcing" and could be reinforced even when the reinforcing speaker was absent!

Language is complex, and the process by which a child acquires it varies considerably from individual to individual. It is acquired, moreover, in an environment much less formal and controlled than the environment of a laboratory. Yet most individuals quickly become members of their respective language communities. Could this phenomenon be adequately explained in terms of operant conditioning? The liberal uses, not to say liberties, which Skinner took with the concept of reinforcement led linguist Noam Chomsky to comment:

> [I]t can be seen that the notion of reinforcement has totally lost whatever objective meaning it may ever have had. Running through these examples, we see that a person can be reinforced though he emits no response at all, and that the reinforcing *stimulus* need not impinge on the *reinforced person* or need not even exist (it is sufficient that it be imagined or hoped for). When we read that a person plays what music he likes, thinks what he likes, reads what books he likes, etc., BECAUSE he finds it reinforcing to do so, or that we write books or inform others of facts BECAUSE we are reinforced by what we hope will be the ultimate behavior of the reader or listener, we can only conclude that the term *reinforcement* has a purely ritual function. (Quoted in Cziko 1995, p. 192)

In explaining the acquisition of language using the tools of operant conditioning theory, key concepts were stretched to the point that they lost all content.

Yet there are phenomena that need to be explained, if not by the concepts open to a Skinnerian behaviorist. First, human parents do teach, and their children do acquire, language. (Obviously, which natural language you speak depends on the contingencies of the community in which you are raised.) Second, human children will happily instruct each other. Third, human children acquire language quickly and can utter old words in new combinations. These impromptu utterances cannot themselves be the result of reinforcement. Moreover, word order and syntax—both the way sentences are constructed and the way they are linked together—are very important. The rules are complex. They are not explicitly taught or reinforced. Although parents may not indeed be able to even state what the rules are, children of four years of age seem to grasp the rules from the standpoint of practical language use. And there is the problem of what Chomsky called *poverty of stimulus:* The language heard by a child is typically not accurate, clear, or adequately structured to permit a deduction of the grammatical rules underlying the behavior (see Chomsky 1957; 1986).

More will be said about human language in Chapter 11, in a discussion of experiments to train apes to use language. For the present, it is important to note that it was critiques such as Chomsky's that caused radical behaviorism to fall out of favor among human psychologists. Exactly how language is to be explained is still highly controversial, but it does not appear to be explicable along behaviorist lines. Nevertheless, if language is to be studied—especially in the context of nonhuman animals—evidential behaviorism requires that we focus on unambiguous behavioral evidence to support hypotheses about language use. Though human language acquisition and use may not be explicable from the standpoint of radical behaviorism, all claims that nonhuman species are genuine language users must be deeply rooted in high-quality behavioral evidence if they are to be scientifically respectable!

But radical behaviorism has not just fallen out of favor with human psychologists; it has fallen out of favor with animal psychologists too, for even though they have a firm commitment to evidential behaviorism, they are not wedded to the behaviorist's stimulus-response (*S-R*) model of explanation. The study of animal behavior has shown the need to introduce *S-S* models of explanation as well as *R-S* models.

To see how an *S-S* explanation might be called for, recall that operants are voluntary responses strengthened by rewards and weakened by punishments. In what is known as *sensory preconditioning* (see Macphail 1998, p. 101), which is one kind of *compound stimulus effect,* animals associate two neutral stimuli, neither of which initially elicits an overt response. In the experimental setting, animals might first experience a series of pairings of two neutral stimuli S_1 (light) followed by S_2 (tone). In the second stage, S_2 is paired with a reward or punishment—say tone followed by food. At the end of this part of the experiment, S_2 elicits a conditioned response, salivation. Finally, S_1 is presented, and though it has never been associated with food, it elicits salivation. Macphail observed:

> Sensory preconditioning is readily explained by the notion that animals form *S-S* associations. In the first stage, the animals learn to associate the light with the tone; in the second stage, they come to associate the tone with the food; so that in the third stage, presentation of the light elicits a representation of the tone, which in turn elicits a representation of food, which ultimately elicits the appropriate response—salivation. (1998, p. 101)

So the occurrence of one stimulus leads to the expectation of another. From the standpoint of the study of animal behavior, the old behaviorist paradigm that all learning is learning about responses to stimuli is shown to be too restrictive. There is more to animal learning than just this.

Another issue to be briefly discussed is this: What is the role of the reward in learning? Do animals learn about the consequences of their responses? Consider the old behaviorist paradigm of *S-R* operant conditioning. The role of the reward in this context is simply to boost the probability of response in the presence of stimuli. By contrast, *R-S* models of explanation have been introduced to explain observations whereby animals seem to learn about the rewards they work for. This is known as *reward comparison*.

An example of this latter phenomenon is *reinforcement contrast*. In experiments first performed by Crespi and also by Zeaman in the 1940s, rats were trained to run down an alleyway for either a small or a large reward. The amounts were then suddenly shifted. Rats that had been running for a large reward now ran more slowly. This is known as negative-reinforcement contrast. Rats that had been running for small rewards, now got bigger ones and tended to run faster. This is known as positive-reinforcement contrast (personal communication from Michael Woodruff, 15 July 2001).

In traditional operant conditioning theory, these phenomena are puzzling because in that context the devalued reward should have no effect on behavior—operant strength can only be modified by procedures in which the response actually occurs. Yet animals behave *as if* they know their responses result in consequences, *as if* they are capable of working for rewards, and *as if* they can sense changes in the value of rewards. Studies of the effect of reward comparison indicate the existence of *R-S* associations.

Lying at the heart of current animal learning theory is the concept of *association* and *associative learning.* When animals experience relations of dependence between perceived events, they form associations between those events. In this regard, Shettleworth observed:

> Associative learning can be described as the process by which animals learn about causal relationships between events and behave appropriately as a result. Associations are hypothetical connections within the animal that represent causal connections between events in the animal's environment. This functional description makes almost perfect sense of the conditions for associative learning. It also reflects the philosophical basis of the study of conditioning in associationism, which suggested that effects should be associated with their causes. Associations have traditionally been thought of as the building blocks of all cognition, but seeing them as allowing animals to represent distinctively causal relationships makes associative learning just as adaptively specialized as, for example, learning about spatial and temporal relationships. (1998, p. 110)

On this view associative learning is to be thought of as an adaptation for coping with causality—it enables organisms to shape their behavior to an environment in which events are linked as cause and effect. A capacity for associative learning is found in many species.

Just as evolutionary psychologists reject the idea of the mind as a general-purpose problem solver, favoring a modular view of mind, so too contemporary learning theorists see learning (in humans or animals) in modular terms (Shettleworth 1998, p. 137). Modularity involves the idea that there are different kinds of learning. The distinct modules are now seen as representing adaptive specializations for the processing of restricted kinds of information in ecologically appropriate ways. Distinct modules may differ from one another with respect to the conditions for learning, the content of learning, and the effects of learning on behavior. There may also be species-specificity with respect to the way information is processed in a given module. In other words, there may be genuine qualitative differences between species with respect to learning abilities and not merely quantitative differences with respect to a hypothetical general learning ability.

So if learning is to be thought of in evolutionary context, when might it be expected to evolve? Shettleworth observed:

> Animals need to know such things as what locally available food is good to eat, where and when to find it, whom to avoid and whom to approach. A capacity for learning will evolve whenever

> specific events of importance for each individual differ across generations but remain the same within generations. The general nature of events to be learned about must be the same in every generation, otherwise no single learning ability could cope with between-generation variation. For instance, as the local structure of the habitat changes, members of each new generation may find food in different places, but there may always be some advantage in being able to learn where food is. The conditions that bring learning about should be reliable correlates of the state of the world that the animal needs to adjust to. This correlation is encoded in cognitive modules, so that particular sorts of experiences bring about particular changes in cognitive state. The effects of learning on behavior must be intimately related to its function. It would be no good to a blue jay to associate the orange and black pattern of the Monarch butterfly with the emetic effects of ingesting it if this association caused the blue jay to attack Monarchs more avidly rather than rejecting them. (1998, p. 106)

But if this is granted, what does it tell us about animal minds? And what is meant by cognition?

This is a matter of some debate by contemporary researchers. Of interest to many people is the issue of animal consciousness. And there is a variety of opinions. Macphail, for example, did not think that contemporary associationism had direct implications for the question of animal consciousness:

> We may now speak readily enough of the "expectations" of animals, but what is meant is simply the formation of S-S or R-S associations. When we say, for example, that hearing a bell leads to a dog's expectation of food, we mean that the activation of the internal representation of a bell leads to the activation of the internal representation of food. It is not clear that the ability to form S-S and R-S associations as well as S-R associations means that animals are any more likely to be conscious. Animals possess devices that detect dependencies between events in their worlds, and form associations between internal representations of those events. But do "internal representations" necessarily imply a mind? This is a question in which contemporary associationists have shown little interest, at least in public. (1998, p. 105)

So one issue is this: Does the capacity for associative learning imply other kinds of cognition, such as consciousness? Another issue to be examined is this: How much animal behavior can be explained simply in terms of associative learning (as opposed to other forms of cognition, such as consciousness)? Current associationists are evidential behaviorists. So are there unambiguous behavioral tests to show that there is more to the explanation of animal behavior than an account of that behavior in terms of associative learning? And what cognitive baggage, if any, does an account of associative learning bring with it?

Finally, what is cognition? We have used the word very broadly to suggest that cognition is present if there is any form of mental life to an organism. That is to say, any advance on *Cartesian minimalism* will involve some form of cognition, but cognition can fall far short of an attribution of consciousness. The attribution of cognitive properties to an organism involves the formation of hypotheses to explain what is observed in the correlation between sensory inputs and behavioral outputs.

For example, if an animal has an internal representation of the world, you might expect behavioral plasticity in response to changes in the environment. As Shettleworth observed:

> Changes in a central cognitive representation are generally expected to be reflected in behavior in a flexible way. For instance, when marsh tits (*Parus palustris*) have stored food in sites in the laboratory, they show that they remember the locations of those sites in two different ways. When they are hungry and presumably searching for food, they return directly to the sites holding hoarded food. In contrast, when they are given more food to store, they go to new, empty, sites. Thus the birds are not merely returning automatically to the sites with food. They seem to have a representation of food in the site that they can act on in a flexible functionally appropriate way. (1998, p. 104–105)

But how should we think of cognition in a case like this? If information is being processed, how is it being processed? What types of cognitive abilities are needed to explain the phenomenon? (Acknowledging cognitive modularity forces us to ask this question.) And to what extent can our attributions of cognition be justified from the standpoint of evidential behaviorism?

Explaining Animal Behavior

Settling exactly what is meant by cognition is an important first step to the elucidation of what is involved in explanations of animal behavior. Shettleworth began with a broad definition:

> Cognition refers to the mechanisms by which animals acquire, process, store and act on information from the environment.

> These include perception, learning, memory, and decision making. The study of comparative cognition is therefore concerned with how animals process information, starting with how information is acquired by the senses. (1998, p. 5)

But as Shettleworth pointed out, there is disagreement over whether words such as *cognition* or *cognitive* should be used in this broad sense. The debate brings attention to the sorts of knowledge humans and nonhuman animals may be said to possess.

Philosophers make a distinction between *knowing how* and *knowing that*—between procedural knowledge and propositional knowledge. The former type of knowledge, knowing how to do something, can be judged by behavioral criteria (it might even be thought of as a stimulus-response connection) and does not require the ability to verbalize what is known. And, indeed, many humans who possess language do not have the ability to put into words some of the things they can clearly do. There seems little doubt that nonhuman animals have varying degrees of procedural knowledge. Such knowledge can be taught by *showing* rather than by *saying*.

But what about propositional knowledge? When we say "Subject *X* knows that *p*," what goes in for the variable *p* is a proposition (a declarative sentence, for example: "Snow is white," "My cave is south of the large tree stump," and so on). Propositional knowledge is verbalized knowledge. This means that knowing that *p* involves an ability to linguistically articulate that *p* (if only to oneself). Such knowledge can be taught by showing, but it can also be taught by saying (or other forms of linguistic articulation, such as writing). Importantly, the possession of propositional knowledge requires language. Although many nonhuman animals have sophisticated communication systems, whether any have language in the sense required to support propositional knowledge is much more controversial. We will return to the language issue in Chapter 11.

Even though most people will agree that animals have procedural knowledge, explaining what is involved in procedural knowledge is quite another matter. And once again, the issue is one of degrees of cognitive sophistication. An explanation of how an animal *knows how* might simply involve a stimulus-response connection or the fruit of some other form of associative learning, or it might involve the attribution of something more complex, such as an internal representation of the external environment—a *cognitive map* of the world in which it lives.

This issue has been discussed most prominently in connection with the ability of animals to get around in their environments—to orient themselves spatially. If spatial orientation is controlled by a cognitive map, then we would explain the ability of animals to get around in their environments through the possession of an internal representation of distances and directions that allows the animal to get from point *A* to point *B*. In this case, the represented system would be the world the animal lives in, and the representing system would be the animal's nervous system, which would also embody rules for correspondence between represented and representing systems.

But how would we know whether an animal was using a cognitive map to get around? The very idea of cognitive maps was suggested as an explanation of the fact that animals sometimes take novel routes from *A* to *B* (ones they have not been trained to take and that do not appear to be explicable in terms of learned chains of stimulus-response connections [Tolman 1948, p. 192]). One explanation of this flexible behavior is that the rats have internal representations of their environment and are able to perform computations based on sensory input and the internal representation to get from *A* to *B* in a novel manner.

The trouble is that the same sort of flexible behavior has explanations that do not involve the attribution of cognitive maps. From the standpoint of the theory of associative learning, flexibility can arise from a capacity for *stimulus generalization*. No repeated stimulus is ever perceived in exactly the same way. Learning from experience involves the ability to generalize from one stimulation event to others that are merely similar in some relevant respects. An alternative explanation of the rats' behavior in taking novel routes is that they are generalizing from experience with familiar local views.

Rats cannot directly tell us whether they have cognitive maps of their environment. So we need an unambiguous behavioral test, but such tests are hard to come by (Shettleworth 1998, chap. 7). From the standpoint of evidential behaviorism, the evidence that supports cognitive maps also supports other kinds of explanation. The problem is not that we know that animals do not have cognitive maps, but rather that we do not have behavioral evidence that rules out other hypotheses that explain the same behavior. For example, consider experimental tests on rats using a water maze. A rat is placed in a tank and must swim to a platform. The platform may be exposed above the surface of the water, or it may be slightly submerged and hidden. The rat is released at novel locations, but nevertheless manages to find the platform.

Does the experiment show that the rat has a cognitive map, or are there other explanations of its ability to find the platform? The test is not unambiguous, because as Shettleworth observed:

> Rats experienced in the swimming task rapidly approach the hidden platform no matter where they are released in the tank. However, this does not necessarily mean they are generating efficient routes from novel locations because rats typically swim all over the tank early in training. Most likely no location or local view in the tank is completely novel to an experienced rat. (1998, p. 313)

Nevertheless, more subtle tests are possible.

In one of these, rats are allowed to learn the *procedural* requirements of the swimming task by learning to approach a platform in a given tank from any direction. The rats were then placed on the platform in a new tank in a novel environment for two minutes before being required to swim for this new platform from novel release points. The rats with prior experience of finding the platform (in a different tank and environment) located the platform in the new environment twice as fast as control animals with different kinds of prior experience. Does this show that a rat with procedural knowledge for the swimming task is able to form a cognitive map of the new environment to generate efficient solutions to the platform location problem? Not necessarily, for as Shettleworth has noted:

> Such results mean that under some conditions rats do not need experience of specific routes or specific views of the environment from their starting place in the water. Passive placement at the goal allows the rats to form a cognitive map of its location. *One might equally well conclude that past experience of how its own movements transform its view of the environment allow a rat to move toward a goal at which it has previously been placed.* (1998, pp. 313–314; my italics)

Thus, from the standpoint of observed behavior, we cannot tell which explanation is right.

Most students of animal behavior are often impressed by the apparent complexity and subtlety of the behaviors they study, but that does not mean that animals solve problems—for example, negotiating their way around the environment—in the same sorts of way that we do—for example, by using internal representations of the external world. The challenge is to find behavioral assays that will rule out lower-level explanations of behavior (for example, those based on stimulus generalization) in favor of richer accounts, involving, perhaps, cognitive maps. As we will shortly see, these issues bear on evidence for cognitive sophistication in nonhuman species.

Cognition and Consciousness

The scientific study of cognition in animals involves the formation and testing of hypotheses about how information is processed on the basis of observations of sensory inputs and behavioral outputs. Yet lurking behind this talk of cognition (broadly conceived as information processing) is the issue of consciousness. We know from our own experience that information processing can be dissociated from conscious awareness. Most of us have had the experience of driving to work (or going home) with our conscious minds occupied by things other than the road and our fellow drivers. The miles go by, yet sensory experience is processed and complex maneuvers are executed without much direct conscious awareness. The fact that nonhuman animals process information, possibly in complex ways, does not logically imply that they do so with conscious awareness of what they are doing.

In the contemporary literature on animal behavior and cognition, there are at least three separate lines of thought:

Cognitive Minimalism

The core of cognitive minimalism is that animals are not conscious. Animals lack feeling consciousness—they do not consciously experience pain, nor, consequently, do they have awareness that they are currently in pain. (None of this implies that they do not react to noxious stimuli, as we shall see.) They lack a sense of self (a subjective view of the world) and a theory of mind (beliefs, hopes, desires, intentions, including beliefs about what they and other organisms believe, hope, fear, and so on). Euan Macphail has recently articulated such a view (1998). On his view, language and consciousness come together, and both are restricted, from the standpoint of evolution, to humans. On this view, nonhuman animals and preverbal children lack consciousness. We will have occasion to examine some of Macphail's arguments in the next two chapters. If animals do not experience pain, and indeed are not conscious in any way, then this would remove them from Bentham's extended utilitarian moral community. It would involve a radical reassessment of the moral status of nonhuman (and preverbal human) animals.

Cognitive Ethology

Cognitive ethology has been explored in Griffin 1992; Allen and Bekoff 1997; Bekoff and Allen 1997; Fouts 1997; and Dawkins 1998. Proponents of cognitive ethology state that animals are conscious.

Their behavior indicates that they experience pleasures and pains; they are self-aware; they have beliefs and intentions; they think, deliberate, and make plans. Good explanations of animal behavior should be couched in mentalistic terms—sometimes the same mentalistic terms we use to describe human behavior. In some forms of cognitive ethology, the issue of animal consciousness is not coupled to animal language; in other forms, the two go together. But as Donald Griffin has recently remarked, the cognitive ethologist's program is enjoying much contemporary interest:

> The scientific investigation of mental experiences is enjoying a productive renaissance. Rebounding after decades of repression by behaviorism and reductionism, neuroscientists have joined psychologists and philosophers in seeking to learn what distinguishes conscious from non-conscious functioning of brains. (2001, p. 4833)

But it is controversial—and those in the sway of traditional behaviorism are apt to dismiss cognitive ethology as mush-headed soft science. To see why, one must look at what the cognitive ethologist is up to.

According to Shettleworth (1998, p. 477), cognitive ethology is concerned with two broad issues: (1) Do animals exhibit behavior that can be taken as evidence of intentions, thoughts, beliefs, hopes, fears, and so on? (2) If they do, then do they have subjective states of awareness similar to those a person would have when engaging in functionally similar behaviors? A positive answer to the first question does not straightforwardly imply a positive answer to the second question, and in assessing these issues later, it will be seen that much hinges on what investigators believe can be legitimately inferred from data. For example, cognitive ethologist Marian Dawkins has recently written:

> [I]f we accept the argument from analogy to infer consciousness in other people on the grounds that they are like us in certain key ways, then it is going to be very difficult to maintain that consciousness should not be attributed to other species if they have at least some of those same key features. What makes us reasonably certain that our fellow human beings are conscious is not confined to what they look like or how they live or even whether we can understand their language. Our near-certainty about shared experiences is based, amongst other things, on a mixture of the complexity of their behavior, their ability to "think" intelligently and on their being able to demonstrate to us that they have a point of view in which what happens to them *matters* to them. We now know that the three attributes . . . are

> also present in other species. The conclusion that they, too, are consciously aware is therefore compelling. (1998, pp. 176–177)

Cognitive ethology is *anthropomorphic* in nature in that it tries to account for animal behavior by attributing human mental characteristics to nonhuman animals—characteristics such as conscious thought, belief, and intention. As we will see in later chapters, Dawkins's conclusion has been found less than compelling by other investigators.

As for the charge that cognitive ethology is soft science, some of the literature in this tradition can certainly convey that impression. Thus Roger Fouts could write of one of the chimpanzees used in the early chimpanzee language studies:

> But for all the outer differences between Washoe and a human toddler there was one thing they had in common: the eyes. When I looked into Washoe's eyes she caught my gaze and regarded me thoughtfully, just like my own son did. There was a person inside that ape "costume." And in those moments of steady eye contact I knew that Washoe was a child, no matter what she looked like and no matter what acrobatics she performed in the top of a tree. (1997, p. 20)

Yet we will also see in the next two chapters that there are serious and genuine experimental issues here, as well as thoughtful arguments concerning the nature of laboratory experiments and the role for field observations and anecdotal evidence.

The issues here are important indeed, for what is at stake are the cognitive properties that confer moral status on nonhuman animals. Indeed, as Marian Dawkins has pointed out:

> For two very powerful reasons, then, what non-human species experience is important to us. If they are conscious, this could change our view of how we should treat them. And it could also change our ideas of the evolution of our own consciousness. It is hard to think of any other subject that touches so deeply on so many important issues. Unfortunately it is also hard to think of anything else that is quite so difficult and intractable to study either. Its intractability comes from its essentially private nature. (1998, pp. 9–10)

Yet in view of this seeming intractability, a third option is open to scientific investigators:

Modest Cognitivism

Proponents of modest cognitivism conclude that there are probably no behavioral assays that can uniquely reveal consciousness in nonhuman species. Regardless of whether animals are conscious, they process information in interesting ways, and this is nevertheless amenable to scientific, behavioral study. This view is implicit in the work of Sarah Shettleworth, who commented:

> [T]he "cognitive revolution" in psychology was not about internal subjective states but about mechanisms of information processing that could be inferred from behavior. That is to say, cognitive psychologists are methodological behaviorists in that they study behavior, but not radical behaviorists in that they do not reject the study of internal psychological processes. (1998, p. 478)

Modest cognitivism really represents the spirit of a healthy skepticism. The modest cognitivist does not rule out animal consciousness but does ask for the provision of unambiguous behavioral evidence. The modest cognitivist also believes that behavioral evidence can be found for information-processing abilities that have varying degrees of complexity yet that fall short of the attribution of consciousness. In methodology, like morality, modesty is sometimes a virtue.

Cognitive minimalism, cognitive ethology, and modest cognitivism, with their very different estimates of the cognitive capacities of nonhuman animals, are nevertheless all rooted in evidential behaviorism and in evolutionary biology. The positions differ with respect to what their respective proponents think may be safely and reliably inferred from behavioral data. Who is right? In the next two chapters it will emerge that there is no straightforward scientific answer to this question. The issue is underdetermined by both evidence and scientific theory, and, moreover, it cannot be settled simply by performing more experiments, for the very issue at hand is how the results of experiments are to be interpreted.

Evolution and Animal Behavior

Since clear thinking about the evolution of animal behavior will be important in the next two chapters concerning controversies about consciousness in nonhuman animals, it will be useful to discuss here some points to bear in mind. In the preceding chapter, it was seen that care must be taken to avoid the *modeler's functional fallacy.* The fallacy is committed when it is assumed, without independent evidence, either that similar structures necessarily achieve similar functions across species lines or that where the same function is achieved, it is achieved in the same way, using the same mechanisms.

For example, a penguin's flipper has the same bones as an eagle's wing. Careful phylogenetic studies indicate that they are both derived from the same ancestral anatomical arrangement. The anatomical traits are said to be *homologous.* What this means is that there has been descent from a common ancestor with subsequent evolutionary modification, and the result is that flippers serve different functions from wings. So similar structures do not necessarily achieve similar functions. By contrast, traits are *analogous* if they have some recognizable similarity but nevertheless arose independently in distinct lineages. The wings of insects and the differently structured wings of birds are analogous, yet they allow the achievement of the same function, flight. The distinction between traits that are homologous and those that are analogous can create serious problems when estimating the significance of similarities and differences between organisms.

In view of this, another consequence of evolution that must be considered is the phenomenon of *recruitment.* Here existing structures, mechanisms, or processes are co-opted in the course of evolutionary time to serve new functions. Thus Lewontin has remarked:

> That is what birds did in developing wings. They recruited what were originally front legs into flying appendages, and in the process gave up some of the functions of those legs, like running and grasping. Penguins then recruited their ancestors' wings to make flippers and had to give up flying. But recruitment does not always involve losing functions, because there can be a chain of recruitment so that all functions are still maintained although carried on by other parts of the animal. The bones of our inner ear, which are used for transmitting sound, were, in our reptilian ancestors, bones of the hinge between the lower and upper jaw. Mammals recruited them for hearing and replaced the hinge bones with an elongation of the jaw bones themselves. So we can hear things that reptiles cannot, and we can still open and close our jaws. It is precisely the recruitment of parts to serve new functions, while still preserving old functions that leads to the greatest errors in using animal models for other species. (1995, p. 13)

It goes without saying that what is true of anatomical traits is also true for behavioral and cognitive traits of organisms. Humans and rats can negotiate mazes, but it does not follow from this that they are doing so using the same underlying (cognitive) mechanisms and processes. This will be important for the concerns in the next chapter, because humans withdraw from noxious stimuli, and so do rodents. Humans typically withdraw to avoid pain. Does this mean rodents do too?

And, because of recruitment, anatomical similarities between human and chimpanzee brains, for example, do not guarantee similarities with respect to cognitive capacities, for example, a capacity for language use. This issue will be taken up in Chapter 11.

In what follows we will see that there are three sources of difficulty when confronting the problem of the explanation of animal behavior. First, what sort of evidence is to count? Are we going to use field observations and/or anecdotal reports, or are we to restrict the evidential base to what can be gleaned from tightly controlled laboratory experiments? Second, what principles are we to use to interpret the significance of data gathered? In particular, what rules are to be used in making inferences from behavioral data to conclusions about the cognitive capacities of the organisms exhibiting behaviors of interest? These first two points are epistemological in nature. They represent concerns over what is to count as a source of knowledge and what rules are to be used in drawing conclusions from evidence.

The third point is ontological. It concerns the implications of causal evolutionary processes for the very nature of similarities and differences between the subjects of behavioral analysis—for our purposes, humans and nonhuman animals: that structural similarities do not imply behavioral similarities; that similar behavioral functions can be achieved through different causal routes; that novel functions can arise by recruitment of existing structures and processes. All these considerations will affect how we confront the first two sources of difficulty. To ignore them is to ignore evolution itself. Indeed, a consideration of evolution's implications for cognitive similarities and differences between ourselves and nonhuman animals provides an important rationale for the consideration of alternative explanations of observed animal behavior. Thus the interest in explanations of animal behavior in terms of (low-level) associative learning, rather than (higher-level) cognitive processing, springs not simply from a desire for explanatory economy but from a desire to see organisms in the light of evolution, and not simply as humans writ dumb (if not necessarily small).

Further Reading

Perhaps the best study of theoretical and experimental work concerning the explanation of animal behavior is S. J. Shettleworth, *Cognition, Evolution, and Behavior* (Oxford: Oxford University Press, 1998). Useful material may also be found in J. Alcock, *Animal Behavior: An Evolutionary Approach* (Sunderland, MA: Sinauer Associates, 1984); G. Cziko, *Without Miracles: Universal Selection Theory and the Second Darwinian Revolution* (Cambridge: MIT Press, 1995); and J. T. Bonner, *The Evolution of Culture in Animals* (Princeton, NJ: Princeton University Press, 1980).

Helpful discussions of issues surrounding cognitive ethology can be found in C. Allen and M. Bekoff, *Species of Mind: The Philosophy and Biology of Cognitive Ethology* (Cambridge: MIT Press, 1997); M. S. Dawkins, *Through Our Eyes Only? The Search for Animal Consciousness* (Oxford: Oxford University Press, 1998); and D. R. Griffin, *Animal Minds* (Chicago: University of Chicago Press, 1992). Skepticism concerning animal cognition and consciousness has been forcefully argued by E. M. Macphail, *The Evolution of Consciousness* (Oxford: Oxford University Press, 1998).

10

The Evolution of Consciousness: A Question of Animal Pain

What has modern science taught us about pain and consciousness? Do animals have conscious experiences? This controversial topic can usefully be broken down to two basic issues: consciousness of pain and other sensations (the issue of feeling-consciousness) and consciousness of self and of others as distinct from self (the issue of self-consciousness). This chapter addresses the issue of feeling-consciousness, and the next chapter is focused on the issue of self-consciousness. Nevertheless, issues about self-consciousness will also surface in this chapter, for there is an important controversy about which type of consciousness is primitive—which came first. Some theorists argue that feeling-consciousness is primitive from an evolutionary standpoint, so that birds and mammals, along with humans, can experience pain and other sensations. Others argue that feeling-consciousness required the prior evolutionary emergence of self-consciousness, and that only humans are self-conscious.

We know from our own experience what it is to have a capacity for feeling-consciousness. Thus Marian Dawkins has commented:

> Our emotions provide us with perhaps the most vivid of our conscious experiences. The way in which a nagging pain can completely preoccupy us or anger temporarily get in the way of our doing or saying anything sensible highlights one of the most important features of consciousness—its attention-grabbing, center-stage quality. If we are in pain or feel an emotion, there is no doubting that we are experiencing something in the here and now. (1998, p. 141)

But do nonhuman animals experience pain and anger too? Can they experience happiness and fear? Do they have conscious experiences? Animal lovers have few doubts about how these questions should be answered. Animal rights activists claim to know the answers with unshakable certainty. But what has science revealed about these matters? And what are the moral consequences of what science has revealed? The issue of pain and feelings needs to be addressed because it has crucial relevance for the moral questions raised by our interactions with nonhuman species.

When animals were viewed as mere machines, questions about animal consciousness had unambiguous negative answers. For the mechanist engaged in vivisection, there was no fundamental difference between damaging an animal and damaging a machine. Machines can be rendered dysfunctional through our interference, but they do not suffer in the process. Radical behaviorism was also silent about questions of consciousness (in both animals and humans), for although it is possible for the radical behaviorist to classify some forms of behavior as pain-avoidance behavior, with painful stimuli being identified with stimuli that are consistently terminated or avoided by the subject (Rowan 1984, p. 77), the mind was a black box not to be opened—observable behavior was all that was scientifically relevant. Pain-avoidance behavior told us nothing about the essentially private, subjective conscious experience of pain (and perhaps there was nothing to be told).

By contrast, the cognitive ethologist committed to evidential behaviorism will use behavior as a means to form and test hypotheses about what is going on in the nonhuman animal mind. Behavior is the starting place for a discussion of animal pain and other sensations. Ideas along these lines have also shaped public policy discussions concerning the welfare of farm animals. For example, the Brambell Committee reported that:

> Nobody can experience the feelings of another individual, however well that person may be known to them. They can be evaluated only by analogy with one's own feelings, from what that person tells us and from one's own observation of his looks, behavior and health. The evaluation of the feelings of an animal similarly must rest on cries, expression, reactions, behavior, health and productivity of the animal. . . . Animals show unmistakable signs of suffering from pain, exhaustion, fright, frustration, and so forth and the better we are acquainted with them the more readily we can detect these signs. Judgment of the severity of their suffering must be subjective. There are sound anatomical and physiological grounds for accepting that domestic mammals and birds experience the same kinds of sensations as we do. (1965, p. 9)

Indeed, it is precisely because some investigators believe that nonhuman animals can have the same sorts of sensations that we do, that such animals are used as subjects in research into the causes of pain and in the search for therapeutic modalities to relieve pain.

In such research, experimental subjects are chosen because (1) they have anatomical, physiological, and biochemical mechanisms similar to those in a human being believed to be able to experience pain; and (2) they behave, when appropriately stimulated, in a way similar to that of a human believed to be in pain. Notwithstanding this, it turns out that there is a serious scientific controversy as to whether creatures such as mice—and indeed our close relatives, chimpanzees—can actually feel pain. The controversy centers on the way scientists interpret anatomical, physiological, and behavioral evidence. So what, exactly, has science revealed about animal pain and feelings? Here it will be useful to begin with the anatomy and physiology of pain.

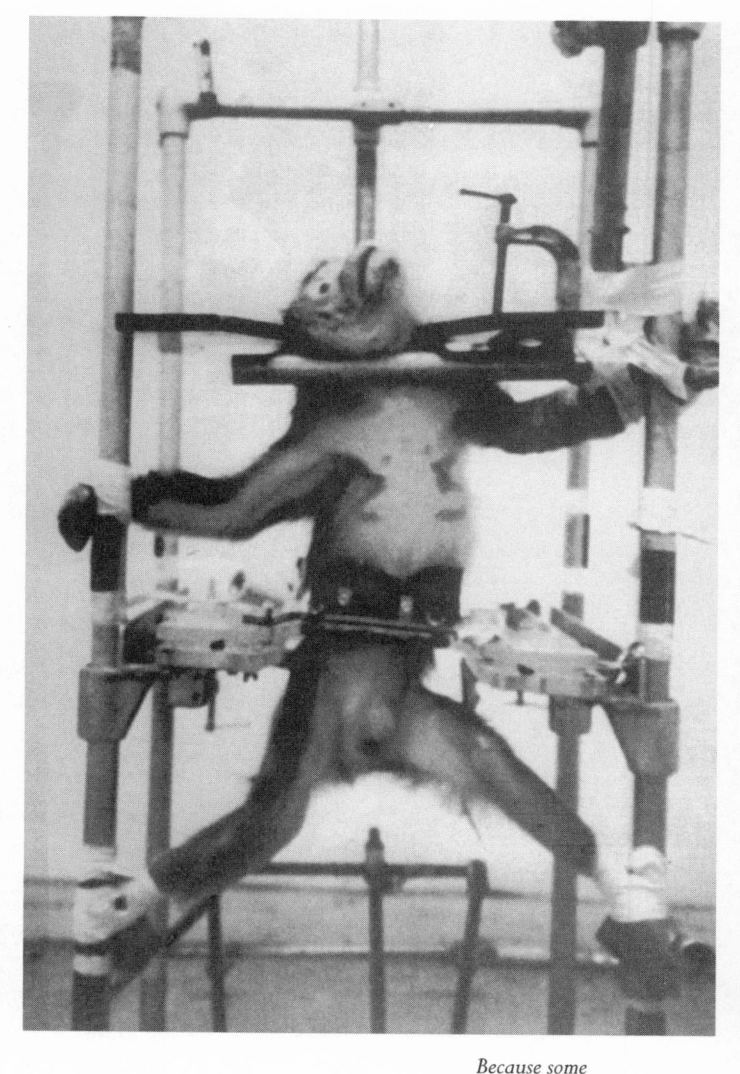

Because some investigators believe that nonhuman animals can have the same sorts of sensations that humans do, such animals are used as subjects in pain research and in the search for therapeutic modalities to relieve pain. (Bettmann/Corbis)

The Physiology of Pain

What do we know about pain? Humans, unlike nonhuman animals, give verbal reports of the characteristics of their painful experiences. They report not just having painful sensations; they also report that they reflect upon their pains. Moreover, they tell us that the perception of pain is influenced by their state of mind. As Rowan has observed:

> Studies in human beings clearly demonstrate that pain perception (not the same as sensation) is governed by subjective phenomena. There are people who can sit through an appointment with the dentist without any local anesthetic, whereas others feel pain (perceive pain) after a couple of healthy jolts of novocaine. Thus pain is a multifactorial phenomenon, consisting on

> the one hand of physiological responses to the stimulation of certain nervous pathways and, on the other, of the central integration and appreciation of these events, including a subjective psychological component. (1984, pp. 75–76)

There are anatomical and physiological similarities between humans and other vertebrates with respect to those parts of the nervous system involved with sensation. There are also behavioral similarities in the presence of noxious stimuli. Are there further similarities with respect to a subjective experience of pain?

The nerve receptors that are believed to be important in the sensation of pain are called nociceptors. These may be contrasted with thermoceptors that respond selectively to thermal stimulation and mechanoceptors that respond to pressures. It is possible that different skin sensations are served by separate sets of sensory nerves. Nerve fibers differ with respect to diameter and whether they are coated with a fatty substance called myelin. The so-called large (diameter) fibers are myelinated, whereas small (diameter) fibers may or may not be myelinated. The effect of myelination is to enhance transmission speed (Macphail 1998, p. 189).

It looks as though pressure stimulations are transmitted by the large fibers, whereas thermal stimulation is transmitted by the small fibers. It has been further supposed that painful stimulation is transmitted by polymodal small fibers that respond to intense stimulation of any kind. Rowan has observed:

> Direct, percutaneous recordings in conscious human subjects have demonstrated that pain sensations are correlated with activity in the small myelinated (A Delta) and unmyelinated (C) nerve fibers. In fact some workers have claimed that activation of "A Delta" fibers is associated with "first," fast, sharp, localized pain, and activation of "C" fibers is associated with "second," slow, burning or aching, poorly localized pain. Research on anaesthetized animals indicates that these same fibers are activated exclusively (or most potently) by stimuli of noxious intensity. These small-caliber fibers with "free" nerve endings appear to be present in all vertebrates. (1984, p. 78)

A veritable host of chemicals will stimulate C-fibers, including, for example, endogenous substances such as bradykinin (a peptide in the blood associated with inflammation), as well as exogenous substances such as acids and insect venom.

It has just been seen that the A Delta and C fibers are the pain fibers. These fibers enter the dorsal (posterior) horn of the spinal cord, where they synapse with neurons. These neurons give rise to two general outputs: those for local reflexes and those that join the anterolateral system (ALS) as part of the lateral spinothalamic pathway to the brain. For the pain system there are basically two ascending pathways that carry information about nociception to the brain, and in particular to the thalamus.

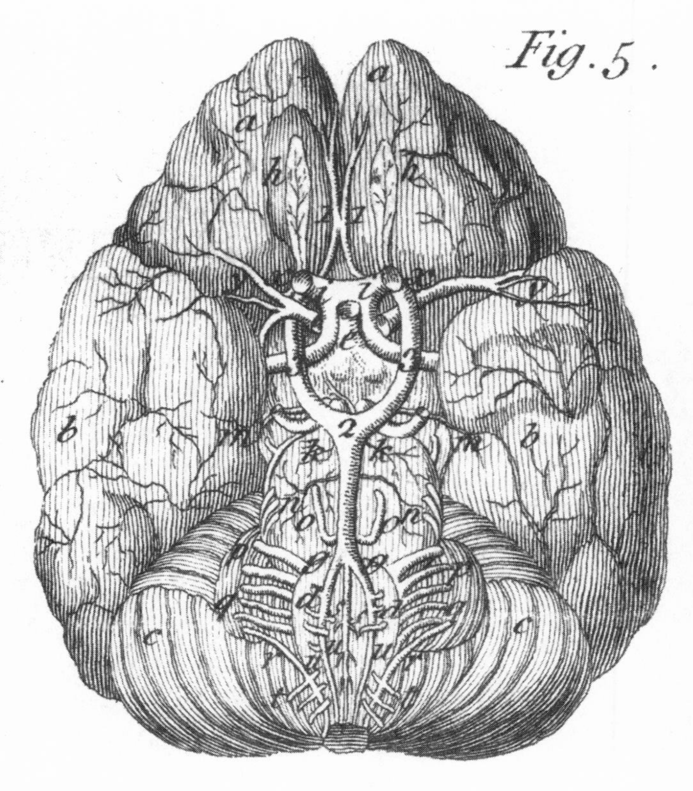

Anatomical drawing of a man's brain and cerebral nerves. The neocortex is that part of the brain containing modules concerned with memory, thought, and language. The neocortex is especially developed in mammals, particularly in humans (our closest phylogenetic relatives, chimpanzees, have about one quarter as many cortical neurons as we have). (Hulton/Archive)

Before discussing these two pathways further, it is worth taking an anatomical interlude to note that the thalamus is a brain structure that is part of the vertebrate forebrain. It is situated above the hypothalamus, and among other things it relays sensory information to the neocortex. The neocortex is that part of the brain associated with higher cognitive functions—for example, the control of voluntary motion and conscious awareness of sensations. It also contains modules concerned with memory, thought, and language. The neocortex is especially developed in mammals—and particularly in humans (our closest phylogenetic relatives, chimpanzees, have about one-quarter as many cortical neurons as we have).

Returning now to the pain pathways to the brain, most A Delta fibers participate in the direct or *neospinothalamic* pathway. This component of the ALS ends in the thalamus. The sites of termination in the thalamus are (1) the ventroposterior lateral nucleus, believed to be the site related to the ability to localize pain (identify where the pain is); (2) the posterior nucleus, believed to be the "ouch" site; and (3) the intralaminar nuclei, believed to be associated with alerting and arousal. This is a monosynaptic pathway that originates from cells in the dorsal horn that receive input from peripheral A Delta fibers and send their axons to the thalamus. This system is most involved in localized, sharp, acute pain.

By contrast, the C fibers mainly terminate on neurons in the dorsal horn that contribute to the indirect, polysynaptic, *paleospinothalamic* component of the ALS. This pathway relays information about pain to the brainstem reticular formation (spinoreticular fibers), which in turn transmit information to the thalamus (reticulothalamic fibers). These latter fibers terminate in the posterior and intralaminar nuclei of the thalamus. This pathway is associated with the sensation of chronic, poorly localized pain. In addition to these pathways leading to the thalamus, a component of the ALS leads from the brainstem reticular formation to the hypothalamus. The hypothalamus provides access to structures in the limbic system that contribute to the emotional aspects of pain response. (I am indebted to my colleague Michael Woodruff for these details concerning the neurophysiology of pain.) In broad outline, the pain pathways described above are the same for all mammals.

It is tempting to say at this juncture that these physiological similarities—especially within mammals—justify us in concluding that (at least some) nonhuman animals feel pain. But this would be a little hasty. We know, for example, from human studies that stimulation of the nociceptors is not enough to generate the conscious feeling of pain. As noted by Rowan et al. (1995, p. 75) and Macphail (1998, p. 7), humans whose spinal cords have been severed will withdraw a foot from noxious stimuli—stimuli that the patient is completely unaware of. In this case there is a nociceptive reflex arc that is mediated by the spinal cord below the point of severance and involving no higher cognitive functions. Withdrawal from noxious stimuli does not require or imply conscious awareness!

If the "cut" in humans is made in the brain, further interesting results can be seen. Rowan et al. drew our attention to patients with prefrontal lobotomies. They noted:

> For some reason, an individual with destroyed frontal lobes of the brain does not experience the agony of pain. If people are pricked with a pin, they will jump (they retain the startle reflex) but if one slowly pushes a pin into their leg or arm with their consent they will merely watch in interest. When asked if it is painful they will respond that the "little pain" is still present but that the "agony" is not. Somehow the removal of the frontal lobes of the cortex removes the affective response to potentially painful stimuli. (1995, pp. 75–76)

These observations suggest that the human experience of pain requires not just nerves but also appropriate processing of noxious stimuli in the brain. And as Rowan et al. went on to point out, some theorists have argued, on the basis of observations of lobotomized humans, that animals with much smaller cerebral lobes may not experience pain at all.

To better appreciate the role for higher cognitive functioning in the subjective experience of pain, it will be helpful to make a brief digression into the neurophysiology of visual experience, and in particular, conscious visual experience. Consider the *phenomenal properties* of conscious visual experience. These are those features commonly characterized as "what appears from a subjective point of view" or what "appears in experience" or is "given in consciousness." Although the exact boundaries of the phenomenal are not important, they include such things as visual shapes, colors, textures, and surfaces; auditorily experienced sound-qualities; sensations of pain; feelings of fear; and so on. There is a strong body of evidence that phenomenal properties are not associated with any particular, localized neural structure but rather with complex patterns of neural activity across diverse parts of the brain. Even a simple phenomenal field contains features whose representations are distributed across a variety of neural areas (Lahav and Shanks 1992).

For example, in primates, the primary visual pathway, running from the retinas to visual and other structures in the cortex, passes through more than twenty distinct visual areas. Electrophysiological studies (recording cells' responses to various visual stimuli) and lesion studies (correlating brain damage with behavioral deficits) show that each one of these areas specializes in the processing and representing of a limited range of visual features.

Thus, in the early stages of the primary visual pathway (in the lateral geniculate nucleus of the thalamus), cells are organized in a two-dimensional map and respond selectively to dots appearing in particular locations in the visual field. In the next area—V1—many neurons respond selectively to lines of a particular orientation in a particular spatial location. From there information bifurcates into several distinct neural structures. Some of the information goes to areas responsible for analyzing various parameters of spatial organization and stimulus-motion—for example, speed and direction of movement (in the MT area). Other information goes to area V4, which processes information about the color of the stimulus. In another cluster of distinct areas (inferotemporal), neurons are found to carry information about highly complex objects such as faces and hands.

Although there is some overlap and redundancy in this system, it has become clear that each visual area carries information only about a narrow range of parameters. Thus cells in the MT area respond to motion, but display no significant responsiveness to the color of the stimulus or to its complex shape. Cells in the inferotemporal areas respond selectively to faces or hands, but not to the spatial location or motion of the stimulus, nor to simple features such as lines or dots, whereas cells in V1 that respond to line orientation are insensitive to complex shapes as well as to dots.

The curious feature of the visual system is this: Despite the fact that distinct structures in the visual system are specialized in the processing of distinct types of visual feature, *there is no known neural structure that integrates all this information*. No single area has been found that consists of neurons that respond selectively to motion, color, dots, line orientation, and complex shapes. The evidence to date strongly suggests that the phenomenal properties of visual experience are not subserved by any particular structure—there is no *central clearinghouse* in the brain where information is integrated. The visual system is modular, and conscious visual experience arises from the concerted action of many, spatially distributed modules.

Moreover, the visual field often contains features that are processed outside the visual system. For example, visual, phenomenal features often contain lexical and semantic information (a set of squiggles on paper may appear as a particular word) or emotional information (a perceived object may appear as beautiful, frightening, or disgusting). Such feelings may also contain various cognitive meanings and associations with memories. All of these phenomena are known to be processed and represented in structures, mostly nonvisual, throughout the brain. And it goes without saying that visual features are often integrated with phenomenal features from other sensory modalities. Thus a visually experienced object may also appear as the source of heard speech and as the same object experienced tactually. Moreover, visual experience is typically accompanied by a stream of subvocal chatter; we talk to ourselves about what we see, and this shapes and colors the experience: Smith is not just seen as a shape in the visual field, for though we may outwardly smile and wave at him, we may be saying internally, "Damn the swine to hell!" Without language, our visual experiences would be very different.

On the basis of these observations about the integrated nature of visual experience and the distributed nature of neural representations, Ran Lahav and I observed:

> In contrast with this distributed picture of neural representations, the phenomenal field constitutes a single unified scene. Phenomenal dots, lines, complex shapes, color, motion, emotional value, cognitive meanings, and non-visual features such as sounds, are integrated in complex ways, often by merging together or "coloring" each other, within a single landscape. It seems, therefore, that phenomenal events are distributed over a

> large multiplicity of areas in the brain. This suggests that phenomenal properties express global patterns of activity across diverse neuronal areas. *They apply globally to large areas of the brain (or to neural activity occurring in them).* (1992, p. 217; my italics)

Conscious visual experience in humans clearly reflects and is colored by higher cognitive processing, including our linguistic capacities. In turn, this involves neural activity in our sophisticated, evolved neocortex.

The study of the neurobiology of conscious visual experience is important for our discussion of consciousness of pain because some very strange phenomena have been observed in patients with certain types of brain damage, and the observations are relevant for the notions of *awareness* and *consciousness.* Of particular interest is the curious phenomenon of *blindsight.* Humans with certain types of damage to the visual cortex (the part of the neocortex that receives visual information from subcortical parts of the brain) report that they are blind—that they are unaware of any visual experiences. If such persons are asked to locate a point of light, they will say they cannot, since they are blind. But when asked to guess its location, they do better than random chance—they are right, say, 80 percent of the time. And some report that though they can see nothing, they have a "funny feeling" about the location of the light (Restak 1994, pp. 129–130). Such patients have an awareness of the light, though not an immediate conscious awareness.

Subsequent research has suggested that the phenomenon arises because of the action of one of the branches of the optic nerve that connects with parts of the brain other than the visual cortex. Restak has commented:

> The most likely candidate is a large branch which goes from the eye to a way-station in the midbrain called the *superior colliculus.* In people with normal vision, this tiny, rounded protuberance activates outside of awareness whenever light is projected anywhere in the visual field. With the evolution of the primate brain, the path from the retina to superior colliculus receded in importance. A monkey blinded by a surgical excision of his visual cortex will learn to see again thanks to the "backup system" provided by the superior colliculus. But if that same monkey is reoperated on so that the superior colliculus is destroyed, he will remain permanently blind. (1994, p. 130)

Some researchers believe that blindsight results from the operation of this subcortical visual pathway. What we see from this phenomenon is that even though consciousness implies awareness, it is possible to respond to stimuli, thus showing some degree of awareness, without being conscious of the stimuli.

This discussion of neurophysiology is relevant to the controversy over pain in nonhuman animals in at least two distinct ways. First, it shows that it is possible for a creature to withdraw from noxious stimuli without experiencing pain at all. Second, and growing out of the discussion of conscious visual experience, the issue is raised as to whether the "higher" cognitive processing of sensory information concerning pain, like that involved in conscious visual experience, requires global patterns of neural activity in the neocortex. The significance of this latter issue will be explored in the next section.

The Cartesian idea that nonhuman animals may not experience pain—indeed may lack both feeling-consciousness and self-consciousness—has recently been reexamined by Euan Macphail, who has contended that the theory of evolution has implications for differences between human and nonhuman animal brains that support this particular claim. His arguments merit serious consideration. Others disagree, seeing in the theory of evolution a host of implications supporting the extension of cognitive charity to nonhuman species, both with respect to feeling-consciousness and self-consciousness. The next two sections will thus be devoted to a presentation of some rival theoretical views about feeling-consciousness in nonhuman animals. This will set the scene for a discussion of some controversies about the nature of evidence. For rival theorists do not just differ over the implications of the theory of evolution; they also differ over the issue of what evidence is relevant and how it is to be interpreted and understood.

Pain, Language, and Consciousness

Macphail's skeptical analysis of feeling-consciousness in nonhuman animals grew out of the role evidently played by higher cognitive processing in human experience of pain, a role he thought was absent in nonverbal animals and preverbal children. Because Macphail has argued for a cognitive discontinuity between verbal humans on the one hand and nonhuman animals and preverbal children on the other, it will be important to pay attention to both a phylogenetic discontinuity between humans and other animals and an ontogenetic (developmental) discontinuity between verbal and preverbal humans. In Macphail's analysis there must be a cognitive *big bang*—a major evolutionary event—to account for the perceived differences between humans and nonhuman animals. This event ought to have consequences for our understanding of the ontogenetic discontinuity between preverbal and verbal humans. Evolutionary theories that involve major leaps and discontinuities are sometimes referred to as saltationist theories. Macphail's theory will be seen later to involve both phylogenetic and ontogenetic saltationism.

How did Macphail make his case? He began with some observations. First, conscious sensation involves activities in the cerebral cortex. Second, the nerves involved in bodily sensation of pain do not send information directly to the cortex. Importantly, information goes to the thalamus first and is relayed from there to the cortex. In the thalamus itself there appear to be cells that respond to light touch, noxious stimuli, and temperature. *However, there do not appear to be any cortical cells specialized for pain* (1998, pp. 190–191). Thus Macphail has suggested:

> This discussion of pain has shown that there is no direct through-line of fibers specific for pain from the periphery, via the spinal cord, to the thalamus, and finally to a pain-specific region in the cortex. Whether a given physical stimulus causes the experience of pain is dependent on what can only vaguely be described as "higher" nervous activity. (1998, p. 196)

As evidence, he noted that sensations of pain are notoriously subject to "higher" influences such as *expectation* and *suggestion*. For example, Leonard Zusne and Warren Jones have observed:

> It is well-known that expectations have a profound effect on the degree of distress that an individual will experience when in pain. Objectively measured, the anticipation of pain can be quite literally worse than the pain itself. . . . The placebo, a physiologically inert substance, can be as effective as a drug if there is an expectation that it will work. The placebo effect has an obvious bearing on the relief of pain in faith healing. Clinical studies show that severe postoperative pain can be reduced in some individuals by giving them a placebo instead of a pain-killing drug, such as morphine. Some 35% of the cases studied experienced relief. On the other hand, only 75% of patients report relief from morphine. (1982, p. 54)

This should serve to caution us that pain and related kinds of suffering are complex phenomena that involve more than simple physiological responses to noxious stimuli.

In this regard, there also seems to be a weak relationship between the objective intensity of noxious effects and the subjective reaction to them. Sportsmen, soldiers, and people in emergency rooms sometimes report little or no pain in the face of injuries—sometimes severe injuries. And there appear to be cultural effects:

> In one study, electric shocks were delivered to the skin of a group of Nepalese porters and of the Western climbers for whom they were to work. Both groups showed comparable thresholds for the detection of the shock: there was, then, no great difference in their general skin sensitivity. But the level of shock required to cause a report of pain in the porters was much higher than that for the climbers. (Macphail 1998, p. 195)

What are the consequences of this dependence of sensations of pain on *higher nervous activity?*

Macphail has observed that there are important differences between sensations of pain (and indeed pleasure) and other sensations. Among these is the observation that although we can have unconscious processing of visual sensations—this was part of the point of the earlier discussion of blindsight—it does not make sense to say that we are in pain but are not conscious of it. There can, of course, be unconscious responses to noxious stimuli—we saw these occur in the spinal patients discussed earlier.

Another important difference between pain (and pleasure) and other sensations (for example, color or motion) is that there appear to be no cells in the cortex specialized for pain. And Macphail observed in the light of this and the observations above:

> But consciousness is generally supposed (and with good reason) to be a state that is dependent upon and mediated by cortical activity. People without neocortical activity, for whatever reason, show no evidence of consciousness however lively their reflexes to stimuli might be. If we assume, as I do, that pain and pleasure are mediated by the neocortex, the implication of the absence of localized activity is that they are mediated by widespread activity—perhaps by global cortical action. (1998, p. 200)

This now takes us to the heart of Macphail's skeptical thesis. Consciousness of pain, like conscious awareness of a unified visual field, seems to require *global patterns of neurological activity in the neocortex.*

To experience pleasure and pain—to have the sensations that confer moral status—an animal must be conscious, and this implies having the right kind of global cortical activity. Comparative anatomy and physiology reveal strong similarities between humans and other mammals with respect to sensory systems. But there are major evolved differences between humans and other mammals in terms of cortical development, complexity, and sophistication and hence, perhaps, in terms of cognitive capacities consequent upon global cortical activity.

Macphail's first observation seems reasonable enough: Animals may not experience pleasure and pain the way we do. We saw in the discussion of visual experience how language shapes what is seen. But Macphail had a more radical conclusion that will seem shocking to many. There is a known correlation between organisms that talk and organisms that we have good reason to believe are conscious. He has argued that there may be more than a correlation. There may exist a causal link. Moreover, he has pointed out that we have not found unambiguous evidence of consciousness in nonverbal animals or preverbal children. There is no definitive evidence of consciousness in nonverbal organisms. Hence we have to consider the possibility that nonverbal organisms do not actually experience pain or pleasure (1998, pp. 201–202). This possibility would not have startled Descartes, but it is likely a conclusion that many will find unwelcome. Unwelcome or not, for Macphail this position was a consequence of evolutionary reasoning. Now common sense suggests that nonhuman animals do indeed experience pleasure and pain and also that pleasure and pain play significant evolutionary roles, so why deny this?

Macphail has noted that you could make the following argument: Sensations caused by the external world are classified by the nervous system as pleasant, painful, or neutral. This reflects the operation of evolutionary processes in that animals that ignored cuts and injuries would be less successful in rearing offspring than those that took heed of the noxious stimuli. But he went on to observe:

> What is odd about this account is that there does not in fact seem to be any need for the experience of either pleasure or pain. The essence of the idea is that the nervous system discriminates between beneficial, dangerous and neutral stimulation, and that the outcome of that classificatory process may activate either the approach or avoidance system. Why should the organism experience pain before the avoidance system is activated? What *additional* function does the pain serve that could not be served more simply by a direct link between signals from the classificatory system and the action systems? (1998, p. 14)

He went on to point out that evolutionary considerations not only allow for the possibility of the evolution of avoidance mechanisms without the conscious experience of pain, *they also lead us to suppose that just this has happened.* For example, most of us accept that lowly critters such as slugs have no "higher" mental life, yet slugs also avoid noxious stimuli, "and if some animals evolved approach and avoidance systems that did *not* involve pleasure or pain, how could we possibly distinguish between them and other animals that did evolve feeling-consciousness" (p. 17)?

Thus the problem that comes into focus is the issue of the evolution of consciousness itself. What does an evolutionary perspective on this issue tell us? Macphail commented:

> [I]f animals are conscious, then consciousness must have evolved: non-conscious elements in some way gave rise to consciousness. But unless consciousness serves some function, it is hard to see how it could have evolved. Organisms that were not conscious would be at no disadvantage to those who were. We cannot tell from the behavior of an organism whether it is conscious or not, and unless we can find a function for consciousness it would seem to be systematically impossible to diagnose conscious from unconscious species. (1998, p. 220)

Where, then do we draw the cognitive line? Macphail's thesis was that we have not seen among the various groups of nonhuman animals an abrupt leap in cognitive capacity that might suggest a transition from nonconsciousness to consciousness. "The only place in which we have seen an unexpected shift in cognitive capacity has been in the transition from non-humans to humans: humans have the capacity for language, and animals do not" (1998, p. 220).

Consciousness, on this view, emerges as a by-product of the evolution of the "higher" cognitive capacities required for language. These in turn reflect the evolution of neurological complexity. From our own experience we know that beings who can talk are also beings who can feel and think. Thus, while feeling-consciousness itself may confer no evolutionary advantage, the ability to use language clearly does. What is it about language that is so special?

Macphail has pointed to the issue of *intentionality.* His thesis, in a nutshell, was this: When we use language to communicate something, for example, "Grass is green," we identify an object and say something about it, and this requires that our minds be able to form internal representations of subject and predicate and relate these in such a way that the latter is *about* the former. This capacity to form *aboutness* relationships is crucial because it is what makes language possible. This raises the issue of the *intentionality* of mental states. As Jacquette has observed:

> To intend an object is to mean it, to be directed in thought toward it and no other object. This is an abstract relation between a mental state and the object that is thought about. When I believe Caesar fought the Gauls, my belief is about Caesar, the

> Gauls, fighting, and the state of affairs which is such that Caesar fought the Gauls. . . . According to intentionalism, all thoughts are directed toward an intended object or objects. But they may be directed in different ways—by believing, hoping, fearing, desiring, doubting, or dreading something about an intended object. (1994, pp. 95–96)

I will return to the issue of intentionality in the next section. For the moment, however, it was Macphail's thesis that the kind of intentionality that comes with the ability to use language is the key to the emergence of self-consciousness.

Concomitant with this ability to see one internal representation as being about another is the ability to distinguish representations of oneself and the world outside oneself—some representations are about you, the subject, and others are about the world beyond you:

> Once the cognitive leap necessary for discriminating between self and non-self has been made—a leap that requires the ability to formulate thoughts "about" representations—the organism has in effect, not only a concept of self, but a "self"—a novel cognitive structure that stands above and outside the cognitive processes that are available to organisms without a self. (Macphail 1998, p. 226)

Macphail's thesis was that the *self* is the fruit of cognitive bootstrapping. And if this is correct, then the emergence of self-consciousness requires the evolution of the cognitive capacities that enable an organism to use language.

And if language and self-consciousness emerge together, what then of feeling-consciousness? Macphail contended that feeling-consciousness was a by-product of self-consciousness:

> The causes of feelings could serve perfectly well if they caused the consequential behavior directly rather than through the intermediary of feelings. It is simply the case that when the self contemplates inputs that are closely associated with behavioral priorities, those inputs are experienced as pleasant or painful. Feelings may make our world much better and much worse, but they are, in the end, functionless epiphenomena. (1998, p. 234)

So far from being primitive, feeling-consciousness emerges as a result of the development of a sense of self, and this in turn emerges as a consequence of the evolution of a capacity for language.

To see what is going on here from an evolutionary standpoint, it will help to reflect for a moment on the nature of adaptations—the quintessential fruits of the operation of natural selection. Elliott Sober has provided the following helpful remarks:

> To say that a trait is an "adaptation" is to comment not on its current utility but on its history. To say that the mammalian heart is (now) an adaptation for pumping blood is to say that mammals now have hearts because ancestrally, having a heart conferred a fitness advantage; the trait evolved because there was selection for having a heart, and hearts were selected because they pump blood. The heart makes noise but the device is not an adaptation for making noise: The heart did not evolve because it makes noise. Rather, this property evolved as a spin-off; there was selection of noise makers but no selection for making noise. (2000, pp. 84–85)

In the course of human evolution, there has been selection for larger, more complex brains with greater degrees of cognitive sophistication and problem-solving ability.

These large, complex brains also turn out to be self-conscious and feeling-conscious. Macphail has argued, in effect, that there had been selection for large, complex, cognitively sophisticated brains. Consciousness, like the "lub-dub" noise made by the heart as it beats, evolved as a spin-off. This is not to say consciousness is useless. Again, as Sober has observed:

> A trait may now be useful because it performs task *t,* even though this was not why it evolved. For example, sea turtles use their forelegs to dig holes in the sand, into which they deposit their eggs. The legs are useful in this regard, but they are not adaptations for digging nests. The reason is that sea turtles possessed legs long before any turtles came out of the sea to build nests on a beach. (2000, p. 85)

Likewise the brain is now useful in part because it has the property of consciousness. But this was not why the brain evolved. Our evolutionary forebears had brains, but if Macphail is right, then prior to the emergence of a capacity for language, those brains were not conscious.

The debate between assimilators and differentiators, between those who see a seamless cognitive continuum between ourselves and nonhuman animals and those who see fundamental discontinuities, has been with us since the rise of modern science. The emergence of evolutionary biology, far from settling the matter, has merely provided another forum in which this debate can take place. Thus, sticking firmly to his guns, Macphail has told us:

> Descartes' answer was that a soul was added by God, and the soul conferred the ability to both think (to talk) and to feel; animals had no soul, and could neither think nor feel. The magic that I propose has obvious similarities: animals are indeed Cartesian machines, and it is the availability of language that confers on us, first, the ability to be self-conscious, and second the ability to feel. But whereas Descartes ascribed language to divine intervention, I, in common with most scientists, prefer to explain language as a capacity that evolved in accordance with Darwin's principle of natural selection. (1998, p. 233)

Macphail's thesis can usefully be labeled as *biocartesianism*. Unlike classical Cartesianism, which attributed higher cognitive capacities and the ability to use language to the possession of a nonphysical soul, biocartesianism attributes higher cognitive capacities and the ability to use language to the possession of a large, complex, physical brain, where the physical brain is viewed as the product of natural evolutionary processes. Both classical Cartesianism and biocartesianism give humans a special place in nature. Only humans who use language are self-conscious and feeling-conscious. These events in the development of the brain, confined to the human lineage, would amount from the standpoint of evolutionary timescales to a cognitive *big bang,* a sudden saltationist leap.

The first important lesson to learn from this is that there is nothing directly in the theory of evolution to justify Darwin's charitable attitudes toward nonhuman animals. Different theorists see the implications of evolution to point in different directions. Darwin emphasized phylogenetic continuity as the basis for his charitable theses about the rich mental lives of animals. Macphail emphasized the other side of the evolutionary hypothesis—modification (of neurological structures) in the course of evolutionary time. For Macphail, these modifications in the hominid line have brought about a fundamental discontinuity between us and other contemporary animals with respect to consciousness of feelings.

It would not be fair to end this presentation of Macphail's ideas without recognizing that he was very cautious. He knew all too well that what he had was a theory about phenomena where definitive experiments have proved hard, if not impossible. And he made a very sensible proposal in the light of this: Because of the enormous moral importance that attaches to the issue of feeling-consciousness, we should be skeptical of the correctness of his theory:

> I hope I have made it clear enough that—to put it mildly—doubt (strong doubt) remains the only sensible attitude. Where

> there is doubt, the only conceivable path is to act as though an organism *is* conscious, and *does* feel. To propose that animals *may* not be conscious can in no way be used as a justification for treating them as though they *are* not conscious. To do so would be irrational, not to say psychopathic. (1998, p. 236)

It should be absolutely clear, then, that Macphail was not trying to justify the cruelties that were observed in the light of the classical Cartesianism.

But the fact remains that *if* nonhuman animals are neither self-conscious nor feeling-conscious, then traditional concerns about animal cruelty will require serious reappraisal. If preverbal humans—both in and outside the womb—are in the same boat as nonverbal animals, claims about whether they are persons with a sense of self and whether they are proper subjects for moral concern will be essentially no different from those same questions asked of slugs or snails. Either way, the stakes are very high, and even if an "is" does not imply an "ought," so that science cannot logically settle moral questions, it would be foolish to pretend that science has no relevance to moral debates concerning such things as animal care, child care, and abortion, to name but a few. All the more reason, then, to consider carefully what science can and cannot say about these matters. This means that we must consider alternative theoretical interpretations (the business of the next section) and also what the evidence says (the business of the final section of this chapter).

Consciousness: Saltationism or Gradualism?

Commonsense intuitions suggest that slugs probably do not have mental lives, whereas rats, cats, dogs, and monkeys probably do, albeit in a limited, prelinguistic kind of way. But we have just seen that some grounds for doubting this conclusion can be raised. Feeling-consciousness, which we might have thought of as something we share with our fellow mammals, could actually be a by-product of self-consciousness—a form of consciousness acquired by humans with the advent of language.

The argument in favor of this conservative solution to the problem of feelings is that feelings serve no function and could not be the fruits of evolutionary processes such as natural selection. Macphail's argument explained feelings away as epiphenomenal by-products of the evolution of structures and processes that may well be unique to our species, certainly to organisms in the hominid line.

But if feelings are to be epiphenomenal by-products, could they not be by-products of the evolution of structures and processes that we share with our fellow mammals and possibly with other animals too? Macphail did not think so:

> Why, then, should we not agree that feeling-consciousness *is* functionless, but suppose that it evolved far back in evolutionary history? The only answer to this objection is that it provides no account at all of any process of which feelings could be an epiphenomenon: we should again be left with the unsatisfactory vacuum in place of a link between feelings and some form of neuronal or behavioral complexity. Language is an instance of undeniable behavioral complexity (with clear adaptive advantages) from which it is possible to derive an account of the origins—via self-consciousness—of feeling-consciousness. (1998, pp. 234–235)

Yet it may be that there are considerations that might lead us to decouple an account of the emergence of feeling-consciousness from self-consciousness.

Recall that Macphail has tried to defend the view that there is on the one hand a phylogenetic cognitive discontinuity between humans and nonhuman animals and on the other an ontogenetic discontinuity between verbal and preverbal humans. Using Darwin's *method of gradation* (discussed in Chapter 6), we might reasonably ask whether, where there are apparent discontinuities, they are not instead extreme cases at the opposite ends of a continuum and separated by a gradation of intermediate cases.

Darwin employed gradualist considerations in his discussion of species differences. Are the apparent discontinuities we see when we look at distinct species absolute discontinuities in nature, or are they the result of many successive small, gradual changes over long periods of time? It was Darwin's contention that the latter was indeed the case—that the big differences we can see between certain species are the result of the successive, gradual accumulation of differences between initially similar populations, differences that become amplified after initial divergence between populations. Can these gradualist considerations be brought to bear on the cognitive discontinuities that Macphail has discussed? Before turning to the issue of phylogenetic discontinuity, consider again the ontogenetic discontinuity between verbal and preverbal humans that Macphail has defended.

When does a human acquire consciousness? Common sense tells us that a child of three is conscious, but that a single fertilized cell is not. So this places some boundary conditions on the problem. What I will call *ontogenetic saltationist* theories of consciousness are theories that require that there be a sudden developmental leap from being unconscious to being conscious—the lights are off, then suddenly they are on. The saltationist needs to give some account of when this leap takes place, and why. Macphail was a saltationist. He was happy to see the neonate as an automaton and identified the crucial ontogenetic period with the emergence of the ability to use language—which leaves nonverbal animals bereft of feelings and comes uncomfortably late in the day for many parents.

The alternative to saltationism is ontogenetic gradualism. For such a gradualist there are no sudden leaps. The lights, so to speak, are on a dimmer switch and gradually come on. Thus Greenfield observed:

> Since the brain develops slowly and gradually, perhaps consciousness does also. It could be the case that consciousness is not an all-or-none phenomenon, but that it grows as brains do. If we accepted that consciousness were a continuum in this way, then it would follow that the fetus is conscious, but conscious to a far lesser degree than the human adult, or even the human newborn. This way of looking at consciousness would also help with regard to the riddle of whether nonhuman animals were conscious. The less sophisticated the brain, the less the degree of consciousness. Hence animals would be conscious, but a chimpanzee would be conscious to a lesser degree than its human counterpart, as the brains of the two species, so similar at birth, then follow different fates. (1997, pp. 108–109)

For gradualists the differences between the species with respect to consciousness are differences of degree, and not kind. For the gradualist there need be no key event, for example, the advent of language, that makes one being conscious and leaves another a mere automaton. If Greenfield was right, considerations similar in kind to the case for ontogenetic gradualism in humans could be extended to make a case for phylogenetic gradualism and a very different view of the cognitive capacities of nonhuman animals from that articulated by Macphail.

Where feeling-consciousness is concerned, ontogeny may indeed recapitulate phylogeny! The early mental life of a neonate may be similar to that of its phylogenetic kin, whereas the uniquely human mental traits blossom in the fullness of developmental time. The gradual emergence of feeling-consciousness may accompany the increasing complexity of the developing infant's brain; at prelinguistic stages of development, the feeling-consciousness may be similar to that of other, adult mammals. As Greenfield noted:

> The brain of the human and that of the chimpanzee are of comparable weights at birth. A vital difference is that the primate

> brain, including that of the chimpanzee, undergoes most of its development within the womb. For the human, much of its development—some might say most—occurs outside the womb. (1997, p. 109)

On this view of things, it is not the ability to speak that shows a human is feeling-conscious, it is the squeals and howls that it shares with its fellow mammals when subject to noxious stimuli.

Once again, though one cannot *infer* an "ought" from an "is," it is clear that the gradualist interpretation of both evolutionary *differences* between species and ontogenetic *differences* between preverbal and verbal humans will be relevant to a discussion of the moral status of infants and nonhuman animals. The concept of "difference" in evolutionary biology and in developmental biology that has been discussed here is one whose contours and characteristics reflect scientific theorizing. What we can learn here is that if theoretical issues in science can be relevant to analyses of important moral issues, it is important to be aware of the importance of differences over theoretical scientific matters and how these matters may (or may not) be resolved. It may well be that scientific issues relevant to the moral status of preverbal infants and nonhuman animals will involve subtle theoretical reasoning about both evolutionary biology and developmental biology.

Dale Jacquette has attempted to work out a fully evolutionary, gradualist theory of consciousness, and we must give it due consideration. Crucial for Macphail's version of biocartesianism was the notion of *intentionality* or *aboutness.* Only organisms with language can have intentional mental states. Nonhuman animals do not have language, and thus do not have intentional mental states.

Jacquette also saw intentionality as being absolutely crucial for an evolutionary understanding of the phenomena of consciousness and self-consciousness. But, unlike Macphail, he saw it as being of crucial importance for something more basic: mere sensation. If an organism senses, it can be said to be aware. And if it senses or is aware, then it senses or is aware *of something.* On this view, something as basic and primitive as sensation involves intentionality. *Qualia* are the immediate, sensibly perceptible qualities of things, such as the greenness of grass. Jacquette's thesis was that even the *qualia* of sentient beings are intentional. And as Jacquette observed:

> [S]ystems need not have language to be intentional. They need only be able to represent something by something else, such as a part of their environment by a neural excitation state. They do not need to be able to put this representation into (anything that

> we would recognize as) words, or think in sentences. Nor do they need to act deliberately or "intentionally" in the sense of reasoned action. (1994, p. 115)

How can these observations be used to defend a gradualist theory of consciousness?

Using this idea of the primacy of the intentional, Jacquette offered a gradualist account of consciousness:

> Consider a phylogenetic gradation of increasingly complicated information processing systems, from simple organisms such as amoebas and paramecia to humans. At the lower end, let us imagine that there are mollusks with a photosensitive eyespot. . . . The eyespot registers discontinuities of light, the reception of which is hardwired into muscle trains that cause the animals to close their shells or move away whenever there is an abrupt interruption of light. . . . If there are mollusks as limited as this, then they are best regarded not as having minds, but as living machines. . . . (1994, pp. 113–114)

But Jacquette added:

> As we move upward along the spectrum from living machines, we find sentient beings, crustaceans, arthropods, fish, reptiles, birds, and lower mammals. Finally we ascend the chain of neurophysiological complexity to the higher mammals, including humans. At this extreme, we find animals that are not merely sentient, but conscious and self-conscious. (p. 114)

In Jacquette's view, intentionality goes all the way down to the level of sensations, and intentionality is also something that emerges in the course of evolution. Thought begins in sentient beings whose brains have the capacity to acquire neurophysiological features that represent the world.

What then of consciousness and self-consciousness? How is the gradual evolutionary emergence of these features of organisms to be explained? Jacquette observed:

> Sensation in primitive thought maps or represents selected parts of the body and limited aspects of the world outside the body, primarily as a survival tactic. But evolution has endowed some species with sufficiently complex neurophysiologies capable of mappings of mappings, and reflexive mappings of mappings of mappings. . . . Consciousness and self-consciousness are special kinds of intentional attitudes that intend or are directed toward the mind's own states. (1994, p. 119)

Jacquette then introduced a hierarchical progression of complexity:

> With sensation, thought makes its first appearance in the world. It is a neurophysiological event representing occurrences within or outside the body in bursts of neural excitations. Neurosystems with the requisite complexity are additionally capable of monitoring sensation. This is another, higher-level intentional event that intends sensation, and has lower-level sensation experience as its intended object. It is consciousness. Since consciousness is not itself a sensation, but a higher-level awareness of sensation, it has an experiential thought content other than its object. Beings capable of this level of intention are aware of their sensations. At the next and highest level, subjects have an awareness of the awareness of sensation and of other thought contents. This is self-consciousness. (p. 120)

On this view, invertebrates such as the octopus and squid may be sentient, as well as fish and reptiles. Many mammals may be feeling-conscious, but not self-conscious. Jacquette was willing to countenance the possibility that self-consciousness can exist without language and that dogs may have a degree of self-consciousness. On Jacquette's approach, feeling-consciousness precedes the emergence of self-consciousness; it is not, as it was for Macphail, an epiphenomenon consequent upon the sudden emergence of self-consciousness. Whether there is any *evidence* for self-consciousness in nonhuman mammals is an issue to be examined in the next chapter.

For the present, however, it should by now be clear that the basic issues surrounding feeling-consciousness in nonhuman animals—involving as they do intuitions about the self, the nature of language, the nature of mental representation and intentionality—are not purely scientific issues. What we have here are scientific issues that are inextricably intertwined with theoretical and philosophical matters, some of them of great vintage indeed, since the nature of the mind has been the subject of human inquiry in many cultures since ancient times. We must now explore in more detail the ways in which the issues surrounding feeling-consciousness are not purely scientific matters. This will be accomplished by considering the nature of behavioral evidence and its role in debates about consciousness.

A Question of Evidence

Is it not just obvious that nonhuman animals feel pain? Perhaps the debates between saltationists and gradualists are simply idle academic debates. If it is indeed obvious that nonhuman animals and preverbal infants feel pain, this is presumably because the evidence is straightforward and compelling. But evidence in science is like evidence in courts of law.

In courts, some forms of evidence are better and more reliable than other forms of evidence. Evidence can thus be divided into categories and graded for quality. Evidence extracted under conditions of duress is considered worthless, as is hearsay evidence. The courts have come to pay special attention to some types of evidence (for example, DNA evidence) even when there has been eyewitness evidence to the contrary. As we have learned more about the influences that shape eyewitness evidence, such evidence has come under critical scrutiny. One of the reasons we no longer hold trials for witchcraft is that jurists came to be critical of the sorts of evidence that had been found to be simple, straightforward, obvious, and compelling at earlier times in our history. In law, evidence is itself the subject of theoretical inquiry. The same is true in science. Scientists do not just test their theories against the evidence; they also theorize about the nature of evidence, grade difference types of evidence for quality, and discuss issues surrounding the interpretation of the evidence they gather. This is especially true of questions concerning the evidence for consciousness of pain in humans and nonhuman animals.

Marian Dawkins has stated the puzzle in a way that takes us to the heart of the methodological issues. Let us suppose we are evidential behaviorists, as most theorists in this branch of science are. She saw the following problem:

> The reason that so many people insist that we can "never know" whether another animal (or another person) is conscious is that there are no critical predictions that we can make. Every prediction that we can think of—that people who consciously experience pain should cry out if we hit them with hammers, for instance—can be matched by a sort of shadow prediction—that people should cry out when they are hit with hammers without actually *feeling* anything. Our predictions seem quite unable to escape from their shadows and the eternal taunt that there is nothing we can do to distinguish "really" feeling from "behaving as if" feeling. This is also the reason why the study of consciousness is often thought to be unscientific. Science thrives on predictions and a theory that makes no predictions can consequently be dismissed as unscientific. (1998, p.168)

This is a version of what is known as *the problem of other minds.* On this view, the hypothesis that an organism *actually* feels is empirically indistinguishable from the hypothesis that it behaves *as if* it feels. The evidence supporting one hypothesis seems to support the other equally well.

But we have seen in this chapter that saltationists such as Macphail accepted, reasonably enough, the existence of human minds with self-consciousness and feeling-consciousness. Why? What behavior enables us to escape Dawkins's skeptical morass of shadow predictions? Macphail commented:

> The answer lies in what we humans do that confirms to each other that we share comparable mental experience—we *talk about* our intentions, our beliefs, our emotions, our feelings. It is because babies do not talk that we find such difficulty in deciding what it is like to be a baby—and even in deciding whether new-born babies feel anything. There is no such difficulty with adults: we share our experiences through language, and one effect of this is that the question of whether minds other than our own exist is of little concern or interest to most of us. The absence of language means that one clear window into other minds is closed for both animals and infants. (1998, p. 18)

To this Dawkins might make the following objection: Talking is simply a form of observable behavior. Why, then, can we not say that there is no empirical difference between talking about one's feelings and talking *as if* one had feelings? If you are really worried about the inference to consciousness from observed behavior, those worries will surely extend to a consideration of what can be concluded from observations of linguistic behavior too.

Macphail might reply that each of us knows that he or she is conscious and has no problem (outside of philosophy class) with extending this idea to other human beings, similarly constituted, and behaving in relevantly similar ways. Yes, there are doubts, but they are not reasonable doubts. For example, raising Dawkins's *problem of other minds* would likely not fly as an effective defense strategy at a trial for someone accused of torturing a fellow human. All other things being equal, when someone says they are in pain—for example, when tortured with a cattle prod—they almost certainly are. (Of course, raising Dawkins's *problem of other minds* in an animal cruelty case will likely not be successful either—though if Macphail is right, this might need to be changed.)

Against this, it might be argued that our nonlinguistic mammalian relatives do feel pain because they are similarly constituted to us—through evolutionary descent with modification they differ by degree and not by absolute metaphysical kind. They also behave in relevantly similar ways. Unlike the spinal patient considered earlier who withdrew his foot from noxious stimuli but experienced no pain, and unlike the lobotomized human with diminished experience of pain, normal humans and most mammals will yelp, squeal, cry out, and whimper or otherwise behave as, for example, a mute adult human might behave (for example, by writhing around), when exposed to suitably noxious stimuli.

Someone who cries out when he wallops his thumb with a hammer is in pain. We do not need the additional behavioral evidence in the form of, "Dammit that hurts like hell." But Macphail has considered this suggestion and found it wanting. He asked:

> Must an animal show all of the non-verbal responses typical of our reactions to pain in order for us to ascribe pain to the animal? Presumably not. Vocalization, for example, does not seem a reasonable criterion for the experience of pain: whatever we may suppose about the existence of pain in fish, it surely is not a good argument to claim that a fish does not feel pain simply because it does not squeal, yelp, or make any other noise in response to noxious stimulation. (1998, p. 6)

To be fair to Macphail, it is not language use per se that is important in this debate. If he is right, the human ability to use language is a reflection of an underlying cognitive discontinuity between humans and other animals, and this cognitive discontinuity reflects in turn a neurological discontinuity in terms of the differential evolution of the neocortex in humans, relative to other mammals. Macphail is thus arguing that there is a relevant neurological difference between animals that can use language and those that cannot, and it is this that justifies us in considering Dawkins's shadow prediction in the case of *nonlinguistic animals* only—perhaps they really do simply behave *as if* they have feelings, without actually having them in fact. Perhaps the ability to talk about one's feelings and one's self really is special— not because it involves language, but because it reflects fundamental, underlying biological differences between humans and nonhuman animals. Perhaps there are relevant biological differences after all.

Although this response to Dawkins's problem of other minds is not entirely satisfying, it has the virtue of focusing attention on the biological similarities and differences between humans and nonhuman animals. And it does so in a way that can help generate rival accounts of cognitive sophistication that are also deeply embedded in evolutionary reasoning. To see what is going on, it will help to focus on preverbal children, such as newly born infants, who yell, squeal, and writhe when subject to noxious stimuli. Vocalization may not be necessary for the attribution of feelings of pain (there may be pathological conditions in which an adult human can not longer vocalize feelings but may nevertheless experience them). Vocalization is not sufficient for the attribution of feelings of pain—many mothers are convinced their newborn babies cry out sometimes for attention and not because they are in pain. But vocalization is nevertheless *relevant* to the issue of the attribution of feelings of pain ("You should have heard the screams when the car caught fire and the occupants could not escape"). If feeling-consciousness can exist without language (so that babies and nonhuman animals can experience pain), we must attempt to meet Macphail's challenge and make sure that feeling-consciousness is grounded in some form of neuronal or behavioral complexity.

Vocalization is not sufficient for the attribution of feelings of pain—many mothers are convinced their newborn babies cry out sometimes for attention and not because they are in pain. (Hulton / Archive)

Moreover, the issue of animal feelings is not entirely restricted to the question of pain. As we have already seen, Darwin argued that there was a continuity between the emotional lives of humans and nonhuman animals. According to Darwin, our phylogenetic kin may experience fear, joy, anxiety, happiness, shame, sadness, and even grief. Much of Darwin's evidence was anecdotal in nature. The anecdotal tradition continues into our own time. Thus as Poole has observed:

> It is hard to watch elephants' remarkable behavior during a family or bond group greeting ceremony, the birth of a new family member, a playful interaction, the mating of a relative, the rescue of a family member, or the arrival of a musth male, and not imagine that they feel very strong emotions which could be best described by words such as joy, happiness, love, feelings of friendship, exuberance, amusement, pleasure, compassion, relief, and respect. (Poole 1998, pp. 90–91)

But this is a long way from the cautious domain of carefully controlled behavioral experiments on animals (or humans, for that matter). To use the language of the preceding chapter, are we going to be *weak* evidential behaviorists and allow anecdotes and field observations to count as a valid source of evidence? Or are we going to be *strong* evidential behaviorists and exclude all such evidence in favor of tightly controlled behavioral assays in the laboratory? So what evidence is to count and why?

To what extent are we actually justified in acquiescing with Darwin's (and Poole's) charitable estimates of the emotional lives of animals? We could extend Macphail's arguments about pain to emotions. Feeling-consciousness requires self-consciousness, and this requires language. In which case the emotional buck may stop with our species.

And to warn us against unwarranted empathy, Macphail suggested that we consider the sea slug *Aplysia californica,* a most unlikely candidate for an emotional life, with a central nervous system consisting of about 20,000 cells (unlike the billions of cells that are constitutive of our own central nervous system). Yet the same behavioral tests that would be taken to reveal the emotional state of fear in vertebrates such as dogs or cats are also seen in *Aplysia*—a critter that is arguably not a candidate for such a rich mental life:

> [A] solution of shrimp extract (a taste stimulus) was released into an *Aplysia*'s tank a few seconds before the *Aplysia* received a strong electric shock to its head. The shrimp taste served, then, as a signal that a shock was imminent. Subsequent tests showed that defensive responses to weak noxious stimuli were strengthened in the presence of shrimp extract. For example, a weak shock to the tail resulted in the *Aplysia* "inking": the minimum strength of tail shock needed to obtain inking was lower in the presence of shrimp extract (in the absence of previous training, shrimp extract did not affect the inking response). One explanation of this finding—the explanation that would be widely accepted for a similar outcome using a mammal—is that as a result of pairing with shock, the *Aplysia* now reacts to the shrimp extract by showing fear, a central motivational state that potentiates *all* defensive responses. (1998, p. 9)

Macphail's point was that few of us would be seriously tempted to say that the slug was neurologically complex enough to experience fear, yet its behavior in the context of controlled experiments was very similar to that of other nonhuman animals to whom we anthropomorphically attribute states of fear.

But not all investigators share Macphail's low estimate of the cognitive possibilities for invertebrates. Thus Sheets-Johnstone has observed:

> To affirm, for example, that scallops "are conscious of nothing," that they get out of the way of potential predators without experiencing them as such and when they fail to do so, get eaten alive (quite possibly) experiencing pain . . . is to leap the bounds of rig-

> orous scholarship into a maze of unwarranted assumptions, mistaking human ignorance for human knowledge. (1998, p. 291)

This is certainly a field where intuitions vary, where the extreme positions are occupied and defended, and where many intermediate positions are possible. Who is right? Better still, where does the evidence point?

In confronting these questions we must bear in mind that the issues may not have a simple scientific resolution one way or the other. For evidence has to be interpreted, and there may be no unambiguous behavioral test, or anecdote, or field observation that will enable us to settle it one way or another. As Rowan has observed in connection with pain:

> But how can we tell if an animal is really in pain, since they do not have the capability to verbally express themselves as we do? Pain is essentially a private matter and none of us can really tell what pain another is experiencing, even when they can describe it in words. (1984, p. 76)

The relevance of the issue of subjective privacy for the study of animal emotions has also been discussed by Allen and Bekoff (1997) and Bekoff (2000). The same could be said of other feelings such as joy or anger. There will always be other interpretations of the very behavior that we may be tempted to think of as decisively settling the matter.

This is precisely where the scientific questions start to get interesting philosophically. Most of us will likely own up to feeling pain, fear, anger, joy, and grief at various points in our lives. We would be unimpressed by arguments to the effect that we must withhold the attribution of these feelings to ourselves simply because the states cannot be studied externally—"from outside"—using unambiguous behavioral tests whose results could not be subject to alternative interpretations.

Similarly, though we cannot perhaps solve the problem of *other minds*—that is, know definitively and beyond all possible doubt that other humans really have minds and are capable of experiencing feelings, their overt behavior notwithstanding—we would not want to deny that it was at least possible that others experience what we experience. Indeed, much of our interaction with our fellow humans is predicated on their having feelings similar to ours. It is this that underlies compassion and empathy. Similarly, then, perhaps we should not deny that animals have feelings, notwithstanding the fact that they cannot be studied objectively and definitively from outside with the aid of unambiguous behavioral tests—tests whose results cannot be interpreted to say that the subjects in question merely behave *as if* they were in pain, or experiencing joy, or grief.

Most participants in the debate about animal pain and animal feelings are evidential behaviorists of one stripe or another. Observable behavior forms the base of data on which conclusions about animal mental lives are to rest. But behavior has to be *interpreted*. And conclusions have to be *inferred* from what is observed. A big part of the issue before us in this controversy is where to set the *interpretative and inferential bar.* It is one thing to set the bar too low—you can end up giving credence to all sorts of silly claims about human *and* animal cognition—from philosophical fetuses to poetic shrimp. But setting the bar too high is equally dangerous—you can end up excluding too much from the court of rational inquiry. For example, with no generally agreed, definitive solution to the problem of *other minds,* you could end up excluding all other humans, let alone nonhuman animals, from the domain of moral discourse. After all, if they do not have feelings, let alone thoughts, notwithstanding their overt behavior—if they only behave as if they had feelings and thoughts—perhaps they have no more moral relevance than rocks. Of course, in dealing with our fellow humans, we readily extend cognitive charity through analogical reasoning, but perhaps because we feel there must be a categorical difference between ourselves and nonhuman animals (a difference deeply ingrained in some of the strands constitutive of Western culture), we have been reluctant to so readily extend cognitive charity to nonhuman animals.

But importantly, where exactly to set the interpretative and inferential bar is no easy matter to be settled simply by performing an experiment or a series of experiments, for the issue before us is that of the very meaning and significance of the experiments we perform. This issue is a theoretical issue that will involve decisions that reflect the current state of scientific theory, the investigator's philosophy of science (broadly conceived), and the broader culture in which inquiry is embedded. The problem, as should now be obvious, is generated by the fact that we are trying to study from the outside of an organism something that can only be directly experienced, if at all, by the organism itself. And this means that all our evidence is of an indirect, circumstantial nature. Consequently, it can be subject to alternative interpretations, with no direct, experiential, phenomenological evidence to settle the matter definitively one way or the other.

An important part of the business of science is to formulate hypotheses about phenomena of interest. These hypotheses should explain our extant observations and should enable us, where possible, to make predictions about the future. In dealing with a given hypothesis, scientists are typically (though not exclusively) interested in whether there is any evidence in favor of the hypothesis, and most importantly whether there is evidence against the hypothesis. But before this inquiry can begin, scientists must have some idea as to the interpretation of evidence and to what may and may not be inferred from such evidence. Only then is the bar set that some hypotheses may pass and others will fail.

As noted earlier, the issue of where and how to set the interpretative and inferential bar in a given branch of scientific inquiry is not simply an evidential matter—something to be settled simply by observation and experiment—since it is the meaning and significance of the results of observations and experiments that are the very issues at hand. The meaning and significance of observations and experiments are very much a theoretical matter. It is something that is informed by the theories we work with. As was apparent in the earlier discussion of behaviorism, behaviorist psychological theory led some behaviorists to refrain from drawing conclusions about the inner mental lives of experimental (human and nonhuman) subjects. In extreme cases, some theorists went so far as to deny the very existence of inner mental lives. Other approaches to the study of psychology, from the old introspectionist tradition that was replaced by behaviorism, to more recent work in cognitive psychology, have been more generous to their experimental subjects in terms of what may be legitimately concluded from observations.

Informed by theory (including the scientific fecundity, explanatory success, and inductive power of theory), the position of the interpretative and inferential bar is not something that is fixed or positioned in exactly the same way for all investigators, regardless of their theoretical perspectives, prior to the onset of their inquiries, nor is it something whose position, once fixed, stays fixed for all time. It is something that evolves and changes with the evolution and development of scientific theory.

The scientific culture we are examining in this book consists of strong evidential behaviorists who tend to rely exclusively on carefully controlled laboratory studies and weak evidential behaviorists who at least give some credence to hard-to-control field observations and anecdotal evidence. But this distinction, with disproportionate value placed on hard evidential behaviorism, can be dangerous. Thus, in an argument for pluralism, Bekoff observed:

> All research involves leaps of faith from available data to the conclusions we draw when trying to understand the complexities of

> animal emotions, and each has its benefits and shortcomings. Often studies of the behavior of captive animals and neurobiological research are so controlled as to produce spurious results concerning social behavior and emotions because animals are being studied in artificial and impoverished social and physical environments. The experiments themselves might put individuals in thoroughly unnatural situations. Indeed, some researchers have discovered that many laboratory animals are so stressed from living in captivity that data on emotions and other aspects of behavioral physiology are tainted from the start. (2000, p. 868)

This is not to say there are no problems with field observations and anecdotal reports. Bekoff's point is that a variety of different sources of evidence needs to be examined, and there needs to be a willingness to interpret and make inferences from this broader body of data.

How, at the end of the day, should we respond to the conflicting scientific estimates that currently exist concerning the cognitive lives of nonhuman animals? Bekoff referred to Jaak Panskepp, who has suggested a version of Pascal's wager. You must answer the question as to whether mammals have internally experienced emotional feelings. If you give the wrong answer, you follow the devil home—that is, the stakes are very high indeed. As Bekoff observed, "how many scientists would deny under these circumstances that at least some animals have feelings? Likely, few" (2000, p. 868). Even biocartesian Euan Macphail urged caution about his cognitive minimalism for fear of being mistaken.

The intuition behind Panskepp's wager is that skepticism about the mental lives of animals is purely academic and that when push comes to shove most investigators would be willing to make concessions. Maybe so. But it must be remembered that this skepticism about the cognitive capacities of nonhuman animals, which, as we have already noted, is deeply ingrained in some of the various strands that are constitutive of Western culture, is now bolstered by serious theoretical reasoning, both scientific and philosophical, as well as by serious methodological concerns about what can and cannot be inferred from behavioral evidence.

Lay persons will naturally appeal to untutored common sense. Surely it is obvious that dogs or other pets can experience pain and a wide range of emotions. But untutored common sense is not necessarily a reliable guide even about things directly observable. The untutored common sense of our ancestors told them that the sun rose in the east and sank in the west, thereby orbiting the earth. Or, to turn to a biological example, in the seventeenth century, Dr. William Harvey told John Aubrey that "after his booke of the circulation of the Blood came out, he fell mightily in his Practize, and 'twas believed by the vulgar that he was crack-brained" (Dick 1978, p. 26). And even today, Mom's untutored common sense leads her to tell her children to stay warm and dry for fear of catching a cold—*Mom medicine* makes no reference to rhinovirus. And lest we appear to be unduly harsh to Mom medicine, recall that until comparatively recently, the tutored intuitions of physicians told them that all stomach ulcers were caused by stress and not, as we now know for many ulcers, by treatable bacterial infections. Could untutored common sense be equally misleading here, where the issue is feeling-consciousness in nonhuman animals? That it is misleading is part of what the scientific skeptic is trying to argue.

How then should reasonable people respond to the arguments of those skeptical of the cognitive capacities of animals? Skepticism is a double-edged sword. If we cannot be certain that animals have emotions and feelings and experience pain, neither can we be certain that they do not. Moreover, the conscious experiences of other organisms (human or otherwise) are subjective and private, and thus hidden from unambiguous scientific resolution in ways in which astronomical relationships, the motions of the blood, and the causes of colds and ulcers are not.

I think that a reasonable person should adopt some ideas found in courts of law when trying to assess the importance of scientific skepticism concerning the cognitive capacities of our phylogenetic kin. When a case goes before the grand jury, the purpose is not to settle questions of guilt or innocence, but to determine whether there is a prima facie case to be heard in the criminal court. The arguments of the cognitive skeptics are sufficiently well formulated and presented to make a prima facie case that animals do not have conscious experiences, do not feel pain and experience emotions. There is definitely a cognitive case to be heard in the court of rational inquiry. In saying that there is a case to be heard, we are not presuming guilt—we are not presuming the cognitive skeptics are correct. Saying there is a case to be heard means that we are confronted with interesting and serious arguments that need to be attended to.

In a criminal court, however, the standard of evidence is different. Here guilt must be proven beyond a reasonable doubt. The charge that nonhuman animals do not have conscious experiences, do not feel pain or experience emotions is a serious charge indeed, and if established beyond a reasonable doubt, would have serious and immediate implications for the claim that they are beings with moral status—beings who should not be treated arbitrarily and capriciously. And it is here that I think the cognitive skeptics run into trouble.

Claims that consciousness only evolved in the evolutionary line leading to modern humans involves a reconstruction of evolutionary history that will be hard to establish and—as with other claims involving historical reconstruction—will almost certainly not be established according to the strict evidential standards demanded by those who will only be satisfied with unambiguously interpreted, tightly controlled laboratory tests.

Moreover, the ambiguities that the skeptics see in experiments and observations that lead cognitive ethologists and others to conclude that animals are conscious and are capable of experiencing pain and emotions are just that—ambiguities. Ambiguity, and the possibility of alternative interpretation, does not constitute evidence, in and of itself, that nonhuman animals have no conscious mental lives. This issue is all the more acute because the arguments of the skeptics often hinge crucially on where the interpretative and inferential bar is to be set in our endeavors to make sense of our data. These arguments, as we have seen, do not simply reflect experiments performed, but are themselves inextricably intertwined with theoretical and philosophical issues and assumptions.

We have not proved that nonhuman animals are conscious. Consciousness is private, subjective, and directly knowable only from within. All our data are public, behavioral, and knowable (and interpreted) from without. By a similar light, we have not proved that animals do not have conscious experiences. Moreover, even though the skeptical arguments raise interesting issues, they have importantly not proved *beyond reasonable doubt* that nonhuman animals do not have conscious experiences. We can attend to the skeptical arguments and nevertheless reasonably doubt their conclusions. This is essentially to adopt the *modest cognitivism* discussed in the preceding chapter. That is one good way in which modesty can be a virtue!

Where does this leave the moral status of nonhuman animals—an issue that has itself been inextricably entangled with the light shed on the nature of nonhuman animals by the sciences? It is clear that we must make decisions under conditions of uncertainty. The cognitive skeptics have a point, but their arguments are neither straightforward nor conclusive. This is especially important since the subjects of those arguments may be conscious, may suffer pain, and may have emotional experiences. That is, they may be beings worthy of moral consideration.

In a case like this, what we need is a *principle of cognitive charity.* Since it is reasonable to entertain doubts about the claim that nonhuman animals, particularly mammals, do not have consciousness, do not experience pain, and do not have emotional lives, we should proceed *as if* they are conscious beings. In the absence of definitive proof to the contrary, we should extend to our phylogenetic kin the benefit of the doubt and should modify our behavior toward them accordingly. Remember that Macphail's skeptical arguments do not just apply to nonhuman animals, but also to preverbal human children. I think most reasonable people would extend the benefit of the doubt.

In this chapter, much has hinged on the ability of humans—and not other animals—to use language. It has been assumed, uncritically, that there is no evidence of language use in other, nonhuman species. But this assumption has been challenged both by theorists and experimentalists, and so, since the issue of language use has been argued to play a critical role in issues about feeling-consciousness and self-consciousness, we must now turn to examine this matter in a bit more detail.

Further Reading

The case against consciousness in nonhuman animals has been made most forcefully by E. M. Macphail, *The Evolution of Consciousness* (Oxford: Oxford University Press, 1998). By contrast, the case for consciousness in nonhuman animals has been well argued by D. Jacquette, *Philosophy of Mind* (Englewood Cliffs, NJ: Prentice Hall, 1994). Issues about the nature of the evidence in these debates have been discussed in useful ways by C. Allen and M. Bekoff, *Species of Mind: The Philosophy and Biology of Cognitive Ethology* (Cambridge: MIT Press, 1997), and M. S. Dawkins, *Through Our Eyes Only? The Search for Animal Consciousness* (Oxford: Oxford University Press, 1998).

Readers interested in learning more about the human brain might examine S. A. Greenfield, *The Human Brain: A Guided Tour* (New York: Basic Books, 1997), or R. Restak, *The Modular Brain* (New York: Simon and Schuster, 1994).

11

Animals through the Looking Glass: Language and Self-Consciousness

In the preceding chapter it was seen that there are serious disagreements about the issue of feeling-consciousness in nonhuman animals. This chapter is concerned with the issue of self-consciousness in nonhuman animals. Humans are feeling-conscious, humans are self-conscious, and humans are genuine language users. Since the time of Descartes, these properties of humans have formed a cognitive trinity, and even today the issue of the relationships between the various forms of consciousness and the capacity for language is a topic of active theoretical and experimental study. In this chapter we will examine some of the controversies that exist in modern science concerning these contentious matters. In the preceding chapter it was seen that *biocartesians* are contemporary scientific theorists who believe that one or more components of the cognitive trinity (usually all three) are confined to the human lineage. That a capacity for language is so confined has been challenged by several theorists, and in particular by Roger Fouts. If it could be shown by the use of unambiguous behavioral assays that (some) nonhuman animals could use and comprehend language in ways broadly similar to human capacities in this regard, this would be powerful evidence that such animals were organisms exhibiting high levels of cognitive sophistication. Given Macphail's arguments reviewed in the preceding chapter, it would be evidence of both the possession of self-consciousness and feeling-consciousness. Clearly there is much at stake in the animal-language controversy.

Language and Nonhuman Animals

The study of nonhuman animals has revealed the existence of remarkably sophisticated communication systems (see Hauser, 1997). But is there any evidence that there are nonhuman animal species that use language the way we humans use language? No lesser figure than

Charles Darwin could remark, "He who understands baboon would do more toward metaphysics than Locke" (quoted in Wynne 2001, p. 120). Much fiction in Western culture, from the Bible with its talking snakes and donkeys, to the sentimental cartoons of the Disney corporation, involves portrayals of nonhuman animals using language the way we use language. Many pet owners swear their animals can almost talk. Some have even trained birds to sing songs and utter short phrases.

Over the last thirty years much interest has been generated by the claims that certain nonhuman primates—notably chimpanzees—could not only be taught American Sign Language (ASL) but could use it creatively, the way we humans—at least young humans—use, say, English. In his book, *Next of Kin: My Conversations with Chimpanzees,* Roger Fouts proffered a very optimistic view of the linguistic capabilities of chimpanzees, comparing the famous chimpanzee, Washoe, to a child in an ape costume (1997, p. 20). Whether or not we agree with the conclusions that Fouts wished to draw, he was certainly correct to point to the importance of the language research involving chimpanzees such as Washoe, for as he observed:

> Descartes and Darwin collided in Project Washoe. If Descartes was correct, then Washoe didn't have a thought in her head and would be unable to name a single object. If Darwin was correct, then Washoe was already thinking and would be able to express her thoughts by manipulating ASL signs like tools. (1997, p. 71)

Perhaps at last there was an opportunity to settle once and for all this great debate between competing scientific estimates of the cognitive capacities of nonhuman animals.

Chimpanzees lack many of the adaptations that in humans allow the production of speech. Not only do they not have a tongue and larynx shaped and positioned for speech, there are also important neurological differences. They are very gestural creatures, however, and it was for this reason that investigators believed that sign language rather than spoken language should be employed in studies of linguistic abilities.

Fouts contended that Washoe had the linguistic abilities of a small child and that her errors in language use were comparable to those of a small child in the early stages of language acquisition (1997, pp. 44–45). Nevertheless, the reported results are very interesting:

> After about ten months she began spontaneously combining words: GIMME SWEET and COME OPEN were soon followed by longer phrases like YOU ME HIDE and YOU ME GO OUT

Chimpanzees lack many of the adaptations that in humans enable speech production. Not only do they not have a tongue and larynx shaped and positioned for speech, there are also important neurological differences. (Hulton / Archive)

> HURRY. She commented on the environment: LISTEN DOG; she asserted possession of her doll: BABY MINE; and she created her own vocabulary when she didn't know a sign: DIRTY GOOD, for her potty-chair. (1997, p. 30)

As was the case in the preceding chapter, so too here. The central issue is not simply what Washoe did, but how it is to be interpreted and understood. Fouts interpreted this sort of behavior as evidence of genuine language use and inferred that chimpanzee behavior showed that Descartes was wrong in his estimates of the linguistic abilities of nonhuman animals. As you might have guessed, Fouts, like all the other participants in this debate, offered arguments that hinge crucially not just on experimental evidence but also on scientific theory, to try to make a solid case.

Fouts made an appeal to the theory of evolution and to molecular studies of humans and chimpanzees designed to establish how closely these two species are related. Darwin's theory of evolution had suggested that humans and chimpanzees were close relatives, but it did not settle how close. As a consequence of the molecular revolution in biology—a revolution that began in the 1960s and has recently culminated in the Human Genome Project—we are now in a position to study molecular similarities and differences between various species to determine their relative degrees of evolutionary closeness. Early studies examined differences and similarities of amino acid sequences in various proteins; now it is possible to examine entire genomes.

If we examine human and chimpanzee DNA, we find a remarkable base-pair similarity of about 98.4 percent. How significant is this? Fouts observed:

> It means that humans and chimps are closer genetically than two hard-to-distinguish bird species like the red-eyed vireo and the white-eyed vireo (they are only 97.1 percent identical). But even more telling is this fact: humans are nearly as close genetically to chimpanzees as a chimpanzee is to a bonobo, a second species of chimp. . . . This fact has led physiologist Jared Diamond to propose that we humans are, for all intents and purposes, a third species of chimpanzee. . . . (1997, p. 55)

Even though we must concede that humans and chimpanzees are closely related from an evolutionary standpoint, this in itself does not establish that Washoe, for example, is a human in a chimpanzee suit! The interpretation of Fouts's observations about a high degree of genetic similarity will have to be examined carefully.

Indeed, geneticist Richard Lewontin has examined the molecular evidence and sees a very different picture to that painted by Fouts. Studies of DNA similarity do indeed support the claim that humans and chimpanzees are very closely related—the evolutionary lineages leading to chimpanzees and humans diverged about 7 to 10 million years ago for approximately 14 to 20 million years of independent evolution. Lewontin observed:

> To get a perspective on the differences that can accumulate in such a period, it should be realized that nearly all the differences between cows, goats and deer, have occurred in the same time period, and this is roughly the length of time separating deer and giraffes! It is often stated that our DNA is 99% similar to that in lower primates, but this figure tells us nothing. The vast bulk of our DNA, about 97%, is of no clear functional significance . . . and it diverges between species at a constant time rate. So the similarity of this part of the DNA is simply another way of saying that we diverged from the other primates 7 to 10 million years ago. In the functional DNA, there is no proportionality between the percent similarity and similarity of function . . . and the claim that we share some percent of our DNA with some other species contains no information about our anatomical or physiological similarity to that species in any particular respect. (1995, pp. 15–16)

Arguably, then, perhaps less hinges on the degree of base-pair similarity than Fouts has led us to believe. So what, if any, are the anatomical and physiological similarities between humans and chimpanzees with respect to language use?

The presupposition of evolutionary biocartesians is that after chimpanzee-human divergence, human ancestors developed adaptations for speech in the form of "a new voice box, a new brain mechanism, or new powers of high-speed auditory discrimination that enables us to develop language" (Fouts 1997, pp. 91–92). In other words, language would be confined to the human lineage. But what if chimpanzees can genuinely use language? What would be the evolutionary significance of such a phenomenon? Fouts was very clear:

Noam Chomsky hypothesized that humans had a language organ located somewhere in the left hemisphere of the brain. (Hulton / Archive)

> [*I*]f *Washoe could learn a human sign language* it meant that the common ancestor of both humans and chimps also must have had the capacity for this kind of gestural communication. And because evolution always uses the materials it has at hand—recruiting existing structures and behaviors to build new ones—early humans must have built up signed and spoken language on the very ancient foundation of cognition, learning and gesture laid down by our common ape ancestor. (p. 92; my italics)

There is a big *if* here that will have to be examined shortly. For the present, we must ask whether this line of evolutionary reasoning is plausible.

Fouts referred to the work of Noam Chomsky, whom we have already met in Chapter 9 of this book. Having rejected behaviorist accounts of language acquisition, Chomsky hypothesized that humans had a language acquisition device—or language organ—located somewhere in the left hemisphere of the brain. This language organ was to be unique to our species, and its sudden appearance in the human lineage was perhaps the *saltationist* consequence of a cognitive "big bang." At any rate, although Chomsky was skeptical that the capacity for language—a cognitive novelty—could evolve by natural selection (Hauser 1997, p. 35), his proposal in fact leaves the origin of language as an unsolved mystery. It was seen in the preceding chapter, however, that an alternative to a saltationist account of an evolutionary novelty was a gradualist theory. Perhaps a gradualist account could shed light on what Chomsky has left as a mystery.

Fouts was willing to consider this possibility, but he argued that a gradualist account of the evolution of the capacity for language that confines it to the human lineage was biologically implausible:

> There simply wasn't enough time, in the brief six-million-year period since humans diverged from our fellow apes, for the evolution of a completely new brain structure. . . . The primate brain did not evolve like an ever expanding house, adding on new rooms as it grew from monkey ancestor to ape ancestor to human. Instead, evolution was continually reorganizing what it already had—taking old structures and old circuits and putting them to use for new mental tasks. *In fact brain research since the 1960s has shown that human language is controlled by a network of independent cortical areas, each of which has an analogous area in the chimpanzee brain.* (1997, p. 94; my italics).

But what exactly are we to make of these similarities between chimpanzee and human brains? Do these similarities demonstrate a common linguistic inheritance for humans and chimpanzees?

Evolutionary biologists pay special attention to the relationship between structure and function. In evolutionary biology, the concepts of *similarity* and *difference* have to be handled with special care. For example, in the context of *convergent evolution,* different structures, systems, and processes can come to serve similar functions, whereas in the context of *exaptation* (the process in which phenotypic features are co-opted for present use via conversion of function), structures serving one function can be co-opted in the course of evolutionary time to serve new functions—so structural similarities do not, of themselves, point to functional similarities. So what do we know of the neurological similarities that Fouts pointed to in the passage quoted above (1997, p. 94)?

Chimpanzees have about one-fourth as many cortical neurons as humans. But does this mean that the chimpanzee brain is simply a human brain writ small? Though Fouts downplayed chimpanzee vocalization, he has emphasized brain similarity. Yet it is interesting to note that roughly the same areas of the brain that are involved in chimpanzee *vocalization* are involved in *speech* and *speech comprehension* in humans. So what are we to make of these analogies between the human and chimpanzee brain? Richard Lewontin made the following observations:

> When the motor areas of the brain are electrically stimulated in monkeys, they produce grunts. When the same area of the brain

is stimulated in humans they produce grunts. When a region called Broca's area is damaged in humans, various speech disorders, aphasias, result that are not motor malfunctions, but have to do with the making and comprehending of sentences. . . . When Broca's area is stimulated in lower primates they move their tongues, lips and mouths, but they do not make any vocalizations. So a region of the human brain that is associated with speech production and comprehension, is nothing but a motor area in lower forms. (1995, p. 16)

Similar remarks can be made about Wernicke's area, whose role in humans concerns the understanding of heard speech. In other primates this area seems to be associated with the differentiation of sounds the individual makes from those made by other individuals. Structural, anatomical similarities do not imply functional similarities.

According to Lewontin, the actual evolutionary story of the evolution of a capacity for language unravels as follows:

> [C]ertain regions of the primate brain were recruited from their simple motor functions to create quite new functions, the functions of speech. At the same time it was possible for these regions to hold on to their old functions, probably because the brain grew so much larger and so had nerve connections to spare. There is no homologue of speech in chimpanzees, certainly not grunting. Speech is a novel function whose anatomical basis was created by the recruitment of parts that could be spared. . . . There is no reason to suppose that a study of the brain or behavior of the gorilla will tell us anything about human language. (1995, p. 16)

So either the neurological analogies Fouts has pointed to are irrelevant, or nonhuman primates do not, after all, have the neurological apparatus necessary for language use.

Fouts's entire case rested on the assumption that chimpanzees working with the gestures of ASL are genuinely using language and genuinely creating sentences. His challenge was to explain how this could be. He has made much of the evolutionary closeness between humans and chimpanzees. But as observed once again by Lewontin, the linguistic abilities of chimpanzees are certainly no better than those of dolphins, creatures extensively studied because, like humans, they have large brains relative to body size. The evolutionary line that leads to dolphins diverged from that leading to humans at least 60 million years ago, for at least 120 million years of independent evolution. Humans are no closer to dolphins than they are to rats (1995, p. 17).

At this point we need to go back to the sorts of experiments that led investigators such as Fouts to suppose that they were really conversing with chimpanzees. No one doubts that the chimpanzees produced interesting and complex behavior. But was it language, or were there simpler explanations of the same phenomenon? Criticisms have been reviewed by Dawkins (1998) and Shettleworth (1998). According to Dawkins (p. 74), there are at least three simpler explanations of the observed behavior than that the chimpanzees were using language: (1) the Clever Hans effect (the chimpanzees were getting subtle behavioral cues from their trainers); (2) failure to specify prior to the experiment which results would be deemed significant and how likely they were to occur by chance (so that the relevance of observed behavior could be properly analyzed); and (3) rules of thumb (simple behavioral rules that collectively give rise to apparently complex behavior).

Commenting on the earlier work of Herb Terrace and his colleagues, who had trained a chimpanzee called Nim using methods similar to those used in the work on Washoe, Shettleworth has observed:

> Terrace *et al.*'s most devastating conclusion came from an analysis of 3.5 hours of filmed interaction between Nim and his trainers. The films revealed that very often Nim's signs were simple repetitions of signs that had just been made by the trainer. The same effect was evident in commercially available films of Washoe. . . . These observations cast doubt on the possibility that even these animals' orderly two-word strings express an implicit knowledge of syntax. (1998, p. 555)

But there were other problems, too, with analogies drawn between the behavior of the chimpanzees and the behavior of young children:

> Children engage in conversation, which means taking turns to exchange (rather than merely repeat) information. Children also use language for more than getting what they want. They talk about the world, apparently for sheer pleasure in naming and commenting on things. In contrast, the signing apes tended to "talk out of turn" and seldom used signs other than as instrumental responses. (p. 556)

So it is not merely that the chimpanzee data had not been properly controlled to exclude Clever Hans effects; the behavior was not as analogous to human behavior as some of the early reports had led us to believe.

Experiments have been performed in which stringent controls

were introduced to eliminate the possibility of the Clever Hans effect. But as Dawkins, herself a cognitive ethologist, has observed:

> [O]ne rather sad fact emerges: the more carefully controlled the experiments are and the more precautions are taken to eliminate the possibility of some sort of chimp-human interaction, the less impressive the performance of the chimps becomes. Certainly they can associate certain objects or actions with gestures and symbols, but then dogs and horses can do as much and we do not instantly rush to label it as "language." (1998, p. 79)

This is a very good example of why unambiguous behavioral tests are so often demanded in this field. It also serves to underscore the differences between *strong* and *weak* evidential behaviorists—between investigators who rely exclusively on tightly controlled laboratory tests and those who are willing to give credence to anecdotes and hard-to-control field observations.

There were other problems arising from inadequate specification of the significance of experimental results. In early chimpanzee studies, we have little data on how often the chimpanzees were in error in their use of language. Investigators wanting to see language use are likely to record successes and discard failures. This is the *fallacy of selective perception,* and Dawkins thought it had been a problem for the assessment of these types of study (1998, p. 83). She has also raised other doubts of a statistical nature. Chimpanzee experiments sometimes yield startling results. But how likely were these results to occur by chance alone? For example, according to Dawkins, in some of David Premack's experiments on problem solving with the chimpanzee Sarah, the number of possible outcomes for the experiment were very limited, and hence the probability of success by chance alone was high (1998, pp. 82–83).

In order to control for inadvertent behavioral cueing by investigators, various experiments have been designed in which chimpanzees interact with keyboards connected to computers. In experiments with the chimpanzee Lana, the words were geometric designs on keys linked to a computer. Lana interacted with the machine to get things she wanted. But as Shettleworth has noted, this arrangement generated special problems:

> The computer was programmed to activate appropriate dispensers upon receipt of grammatical strings like, "Please machine give apple" and "Please machine give drink." Not surprisingly, what Lana learned mirrored the contingencies built into the system. Her behavior could be accounted for as associations

> between actions, people, or objects and symbols that could be plugged into six stock sentences such as "please (person) (action)." Terrace and others subsequently demonstrated, in both pigeons and monkeys, the kinds of sequence learning that might underlie learning such a stock sentence and studied some of its properties. . . . Nevertheless, even those who promoted the Lana project at the time now agree that any training regime in which "words" are used primarily as operants to obtain food and activities does not promote genuine linguistic competence, even if chimpanzees might be capable of it. (1998, p. 558)

The consensus among experimental psychologists is that at present there is no convincing data to support the claim that chimpanzees and other nonhuman primates have a capacity to create sentences that is similar to the human capacity.

Yet if chimpanzees cannot create sentences, can they perhaps understand them nevertheless? A curious study in this regard concerns a male Bonobo chimpanzee called Kanzi. Kanzi apparently has demonstrated the ability to understand human speech. His ability to carry out instructions in carefully controlled environments appears to be similar to that of a two-year-old child tested under similar conditions. The instructions were in the form of sentences such as, "Get the telephone that's outside." The subjects had to perform several actions with the objects under study, and they were exposed to novel sentences. Kanzi, who learned to use lexigrams by observing his foster mother use them, also demonstrated skills in their use that were comparable to those of a one-and-a-half-year-old human (Shettleworth 1998, p. 561).

But what are we to make of this? Unlike the child tested alongside Kanzi, Kanzi's grasp of English did not blossom into the full-blown linguistic competence characteristic of a human language user. Some investigators, notably Macphail (1998, p. 124), were skeptical of Kanzi's achievements. For although the Kanzi study did not obviously fall prey to the objections that had plagued the earlier work on talking apes, there were nevertheless reasons for concern. For example, as Clive Wynne has recently observed, it is hard to assess Kanzi's grasp of grammar since the average length of his utterances on the keyboard is around 1.5 pushes of the keys, and moreover:

> A critical test for Kanzi's comprehension of sentence structure involved asking him to respond to an instruction such as "would you please carry the straw?" Sure enough Kanzi picks up the straw. But there's a weakness here. Although it could be that grammar conveyed the correct meaning of the test sentence, it

> could equally be just the circumstances that made the requested action obvious . . . after all, a chimp may carry a straw; a straw cannot carry a chimp. (2001, p. 122)

Once again, we are plagued with problems pertaining to the use of unambiguous behavioral tests.

The Kanzi study raises an important issue about the early stages of language acquisition, however. As Shettleworth has observed, for some investigators

> any demonstration that linguistic output, whether it be vocal, pressing symbols, or responding to spoken commands, can be explained as simple discrimination learning is taken as showing that the subjects are not doing what a young child would do in similar circumstances. Yet simple associative learning may well explain some of the young child's early responses to speech, and some of the child's early sentences may be no more complex than Lana's stock sentences. One may note that Kanzi acquired his comprehension of spoken English only after intensive exposure and an extraordinary amount of attention from human companions, but such experience is the norm in young children. (1998, p. 563)

Kanzi seemed to have the same linguistic capacity as a child under two. What might this tell us about human and nonhuman animal language use?

The Evolution of Language

As we have seen in the earlier discussion of the rise and fall of radical behaviorism, Noam Chomsky (1957; 1986) provided a powerful critique of behaviorist accounts of language acquisition. Chomsky's work on the nature of language contained the following central claims:

1. For each human language there is a *generative grammar.* This consists of a finite set of rules that enables a competent speaker of that language to generate, from a suitable finite stock of components, any sentence of that language, including sentences he or she has never been trained to utter.
2. Human children have the capacity to learn any particular human language, and hence grasp its generative grammar. This cognitive capacity that enables a child to acquire its mother tongue is known as the *universal grammar.*
3. The universal grammar has an innate component.

Chomsky believed that these three points provided an explanation for the rapid acquisition of language by young humans—for the processes that account for the significant developmental differences between Kanzi and children *over* two years of age (Maynard Smith and Szathmáry 1999, p. 149).

Subsequent research seems to support some of Chomsky's contentions in this regard. Thus, Maynard Smith and Szathmáry have recently observed:

> During the last two decades, the conviction has grown that language has a strong genetic component. In some sense, our language capacity must be innate; we can talk and apes cannot, and the reason is that we are genetically different. Yet there are two ways in which language might be innate. We may be just generally more intelligent that apes, and the ability to talk is just a by-product of this fact. Or, and this seems more and more likely, there is a specific "language organ" in our brain analogous to a "language chip" in a computer; this organ is to some degree hard-wired, in that some of its neural connections are set correctly without external stimuli. (1999, p. 149)

We have already seen that the study of the brain has revealed the existence of modularity with respect to the ability to use and comprehend language: damage to specific regions of the brain correlates with specific types of dysfunction with respect to language use and comprehension. (See also Restak 1994, chap. 5.) We have also seen, at least for some of these regions of the human brain, that the corresponding regions in the nonhuman primate brain are not associated with the use and comprehension of language. We can expect neuroscientists to continue to teach us much about the neurological basis of the human capacity for language.

The study of the *evolution* of a capacity for language is not in such good shape. We have it and apes do not, which suggests that the story of the evolution of this cognitive capacity is confined to the lineage that leads to modern humans. But where, exactly, is hard to say. People were certainly speaking before they were writing, so we cannot expect a written record of the emergence of the abilities we associate with the capacity for language. Moreover there is no relevant fossil record; we cannot tell from skulls alone, which of them held the neurological adaptations for speech and which did not. Stephen Pinker (1994, p. 334) compared the evolution of language to the evolution of the elephant's trunk. The trunk is clearly of adaptive value to the elephant, yet it does not fossilize. This is an area of science that is almost inevitably long on theory and short on facts.

Is there any circumstantial evidence that might help to pinpoint the origins of language? Maynard Smith certainly seemed to think so. In a recent interview he observed:

> It is a very odd fact that until relatively recently, although humans had become erect and physically like us, and their brains had gotten bigger, and so on, they were incredibly conservative in the tools they use. The same kind of hand ax was made for a quarter of a million years. It is an unbelievable fact but it's true. Then suddenly, about 50,000 years ago, people became very inventive. They refined fish hooks, needles, painting on the walls, burial of the dead, figurines and spears. All sorts of new things suddenly started happening. What on Earth happened 50,000 years ago to make this possible? I think most of the people who have thought about this seriously have come to the conclusion that the only answer that makes sense is language—it helps you to think. Just try thinking without words. (Interview in Campbell 1996, p. 397)

If Maynard Smith is right, this is a matter of great importance for who and what we humans are. For fossils with identifiably modern human features have been found that are around 120,000 years old (Park 1996, p. 240).

In this book we have seen that scientists, philosophers, and laypeople talk as if the capacity for language is what sets us apart from the rest of the living world and from nonhuman primates in particular. They talk as if this capacity is almost the essence of *humanness.* Yet an evolutionary perspective teaches otherwise. If the capacity for language evolved, then without knowing exactly when, the following remarks seem relevant. If the hominid species that preceded the evolutionary emergence of our species—say *Homo habilis* or *Homo erectus*—could speak, then language would not be a characteristic confined to our species. Yet if, as seems likely, the capacity for speech is confined to our species, it almost certainly did not appear, conveniently and coincidently, at the exact time our species emerged as a good and true species in its own right—so that all and only those individuals in the modern human lineage had the capacity for speech. And here is the point of all this: If Maynard Smith is right, humans were around for a long time before the capacity for speech appeared. Sophisticated language use may differentiate humans from apes today; it may not always have done so.

And it must be noted that if a capacity for language did make a sudden appearance, say 50,000 years ago, then there was almost certainly a long history of evolutionary events preceding it. In addition to

Elementary school students working on a language exercise. It is theorized that humans' use of sophisticated language is what may differentiate us from apes today. (Hulton/Archive)

events in the evolution of the brain, the capacity for language required the coevolution of anatomical structures in the throat and mouth—and there must presumably have been strong selective pressures at work, because the adaptations in the throat and mouth that are superbly designed for the production of the sounds constitutive of speech have come at the expense of compromises in other functions. Our short palate and lower jaw are less efficient for chewing than that found in nonhuman primates. There is also less space for teeth—resulting in a tendency for humans to have problems with wisdom teeth. And as Darwin himself noted, our food has to pass over the orifice of the trachea, with an attendant risk of choking—an arrangement unlike that of other mammals who have separate pathways for breathing and feeding (Cziko 1995, p. 182).

Moreover, if there is an innate capacity for language, as Chomsky and others have claimed, it is a capacity that requires appropriate sensory stimulation for its normal development. To illustrate what is meant by this claim, consider the visual system—a system that can be studied in humans and other mammals. Cziko has noted:

> Cats who have one eye sewn shut at birth lose all ability to see with this eye when it is opened several months later. The same

> applies to humans. Before the widespread use of antibiotics, eye infections left many newborn infants with cloudy lenses and corneas that caused functional blindness, even though their retinas and visual nervous systems were normal at the time of birth. Years later a number of these individuals underwent operations to replace their cloudy lenses and corneas with clear ones, but it was too late. Contrary to initial expectations, none of these people was able to see after the transplant. (1995, p. 62)

The point of this observation is to show that normal brain development hinges crucially on there being appropriate sensory stimulation. The visual system is not simply genetically hardwired to unfold into a functioning system—if it were, the postoperative kittens and the humans should have been able to see after the sutures were removed. As Cziko put it, "The genome provides the general structure of the central nervous system, and nervous activity and sensory stimulation provide the means by which the system is fine-tuned and made operational" (p. 62). In other words, you do not simply inherit a functioning visual system.

There is evidence from the study of abused children to suggest that something similar happens in humans in the case of language. Much interest has focused on the case of Genie, an abused child who was locked away and deprived of linguistic stimulation for the first thirteen years of her life. Though later exposed to English, her outputs remained ungrammatical and primitive—not unlike the apes examined earlier in this chapter. As with the kittens above, it appears that there is a developmental window of opportunity to acquire normal linguistic functioning, and this requires sensory stimulation of the right kind (Maynard Smith and Szathmáry 1999, p. 152). We do not inherit a fully functioning language-processing system. We inherit a less complex system that requires environmental stimuli for its proper development and maturation.

The Gap between Humans and Apes

In their theoretical discussion of the evolutionary transition that resulted in our capacity for language, Maynard Smith and Szathmáry examined some rudimentary communication systems that, borrowing from the work of linguist Derek Bickerton (1990), they referred to as *protolanguage.* This is claimed to be the language of apes, that of children under two years of age, that of children raised under conditions of sensory deprivation, and that of speakers of Pidgin. The outputs of these rudimentary communication systems are certainly similar in nature. A child under two might say "red book" or "go store." A chimpanzee raised by humans might say "drink red" or "tickle Washoe." The abused child Genie might say "want milk" or "Mike pain" (Maynard Smith and Szathmáry 1999, p. 194). Much more controversial is whether these outputs are brought about by similar cognitive mechanisms and processes.

For example, consider the contrast between Pidgin communication systems and the Creole languages they give rise to:

> Pidgin is a means of communication that emerges when adults with no common language come into contact . . . for the present it is sufficient to say that it is a limited means of communication without grammar. Creole languages emerge when children grow up in such communities, where the main linguistic input is pidgin. Such Creole languages are proper languages with a fully developed grammar. (Maynard Smith and Szathmáry 1999, p. 155)

But it must be borne in mind that speakers of Pidgin were already proficient language users before they had to resort to Pidgin as a communication system. This is not the case with apes or children under two, for example.

Nevertheless, Maynard Smith and Szathmáry identified the central characteristics of the rudimentary communication system referred to as protolanguage as follows (1999, p. 161):

1. The use of words as Saussurean signs. The idea here is that concepts mediate the relationship between words and the objects they stand for. To use their example, the sight of a leopard elicits the concept of a leopard, which in turn elicits the word *leopard*. Alternatively, the word elicits the concept, which enables the hearer to imagine the sight of a leopard.
2. The lack of purely grammatical items such as *if, that,* and *when,* that is, words that do not refer to anything.
3. The absence of syntax. For example, rules that enable the formation of a sentence in English as the conjunction of a noun phrase and a verb phrase; rules that allow the formation of a noun phrase as the result of combining a determiner such as *the* with a noun; and rules that allow the construction of a verb phrase as the result of combining a verb and a noun phrase.

Bickerton himself introduced the idea of protolanguage to help solve the problem of the origin of language. Protolanguage is thus to be thought of as an intermediate communication system standing between speechlessness and a capacity for speech. Bickerton commented:

> If there indeed exists a more primitive variety of language alongside fully developed human language, then the task of accounting for the origins of language is made much easier. No longer do we have to hypothesize some gargantuan leap from speechlessness to full language, a leap so vast and abrupt that evolutionary theory would be hard put to account for it. We can legitimately assume that the more primitive linguistic faculty evolved first, and that contemporary language represents a development of the original faculty. Granted, this assumption does not smooth the path, for the gulf between protolanguage and language remains an enormous one. But at least it makes the task possible, especially since the representational systems achieved by some social mammals amounts to a stage of readiness, if not for language, at least for some intermediate system such as protolanguage. (1990, p. 128)

Maybe so. Nevertheless, there is controversy concerning the extent to which nonhuman animals are capable of protolanguage. Do the linguistic outputs of apes and other animals unambiguously support the hypothesis that they are using words as Saussurean signs? In particular, is there any evidence that concepts are used to mediate the connection between words and things? Vervet monkeys have distinct alarm calls for pythons and eagles—and behave differently, contingent upon the call they hear. But the possibility exists that such a monkey may hear the call and respond without knowing why it does so. It may have learned to respond appropriately without forming the concept of snake or eagle in its brain. These are complex and controversial issues. They are reviewed extensively in Hauser (1997) and Shettleworth (1998). But a possibility to be borne in mind is that protolanguage as characterized here may involve cognitive capacities that are confined to the lineage leading to modern humans.

Even so, if it should turn out that something like protolanguage is a halfway house between speechlessness and speech, an important part of the evolutionary account of the emergence of our sophisticated linguistic abilities may well involve a role for *intermediate grammars* that extend the range of what can be said beyond the limited domain of protolanguage. Maynard Smith and Szathmáry have proposed a number of ways in which protolanguage could have been extended—for example, by the addition of items for negation, pronouns, verbal auxiliaries (for example, *can* and *must*), expressions for temporal order, and quantifiers such as *many* and *few* (1999, p. 164).

To this end, Maynard Smith and Szathmáry drew a parallel with the evolution of a complex adaptive structure such as an eye. You do not need to have everything at once. In the land of the blind, a few light sensitive cells are better than none at all. Light sensitive cells in a pit enable detection (by shadow) of direction of light source; as the mouth of the pit gets smaller, you get a pinhole camera, and so on. Similarly, in the land of the speechless, a few concatenated words may be better than none at all, and gradual additions of grammatical richness, by slight extensions of the domain of communication, may confer a host of subtle advantages. But we do not know, and the evolution of our linguistic capacities remains an unsolved puzzle in science.

The Question of Self-Consciousness

It now looks highly doubtful that apes are capable of using and comprehending language in ways that are more sophisticated than humans under two years of age. Nevertheless, can the issue of self-consciousness in (some) nonhuman animals be decoupled from the issue of the possession of sophisticated linguistic abilities? In particular, is there any *evidence* that nonhuman animals exhibit self-consciousness? Like the issue of language, this is a highly contentious issue in science. As with our previous discussions, we must address issues of theory as well as evidence, since the former affects the interpretation of the latter.

For some theorists such as Macphail, the issue of self-consciousness in nonhuman animals is answered in the negative. Self-consciousness is something that comes with language, and language is something confined to the human lineage. By contrast, theorists such as Jacquette favor a gradualist evolutionary account of self-consciousness that has the phenomenon decoupled from the evolution of our species-specific linguistic capacities. Thus he remarked:

> At the highest known level, self-consciousness occurs as an intentional act directed toward conscious acts that normally confers on them a unity we refer to as the self. Here we find dogs, dolphins, higher apes and man. As before this kind of consciousness admits of degree. . . . All that is necessary for consciousness and self-consciousness is the occurrence of mental states that take other mental states as their intended object in the case of consciousness, or the unity of consciousness in the person, self, or ego, as an intended object in the case of self-consciousness. (1994, p. 120)

As with all cases of differing theoretical estimates concerning what is possible and what is not, we must try to see where the evidence lies and how participants in this debate are interpreting the data.

Before proceeding further, however, it is appropriate here to give a word of evolutionary warning. I have discussed Macphail and Jacquette in some detail to show the extremes that still exist on these contentious issues that emerged with the rise of science over 300 years ago. We must now examine the evidence for self-consciousness in nonhuman animals. Jacquette considered the possibility, and Macphail rejected it. But if we cannot find evidence of self-consciousness in nonhuman animals, even our closest phylogenetic relatives, the chimpanzees, it does not follow that Macphail is vindicated. It may well be that the issue fizzles out in empirical underdetermination. Humans descend in evolutionary time from other, now extinct, hominid species. If consciousness and self-consciousness emerged after divergence from our common ancestor with the line leading to modern chimpanzees, we simply may not be able to settle, using an appeal to extant evidence, who is right. All that will be left will be theoretical speculation—speculation that in the nature of the case will only be loosely constrained by observational evidence. We have language, we are feeling-conscious, and we are self-conscious. Without live evolutionary ancestors in the hominid line available for study, the connections between consciousness and language, on the one hand, and consciousness and brain complexity, on the other, may be impossible to unravel. Every evolutionary biologist knows that evolution is a historical science, like geology. This means that some important evolutionary events that we would like to know about are simply lost to history, and thus some debates may simply remain both unsettled and unsettleable in principle.

Self-Consciousness and the Mirror Tests

Before proceeding to the issue of self-consciousness in nonhuman animals, it will be helpful to say something about the issue with respect to humans, since this will bring out what the animal tests are supposed to be revealing. In his review of animal communication systems, Hauser made the following observations:

> A general consensus in the literature is that at approximately four years of age, the child begins to use words that reflect a more sophisticated understanding of own and other minds (e.g., "I *think, remember, believe,* that X," "I don't *think* she *believes* there's a tiger on the couch") . . . children begin to use their representational abilities to predict and explain the behavior of other individuals and, in so doing, are in the process of developing a "theory" of other minds. Children at this stage pass false-belief

> tests that younger children fail and show an understanding of how specific kinds of knowledge influence beliefs and intentions. (1997, p. 601)

Thus, to use Macphail's example, if you ask a three-year-old child if he knows what is in a box, he will answer "yes" if he has seen inside and "no" otherwise. If you ask what some other child knows, his answer does not take into account what the other child has seen:

> Above four years old children answer the questions about the other child's knowledge correctly. They now take into account what the other child has seen in assessing what that child will know—they appreciate that other individuals have a different visual perspective from their own. Experiments of this kind are taken to demonstrate the gradual emergence of a theory of mind in children. (p. 185)

In older children, (falsifiable) expectations are formed to allow prediction of the behavior of others.

To have a theory of mind and use it in the formation of expectations about the behavior of other individuals is to adopt what Daniel Dennett has called the *intentional stance* (1987, p. 49). Suppose we are explaining the behavior of a fellow human. We take the intentional stance if we decide that the best explanation of the behavior we observe involves our crediting that person with beliefs, hopes, fears, desires, even rationality. But there is no reason to restrict this to considerations of human behavior—we might take the intentional stance toward space aliens, were we to encounter them, or toward a suitably complex computer. The intentional stance, as Jacquette has observed, involves first-person intentionality in that the subject taking it can have no concept of the beliefs, hopes, fears, desires, and so on of other systems except from first-person acquaintance (1994, p. 116). Do we have reason to take the intentional stance toward the behavior of nonhuman animals, and is there evidence of nonhuman animals taking the intentional stance to each other?

In order to see whether apes, like humans, have self-awareness and a theory of mind, a number of experiments have been performed. First among these are the famous mirror self-recognition (MSR) tests. Many animals when confronted with a mirror image of themselves treat the image as though it were a conspecific. In a famous series of experiments due to Gallup (1970), chimpanzees lacking prior experience with mirrors were exposed to mirror images of themselves. At first they treated the image as if it were a conspecific, yet over a few days the social responses diminished and self-directed responses were observed—including grooming of body parts otherwise visually inaccessible, removing food from between teeth while watching in the mirror, making faces, and so on (Shettleworth 1998, p. 489).

In order to test the hypothesis that this behavior in response to the mirror image reflected self-awareness, Gallup devised the *mark test:*

> The chimps were anaesthetized and marked on one eyebrow and on the top of the opposite ear with an odorless, nonirritating red dye. When they had recovered from the anaesthesia, the animals were watched for thirty minutes without a mirror; there was virtually no behavior directed at the marks during this time. But when the mirror was reintroduced, the animals behaved . . . [by] touching and rubbing the marks, sometimes looking at their fingers or sniffing them in between touches. (Shettleworth 1998, pp. 489–490)

It is now generally recognized that chimpanzees, bonobos, and orangutans can recognize themselves in mirrors (Wynne 2001, p. 122). It has recently been reported that bottlenose dolphins also pass the mark test (Reiss and Marino 2001, p. 5937). Gorillas, monkeys, and other mammals, such as elephants, do not seem to pass the test.

The recent work on dolphins is of particular interest since it suggests that self-recognition is not confined to the great apes and humans. The line leading to humans and chimpanzees diverged from that leading to dolphins over 60 million years ago. Thus, though dolphins, chimpanzees, and humans show a high degree of encephalization and neocortical expansion, the brains of dolphins are very different from those of chimpanzees and humans with respect to both cortical cytoarchitecture and organization. On the basis of these observations, Reiss and Marino observe:

> The present findings imply that the emergence of self-recognition is not a by-product of factors specific to great apes and humans but instead may be attributable to more general characteristics such as high degree of encephalization and cognitive ability. Hypotheses about the evolution of self-recognition have, to date, focused on primate characteristics. Our findings show that self-recognition may be based on a different neurological substrate in dolphins. (2001, p. 5942)

The authors of this dolphin study see self-recognition in dolphins as an example of convergent cognitive evolution—the same cognitive capacity has evolved but is dependent on different neuroanatomical characteristics and evolutionary history. Reiss and Marino are cautious about the broader implications of their study. They draw no conclusions about general self-consciousness and the issue of a theory of mind (p. 5942). But it is precisely here that the original MSR tests got into trouble.

The issues are raised not by the data but by the interpretation placed upon them. Thus Gallup observed:

> Recognition of one's own reflection would seem to require a rather advanced intellect. . . . Moreover, insofar as self-recognition of one's mirror image implies a concept of self, these data would seem to qualify as the first experimental demonstration of a self-concept in a subhuman form. . . . Over and above simple self-recognition, self-directed and mark-directed behaviors would seem to require the ability to project, as it were, proprioceptive information and kinesthetic feedback onto the reflected visual image so as to coordinate the appropriate visually guided movements via the mirror. (Quoted in Shettleworth 1998, p. 491)

But the critics of the MSR tests have focused on just this issue: Exactly what can be inferred by way of cognitive sophistication from what should perhaps be referred to more neutrally as *mirror-guided body inspection* (*self-recognition* being a theory-loaded term)?

Does mirror-guided body inspection imply self-awareness? Shettleworth has observed that it is unlikely that chimpanzees behave as they do in front of mirrors because they have a humanlike sense of self—involving, perhaps, a sense of personal uniqueness, perhaps a sense of one's own mortality or place in the universe (1998, p. 491). Wynne has observed that autistic children, who have an impaired sense of self-awareness, nevertheless develop the ability to recognize themselves in front of mirrors in a way comparable to that of normal children (2001, p. 122). This is particularly significant because some investigators think that part of the deficit in autism is a failure to develop a theory of mind (Macphail 1998, p. 185). And in the case of chimpanzees, Macphail has observed:

> One possibility is that a (self-aware) chimpanzee sees the mark on, say, his eyebrow, and, knowing that the mark is on his body, moves his arm and fingers so as to touch the area. A second possibility differs only in that the chimpanzee might use the mirror to guide his arm towards the mark. But a third possibility is that the (non-self-aware) chimp sees the mark—a novel stimulus—and simply uses the mirror to guide his arm movements so the his finger contacts the novel stimulus. (1998, p. 180)

So once again, it is just not clear what we should infer from the observed behaviors in the context of MSR tests.

Are there other ways in which we might discover that nonhuman animals have a theory of mind? Suppose we observe one chimpanzee—Bonzo—regularly eating a banana only when out of sight of another chimpanzee—Mr. Jiggs. One hypothesis is that Bonzo *believes* that Mr. Jiggs wants the banana and that he will take it from him if he sees it. This hypothesis explains the behavior of Bonzo by assuming he has a theory of mind that enables him to anticipate the behavior of another chimpanzee on the basis of what he believes the other chimpanzee wants, desires, or hopes for. Another hypothesis explains the same phenomenon by noting that on several previous occasions Bonzo's eating a banana near Mr. Jiggs has resulted in Mr. Jiggs's stealing the banana. In this case, the secretiveness exhibited by Bonzo is the fruit of nothing more than associative learning.

In order to test whether chimpanzees have a theory of mind and explain the behavior of others on the basis of what they know others have seen, a number of experiments have been performed. Shettleworth described one of the experiments as follows:

> In the experiment with chimpanzees there were four food containers, each provided with a handle that the animal could pull to get the food. As each trial began, the containers were hidden behind the screen, and the animal watched as one experimenter baited a single container in view of a confederate (the Knower), while a second confederate (the Guesser) was out of the room. Then the Guesser returned, and Knower and Guesser each pointed to a container as the chimpanzee was allowed to make a choice. A creature whose theory of mind encompasses the understanding that seeing conveys knowledge would obviously choose (correctly, in this case) the container indicated by the Knower, whereas one that does not would choose randomly. (1998, p. 501)

But once again, though the chimpanzees eventually began selecting the container pointed to by the Knower, *they required many hundreds of trials before doing so consistently.* Thus Wynne observed, "This pattern suggests only that the apes involved gradually learned to associate one stimulus (the knower) with the reward, not that the animal is treating the trainers as people with minds"(2001, p. 122).

The last group of approaches aimed at uncovering whether or not nonhuman animals have a theory of mind involves looking for evidence of deceit—evidence of behavior that reveals an intent to mislead. Since we cannot ask nonhuman animals what their intentions are, we must look for behavioral evidence. Four behavioral criteria have been suggested, which if satisfied, should lead us to classify behavior as deceptive (Dawkins 1998, p. 132):

1. The instance of deceptive behavior should be part of an individual's normal, nondeceptive behavioral repertoire. (It must be behavior that would normally not rouse suspicion.)
2. The behavior must be rarely used in its deceptive role. (Cry "wolf" too often and no one listens.)
3. The behavior must be used in a way that is likely to be misinterpreted by others who take it at face value.
4. The deceiver must have something to gain.

Given the characteristics of the criteria for the identification of deceptive behavior, almost all of the evidence of deception is likely going to come in the form of anecdotes based on field observations. And there are many interesting anecdotes that seem to flesh out the idea that animals think ahead and deliberately try to deceive each other. Deceit? What could be more human, and what could be more indicative of cunning thought?

For example, Hans Kummer has reported that he was watching a troop of baboons who were resting. Over a period of twenty minutes, one of the females gradually moved about two meters to end up behind a rock, where she began to groom a subadult male. Dawkins commented of this case:

> If the dominant male of the group had seen this, he would have attacked both of them, but from where he was sitting, he could only see the female's tail, back and top of her head. He could not see her front or hands and he could not see the subadult male, which had bent down behind the rock. In other words the adult male could see where the female was but he could not see what she was doing. (1998, p. 133)

The problems here are, of course, all of the problems that are associated with uncontrolled anecdotal evidence.

Deception, by its nature, must be rare, which means that much of the evidence will be of "one-off" incidents. The anecdotes involve interpretation of the behavior in the light of our experience with humans (taken antecedently to have a theory of mind and to be deceptive critters, so that they are seen to do what we might do—or what we could imagine other humans doing—under similar circumstances). But the subadult male baboon need not have planned an illicit liaison with the female. He may have been behind the rock by chance when the opportunity for grooming arose. As Shettleworth has observed:

> It is usually impossible to tell whether functionally deceptive behavior in the field reflects situation-specific learning about behavioral contingencies or inferences by the perpetrator about the intentional states of its victim. A stable group of long-lived animals provides many opportunities for its member to learn from interactions with each other. The field worker witnessing one incident in unlikely to have access to all the relevant past history of the individuals involved. . . . The great mass of anecdotes that has been collected and classified includes reports from many different observers about many individuals of many species. A single individual practicing different kinds of deceptive acts might be convincing evidence of a theory of mind, but this is not the same thing as different acts by different individuals. (1998, p. 507)

For Shettleworth, then, ten leaky anecdotal buckets hold no more evidential water in the long run than one. For Dawkins, by contrast, the anecdotes and field observations of deceptive behavior *could* be explained as learning, as the result of chance encounters, and so on, but she thinks that the *simplest* explanation in these cases is that the animals are actually thinking (1998, p. 135). This is but one place out of many where disputes between strong and weak evidential behaviorists come to the fore—between those who demand tightly controlled experimental studies and those who rely to varying degrees on anecdotes and field observations.

And at least part of the dispute here concerns the invocation of the *simplicity* criterion. For Dawkins, low-level cognitive explanations of the observed instances of deception are possible but overly complex—an account of the observed deceptive behaviors in terms of associative learning might indeed be long and cumbersome: "This had to happen, then this had to happen, which had to be reinforced so this had to happen," and so on, whereas "X intended to deliberately deceive Y" is simpler by far. But one investigator's *simplicity* is another investigator's *complexity*. Shettleworth might argue back that Dawkins has only gained simplicity of behavioral explanation at the expense of attributing an unwarranted cognitive complexity to the animal subjects.

I conclude this chapter by observing that there is little good, that is, unambiguous, behavioral evidence to support the claim that nonhuman animals are capable of language acquisition, use, and comprehension that transcends that exhibited by humans two years of age.

And even giving the extant data a charitable reading, the linguistic accomplishments of chimpanzees and bonobos are slender indeed. The case for self-consciousness and the existence of a theory of mind is hardly better, fizzling out as it does in empirical underdetermination: The evidence that is taken to support self-consciousness or use of a theory of mind can also be explained by alternative hypotheses involving neither. Perhaps there is another lesson here. For although the light of contemporary science has not definitively illuminated important cognitive questions concerning nonhuman animals, the scientific study of animal behavior contains many lessons for those who would study the nature of science. As did the preceding chapter, the present chapter raises many intriguing questions about the design of experiments, the interpretation of data, and, importantly, what can and cannot be inferred from data.

A lot of the dispute in this perplexing field of the study of consciousness is in fact generated not strictly by factual data but by disagreements over the setting of the interpretative and inferential bar. Set the bar too low, and too many species will jump over it. You end up with self-conscious slugs and reflective termites. Set the bar too high, and other humans might be excluded. But between these extremes, there is a lot of room for difference of opinion. There probably are no unambiguous behavioral tests that can settle once and for all whether any organism definitely has private, subjective states of conscious awareness.

We only have externally observable behavioral data, and we are forced to formulate hypotheses about what is going on inside the organism to explain that data. The publicly observable behavior is thus one step away from the hypothetical subjective, private states of consciousness we wish to investigate. Theories about the inner mental lives—or lack of such lives—of organisms are thus underdetermined by the body of available evidence. The same data that lead one theorist to postulate consciousness in an organism can be reinterpreted by other investigators in ways that involve no such cognitive attribution.

The scientist may say, "All *we* know is that creatures who use language are also creatures who are feeling-conscious and self-conscious." But the philosopher will point out that the use of the word *we* in this statement had better involve the *Royal we,* in the sense that it is I, the person who uses language, who knows through direct introspection that he is feeling-conscious and self-conscious. You may tell me you are in pain, and you may tell me you are self-conscious. But the motions of your lips and tongue are just further examples of external, publicly observable behavior that underdetermines which, of at least two mutually incompatible hypotheses about what is going on inside you, is correct. One such hypothesis is that you really are conscious; another is that you only behave *as if* you are conscious; in reality you have no subjective conscious states of awareness whatsoever. This is the problem of other minds again. It is as experimentally intractable as the problem of other animal minds. Our practical solution to this problem is to reason about other humans by analogy. But we have also seen that extending this eminently practical solution to nonhuman animals is a matter fraught with difficulties that are simultaneously evidential, theoretical, and interpretative.

Could there be a technological solution to this problem raised by the private and subjective nature of consciousness? What of brain-imaging technologies? Can these take us within the organism? The answer is that they can take us inside, but not far enough. We know there is no single, localized, neuroanatomical structure—something analogous to Descartes's view of the pineal gland as the hypothetical localized seat of consciousness—that is present in humans and absent in chimpanzees. Imagine a comparative imaging study of a human and a chimpanzee. When the human becomes conscious of something in her environment, a pattern of neurological events is revealed by the imaging device.

When a chimpanzee is exposed to the same stimulus, a similar pattern of brain activity is revealed. Because of the consequences of evolutionary processes, in particular, the recruitment of existing structures to serve new functions, it cannot be concluded definitively that the human and the chimpanzee are having the same conscious experience or indeed a conscious experience at all. It is a lesson of evolutionary biology, that similar structures and processes do not necessarily serve the same functions in distinct species of organism. They may be like the wings of a bird and the flippers of a penguin.

Suppose the organisms in the imaging study differ with respect to patterns of brain activity when exposed to the same stimulus. It is hard to know what to conclude. Another lesson of evolutionary biology is that through convergent evolution, similar functions can be achieved by different structures and processes. How do we know, when dealing with a dolphin, for example, that differences between them and us with respect to patterns of neurological activity imply differences with respect to cognitive function? The answer is that we do not. Brain-imaging data will be new and exciting data, but there is no reason to believe they will offer definitive resolutions to these puzzles about consciousness that arise where philosophy and science collide.

Concerning our knowledge of the cognitive abilities of our closest phylogenetic kin, I will give the last word to Clive Wynne:

> In truth, scientists do not yet know much about the soul of the ape. . . . It is clear that these animals do not show self-awareness in anything like the ways a human does even when given the tools to do so. But if their minds are not like ours, what are they like? I don't know. . . . The only thing I'm sure of is that we should view apes as worthy of wonder for being what they are—not merely as reflections of ourselves. (2001, p. 122)

Further Reading

Issues raised by the ape-language controversy, as well as issues surrounding self-consciousness in nonhuman animals, are given an excellent review in S. J. Shettleworth, *Cognition, Evolution and Behavior* (Oxford: Oxford University Press, 1998). The case against self-consciousness in nonhuman animals has been forcefully made by E. M. Macphail, *The Evolution of Consciousness* (Oxford: Oxford University Press, 1998). The case for consciousness in nonhuman animals has been argued forcefully by M. S. Dawkins, *Through Our Eyes Only? The Search for Animal Consciousness* (Oxford: Oxford University Press, 1998).

12

Conclusion: The Questions that Remain

At its most basic level, this book has been concerned with controversies concerning the similarities and differences between human and nonhuman animals and their relative places in nature. Although debates over these issues are probably as old as humankind itself, the discussion here began with an examination of influential medieval views, and it was shown that these perspectives derived in turn from the philosophical traditions of ancient Greece. The theoretical perspectives of medieval thinkers provided important parts of the foundations upon which later investigators would build the edifice constitutive of modern science. Nevertheless, the rise and subsequent development of modern science have transformed our understanding of the basic issues considered in this book in ways that medieval theorists probably could not have comprehended.

Although experiments on animals played a crucial role in the rise of modern science in the Renaissance, the gradual evolution of modern science over the past 300 years has had enormous implications for our understanding of our place in nature and that of nonhuman animals. In this story, much interest has been focused on the cognitive capacities of nonhuman animals and on how similar those capacities are to, and how different from, our own. And inextricably intertwined with the debates over these issues are debates over the moral status and worth of nonhuman animals. We are not disinterested observers of this debate over the cognitive, and hence moral, status of nonhuman animals. Indeed, science has tried to shed light on the question of animal cognition itself.

The conviction that there are cognitive similarities between ourselves and nonhuman animals has had enormous implications for the characteristics of moral debates concerning our interactions with nonhuman animals. One of the reasons we do not permit nonconsensual experiments on human beings is that we believe that humans can suffer pain and be aware that they are suffering pain. All other things being equal, we recognize that it is morally wrong to inflict pain on our fellow human beings. To the extent that nonhuman animals can be shown, by the rigorous standards of scientific inquiry, to be cognitively similar to us in this regard, to that extent we would have scientific knowledge *relevant* to the debate over the moral status of nonhuman animals. It is perhaps all the more surprising that these important issues continue to be sources of genuine controversy among scientists.

For the past 300 years scientists have been using animals as experimental subjects to advance human knowledge. It is undeniable that this experimentation has conveyed much of interest about animals. But the same experiments have also shown that there are not just similarities; there are also important differences between humans and nonhuman animals. Mice are not humans writ small. This conclusion is reinforced by the theory of evolution, whose central legacy is that even though we are all animals, human and nonhuman animals are similar in some respects and different in others. The exact similarities and differences depend on the animal with which the human is compared. There are not just biochemical, physiological, anatomical, and behavioral similarities and differences; there are also cognitive similarities and differences.

But it has been argued in this book that it is precisely here that the issues become very interesting. For scientists do not just gather data—they analyze them, they interpret them, and they make inferences from them. In short, they theorize about them. And in the last two chapters, it has been seen that the same data can be analyzed and interpreted differently—so differently as to lead investigators to come up with radically different estimates of what can be safely inferred by way of nonhuman animal cognition and cognitive sophistication.

Untutored common sense may suggest simple and straightforward answers to questions about the cognitive abilities of nonhuman animals. Of course they can experience pain, of course they can suffer, of course they are conscious, of course they are self-conscious, and so on. But science has shown untutored common sense to be erroneous many times in the past—even about things that seem obvious (recall how obvious it was to our ancestors that the sun rose in the east and sank in the west, orbiting the earth, the natural center of the universe).

I have argued that scientific studies of animal cognition have involved debates about what sort of evidence shall count—field studies and anecdotes versus tightly controlled laboratory studies. But even where there is agreement over which evidence is to count, there are differing estimates concerning the interpretation of this evidence and what may be inferred from it. Differences over the interpretation of evidence and over what can be reliably inferred from it are not issues that can be settled simply by performing yet another experiment. These differences are deep intellectual differences that reflect in part the scientific theories held by investigators, in part the way they interpret those theories, and in part the philosophical presuppositions of investigators. This is all the more so when we are trying to draw conclusions about subjective, private, internal states of conscious awareness on the basis of external, public behavioral data.

The great twentieth-century geneticist Theodosius Dobzhansky could remark that "Nothing makes sense in biology except in the light of evolution" (1973, p. 125), but he might usefully have added, "once its interpretation and significance have been agreed upon." For example, in these debates we have seen that although the theory of evolution has had enormous implications for the analysis of the relationships between humans and nonhuman animals and their relative places in nature, these debates have nevertheless not been definitively settled by the theory of evolution. Philosophical views about the nature of consciousness and its connection to a capacity for language have been shaping these debates since before the seventeenth century. (And we have also seen how animal science in the seventeenth and eighteenth centuries reflected deeply held religious presuppositions—animals and humans were the mechanical result of intelligent supernatural design, a view that survives to this day in various species of so-called creation science.) The point here is that science does not take place in a cultural and intellectual vacuum. In particular, theories, both scientific and philosophical, really do matter. They shape the way investigators see the world and in particular how they understand and render luminous the data they have gathered. In this sense the interpretation and analysis of data are always contaminated by the theoretical beliefs of the investigator. Such theoretical beliefs also shape the interpretation of particular theories, such as the theory of evolution. Does this mean, then, that in science evidence is not important? Absolutely not.

Admitting that we sometimes have cases where we have competing theories that offer differing estimates of the significance of data, and that we also have cases where a given theory can be subject to divergent interpretations by different investigators, does not detract from the importance of evidential considerations. In particular, we have seen cases where investigators working under the aegis of the theory of evolution nevertheless differ over its implications for issues of language and consciousness and other cognitive capacities in nonhuman animals. Science thrives on these differences, for it is through their resolution, based on a combination of theory and evidence, that science makes headway against the tides of ignorance. In this sense, though the issues are complex and are shaped by both scientific and extrascientific factors, it is false to say that where science is concerned, *evidence is irrelevant.* Indeed, nothing could be more important, in the complex process of separating the intellectual wheat from the chaff, than carefully gathered evidence. The growing consensus that chimpanzees are not capable of sophisticated language use and comprehension is a good example of the central importance of *evidence* as a factor shaping the resolution of a scientific controversy.

The scientific study of feelings and consciousness in nonhuman animals should give pause, however, to anyone who thinks that the refutation of hypotheses by evidence is a simple and straightforward matter. The importance of evidence lies in the fact that it constrains our theorizing about the world—it constrains in particular, the sorts of explanations we can offer for phenomena of interest. But it rarely, if ever, determines unique theories about, and explanations of, phenomena of interest. Evidence is thus relevant in science not because it picks out unique theories, but because it provides checks on theorizing.

Nearly half a century ago, the importance of this underdetermination of theory by evidence was made forcefully by the philosopher W. V. Quine:

> The totality of our so-called knowledge or beliefs, from the most casual matters of geography and history to the profoundest laws of atomic physics or even of pure mathematics and logic, is a man-made fabric which impinges on experience only along the edges. Or, to change the figure, total science is like a field of force whose boundary conditions are experience. A conflict with experience at the periphery occasions readjustments in the interior of the field. Truth values have to be redistributed over some of our statements. . . . But the total field is so underdetermined by its boundary conditions, experience, that there is much latitude of choice as to what statements to reëvaluate in the light of any single contrary experience. No particular experiences are linked with any particular statements in the interior of the field, except indirectly through considerations of equilibrium affecting the field as a whole. (1963, pp. 42–43)

As we have seen, the same experience of squeals and howls, of self-recognition in mirrors, of deception in the field, and so on that was the evidence that led some investigators to attribute feeling-consciousness and self-consciousness to nonhuman animals was the same evidence that led other investigators to very different conclusions. And evidence of the disputes that can arise over which readjustments to make, in the light of new evidence, is seen nowhere more clearly than in the disputes between those who value evidence in the context of strictly controlled laboratory experiments and those who view such experiments as unnatural contexts for the study of animal cognition, and who in turn value the evidence derived from field study and anecdote. Differences over theory can sometimes boil down to differences over which sources of evidence are to be trusted. This is just another way of saying that in science, evidence itself is sometimes the object of theoretical inquiry. Differing theoretical estimates of the nature of legitimate evidence underlie a significant part of the dispute between strong and weak evidential behaviorists.

So science has not provided nice, clear-cut answers to the questions with which this book has been concerned. But science has transformed our understanding of the issues. Animals in the light of science are very different from animals in the light of medieval Christian theology. But perhaps of equal importance are the lessons we learn about the nature of science itself from careful observations of its investigations into the nature of animals. The science of animals may tell us as much about science as it does about animals themselves.

Looking to the Future: Organisms in the Light of Science

In this book I have examined controversies generated by the question as to whether nonhuman animals have morally relevant cognitive properties. There are scientific controversies concerning animals, and organisms generally, however, that I have not covered in this book but which deserve mention because of their potential impact on our species. Moving beyond the issue of animal cognition, there are issues about our relationships to other species that should concern us merely from the standpoint of a rational consideration of human self-interest.

These issues arise because there is more to science than a disinterested pursuit of truth about nature. I have argued in the early parts of this book that an important feature of the rise of science was *machine thinking.* This involved not just seeing nature in mechanical terms but also seeing it as something that could be studied with the aid of machines. One consequence of this was that the study of nature became increasingly coupled to the technological fruits of the engineer's labors.

But especially since the dawn of the Industrial Revolution in the eighteenth century, this coupling of science and technology has led to a scientific interest in the control and manipulation of various aspects of the natural environment. The nineteenth and twentieth centuries provide many examples of the harnessing of scientific and technical knowledge in physics and chemistry to serve a wide variety of industrial purposes. As we move into the twenty-first century, science itself is starting to reveal some of the (perhaps unintended) consequences of our attempts to manipulate and control the organic environment. It turns out that there is a dark side to the technological fruits of the biological sciences that scientific studies themselves have started to reveal and explain. I will briefly mention some of these below, as food for further thought.

Cattle Farming in the Light of Science

Though the biomedical sciences are an important arena in which to discuss controversies about the use of animals, science has had important implications in other arenas, such as agriculture.

Science has undoubtedly enabled us to boost crop yields and the production of animal products (milk and beef, for example). But there have been costs associated with these scientific practices. For example, consider bovine spongiform encephalopathy (BSE), otherwise known as *mad cow disease,* which resulted from the unforeseen consequences of nutritional technologies—in particular, the use of high-protein foods that are based on animal products.

In the United Kingdom, by effectively turning dairy cattle into cannibals—by using meat-and-bone meal consisting of the dried, cooked remains of dead animals, including cattle with undiagnosed diseases (Rhodes 1997, p. 175)—conditions were created in which BSE could flourish. Meat-and-bone meal was used extensively in dairy farming, and as Rhodes has observed:

> Beef cattle are also sometimes fed meat-and-bone meal during the finishing phase. Calves are fed meat-and-bone meal to maximize their growth. Dairy cows are slaughtered for meat when their milk production drops after three or four pregnancies, and since their meat is tougher than the meat of young beef animals, most of it is ground into hamburger or chopped for meat pies, a British staple. Cattle brains go into hamburger as well. . . . High quality animal protein is a relatively rare commodity in the world; not surprisingly, every part of beef and dairy cattle is eaten, recycled back into the next generation of animals, or processed into commercial products. (p. 175)

Mad cow disease resulted from the unforeseen consequences of nutritional technologies. This picture, taken January 22, 2001, shows two cows on a farm near Merseburg, Germany, where another case of mad cow disease was confirmed January 24, 2001. (AFP / Corbis)

The economic consequences of BSE were enormous. Whole herds had to be slaughtered. The implications for human health are still under investigation—but are potentially enormous.

Prions (abnormal protein molecules rather than bacteria or viruses) are the infectious agents indicated in BSE. They can cross species barriers and are believed to be associated with the appearance of serious and invariably fatal neurological disorders, such as variant Creutzfeldt-Jacob disease (vCJD) in humans—a disease that has killed more than ninety Britons, with a much larger number believed to be incubating the disease. If the average incubation time is twenty-five years, some worst case estimates suggest that Britain could be facing 200,000 deaths per year from vCJD by 2015 (Rhodes 1997, p. 222).

This will be the first hint that some practices based on animal science may in fact be quite dangerous—with consequences for animals and humans alike. A discussion of human self-interest clearly requires a discussion of the responses of organisms to the things we do to them.

Raising Crops in the Light of Science

The second case study concerns the consequences of scientific farming practices as they apply to crops. In the green revolution in India in the 1960s and 1970s, farmers were able to dramatically boost crop production by use of high-yield varieties (HYVs). These varieties were hybrids derived from crosses of pure-breeding genetic strains. The hybrids required chemical fertilizers, and since they lacked the natural, evolved resistance to disease and pests that was found in indigenous strains, the seeds were sold to farmers along with pesticides and herbicides (Goodwin 1996, p. 225).

There were short-term gains, but in the long term the use of fertilizers, pesticides, and herbicides causes environmental degradation including polluted drinking water and deterioration of soil quality. And as Goodwin has remarked:

> [B]ecause the HYVs are hybrids, their seed does not breed true. The farmers are not able to use part of one year's crop to sow the next, and must obtain seed directly from the supplier. The overall result is a pernicious cycle of increasing ecological damage, dependency of the farmer, dislocation of normal integrated farming practices, and debt. Farmers become enslaved to "scientific" methods of production that are intrinsically unsustainable, and new technological "fixes" are required to sort out new problems. (1996, p. 226)

And because pesticide and herbicide resistance genes arise as the result of mutations and spread quickly in populations of interest, ever higher levels of pesticides and herbicides are needed to combat "weeds and weevils."

Currently scientists are developing transgenic varieties of crops that are artificially modified to contain genes that produce enzymes that destroy the herbicides and pesticides used to raise them—herbicides and pesticides that are lethal to weeds and pests. Ciba-Geigy, for example, has designed soybeans resistant to their Atrazine herbicide (Goodwin 1996, p. 227). Dupont and Monsanto are following similar paths. But as Goodwin has commented:

> The problems associated with this strategy are so obvious that it is difficult to believe it is seriously entertained after the experience of the green revolution. The herbicide- and pesticide-resistant genes introduced into crop plants are unlikely to stay in these species, and will be transferred by viruses, bacteria and fungi to other species so that "weeds" will themselves become tolerant, necessitating higher doses of herbicides to destroy them. The result will be more toxicity in the soil and drinking water, and greater ecological destruction and ill health. (p. 227)

The economic and environmental consequences of these new farming technologies could well be enormous. In effect, agricultural practices based on these technologies will be enormous experiments on humans and other organisms alike.

These agricultural cases highlight the fact that the biological world is an evolving world that responds and adapts to our practices.

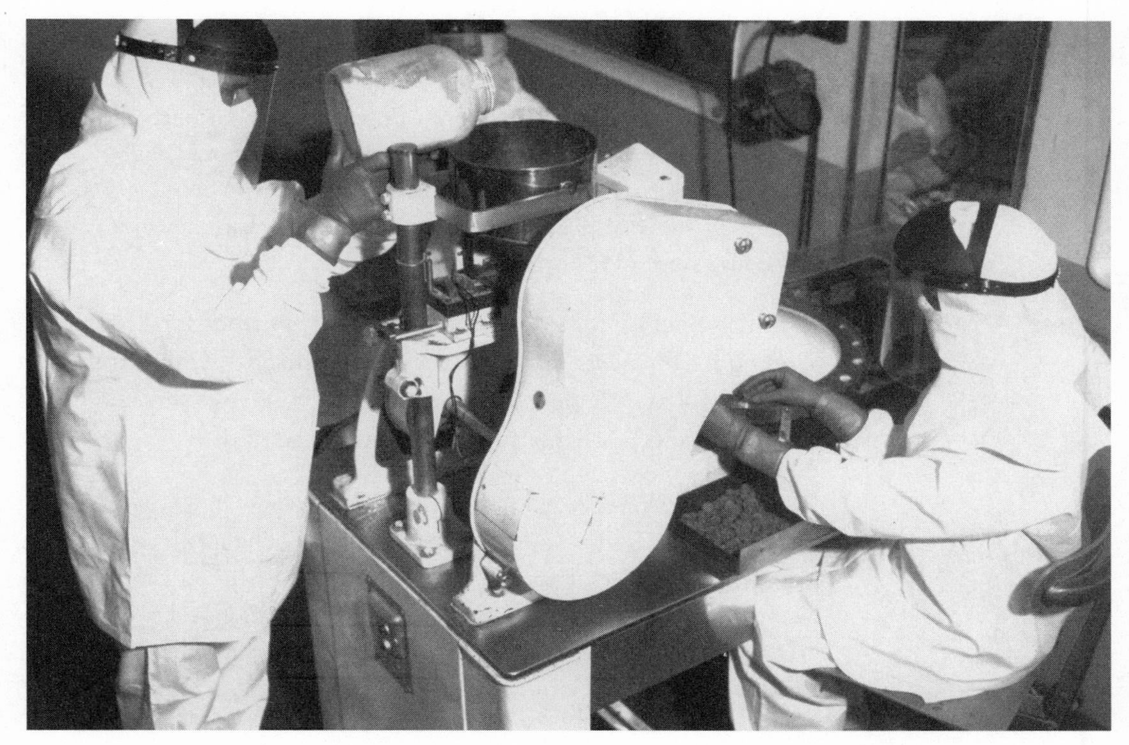

Medical scientists manufacturing antibiotics at Pfizer's factory. Antibiotics have been used extensively in human medicine to cure disease as well as to boost the production of milk and beef. This has created environmental circumstances conducive to the rapid evolution of drug-resistant strains of bacteria—bacteria that then pose a serious threat to human health. (Hulton/Archive)

There are good scientific reasons to believe that we should be cautious in applying the technological fruits of science to other organisms. This need for caution can also be seen in the next case study.

Microbes in the Light of Science

Organisms evolve in response to changing environmental circumstances. In particular, organisms are known to co-evolve in response to environmental changes resulting from human cultural evolution. In this context I will make a case for the claim that human fates (and interests) are intertwined with the evolutionary responses of other organisms to the technologies that are the quintessential fruits of the advance of science.

Consider the evolution of antibiotic resistance in bacterial species. Antibiotics have been used extensively in human medicine to cure disease and in the context of agriculture to boost the production of milk and beef. This practice has created environmental circumstances conducive to the rapid evolution of drug-resistant strains of bacteria—bacteria that then pose a serious threat to human health. For example, drug-resistant tuberculosis is now a major public health issue in many of the larger cities in the United States. Thus, Nesse and Williams observed:

> Perhaps most frightening of all, one third of all cases of tuberculosis in New York City are caused by tuberculosis bacilli resistant to one antibiotic, while 3 percent of new cases and 7 percent of recurrent cases are resistant to two or more antibiotics. People with tuberculosis resistant to multiple drugs have about a 50% chance of survival. This is about the same as before antibiotics were invented! (1995, p. 54)

Similar stories can be told of many of our ancient microbial foes.

Consider *Staphylococcus aureus,* a major cause of wound infection in hospitals. Today, 95 percent of strains of *S. aureus* show signs of penicillin resistance. Nesse and Williams commented:

> The drug ciprofloxacin raised great hopes when it was introduced in the mid 1980s, but 80 percent of staphylococcus strains in New York City are now resistant to it. In an Oregon Veterans' Administration hospital, the rate of resistance went from less than 5 percent to over 80 percent in a single year. (1995, p. 53)

In addition to drug resistance, strains of various bacterial species are now known that metabolize some antibiotics as a food source; other strains have been identified that can tolerate soap and disinfectants.

These cases illustrate the long reach of evolutionary principles. Mutations arise by chance that enable a bacterium to tolerate a clinical or subclinical dose of antibiotics. This ability confers a selective advantage in an environment contaminated with antibiotics. Less tolerant bacteria may be eliminated, along with the genes they carry, but the mutant bacterium will pass its useful genes to its offspring. Soon populations of bacteria will emerge that are tolerant to antibiotics. The useful gene may also be passed laterally to bacteria in other species through the mechanism of plasmid transfer, so that drug resistance, once it arises by chance, can spread far and wide.

In both these cases, the public will ultimately come to pay an awful price for the ways in which we have ignored predictable evolutionary consequences of our actions. There is no reason to believe that these cases cited here are isolated cases.

Ecology in the Light of Science

It can also be argued that a selfish human concern for the welfare of animals (and organisms generally) can also be defended in the context of our growing understanding of ecosystems in the light of the ecological sciences. The issue here is not a trendy concern for fashionable

A bulldozer clearing trees in the Borneo rain forest to make way for logging and agriculture. Concerning species lost by reduction in forest area alone, E. O. Wilson has conservatively estimated that we are losing about 27,000 species per year and concludes that we are in the middle of one of the great mass extinctions in world history. (Wayne Lawler; Ecoscene / Corbis)

endangered species. Instead, it concerns the rate at which species are being lost as a result of widespread habitat destruction and environmental damage consequent upon human activity.

In 1989, tropical rain forest was being lost at the rate of 142,000 square kilometers per year, equivalent to losing an area the size of a football field each second (Wilson 1992, p. 275).

Concerning species lost by reduction in forest area alone, E. O. Wilson has conservatively estimated that we are losing about 27,000 species per year (or about three per hour). His conclusion was that we were in the middle of one of the great mass extinctions in world history (1992, p. 280; see also Ward 1994).

Though ecosystems are resilient, there is no reason to believe that ecological damage on this scale will not have serious consequences for our own species. The economic consequences of unrestricted human exploitation of the environment are already with us—viz the decline of fisheries in the United States and Canada, as well as in other regions, for example the North Sea, where cod, the main ingredient of British fish and chips, is now severely threatened.

A case can be made for the need for rational public policies, based on our growing ecological knowledge, for the welfare of the biosphere as a whole. This must inevitably involve control of the growth of human populations—for we know that expanding populations bring about habitat destruction. Ultimately human welfare is inextricably intertwined with the welfare of other species. Enlightened self-interest alone, shorn of lofty moral reasoning and ecological mysticism, should lead us to this conclusion.

Science itself reveals a need for caution when it comes to the unrestricted economic exploitation of the environment using the technological fruits of science—technological fruits that are enabling us to destroy and alter the environment at ever-increasing rates that are arguably unparalleled in world history and certainly unknown to our ancestors as recently as 100 years ago.

Further Reading

Biological concerns about some of our technology-based agricultural practices are reviewed in B. Goodwin, *How the Leopard Changed Its Spots: The Evolution of Complexity* (New York: Harper Torchbooks, 1996), and R. Rhodes, *Deadly Feasts: Tracking the Secrets of a Terrifying New Plague* (New York: Simon and Schuster, 1997). The implications of evolutionary processes for human health and well-being are given excellent coverage in R. M. Nesse and G. C. Williams, *Why We Get Sick: The New Science of Darwinian Medicine* (New York: Vintage Books, 1995). Issues about the global ecological implications of human activity are given good discussion in P. Ward, *The End of Evolution: A Journey in Search of Clues to the Third Mass Extinction Facing Planet Earth* (New York: Bantam, 1994), and E. O. Wilson, *The Diversity of Life* (New York: W. W. Norton, 1992).

Documents

From *Summa contra Gentiles*

Medieval thought about humans and their relation to nonhuman animals is very complex, and many contrary opinions were defended by the philosophers of the period. Yet Thomas Aquinas (1225–1274) is generally recognized as one of the towering intellectual figures of medieval philosophy, and his views are included here because of the enormous influence that his ideas have had upon subsequent debates. The selection included here comes from Book 3 of his great work Summa contra Gentiles, *which was written sometime between 1259 and 1264. The subject discussed in the selection is a fragment of a much larger discussion of the relation of God to his creatures. In the passage below, it can be seen that the relationships between humans and nonhuman animals are matters that are taken by Aquinas to be inseparable from a discussion of the relative places of these beings in the grand created order. The rise of science did not initially banish appeals to the supernatural that we see in the work of Aquinas and other medieval thinkers. It was left to no lesser a figure than Charles Darwin to set things to right, to offer a scientific, naturalistic account of the relationship between humans and nonhuman animals, and thus to remove appeals to the supernatural from the domain of legitimate scientific discourse.*

112. Of God and His Creatures

That Rational Creatures are governed by Providence for their own sakes, and other Creatures in reference to them THE very condition of intellectual nature, whereby it is mistress of its own acts, requires the care of Providence, providing for it for its own sake: while the condition of other creatures, that have no dominion over their own act, indicates that care is taken of them not for themselves, but for their subordination to other beings. For what is worked by another is in the rank of an instrument: while what works by itself is in the rank of a prime agent. Now an instrument is not sought for its own sake, but for the use of the prime agent: hence all diligence of workmanship applied to instruments must have its end and final point of reference in the prime agent. On the other hand all care taken about a prime agent, as such, is for its own sake.

2. What has dominion over its own act, is free in acting. For he is free, who is a cause to himself of what he does: whereas a power driven by another under necessity to work is subject to slavery. Thus the intellectual nature alone is free, while every other creature is naturally subject to slavery. But under every government the freemen are provided for for their own sakes, while of slaves this care is taken that they have being for the use of the free.

3. In a system making for an end, any parts of the system that cannot gain the end of themselves must be subordinate to other parts that do gain the end and stand in immediate relation to it. Thus the end of an army is victory, which the soldiers gain by their proper act of fighting: the soldiers alone are in request in the army for their own sakes; all others in other employments in the army, such as grooms or armourers, are in request for the sake of the soldiers. But the final end of the universe being God, the intellectual nature alone attains Him in Himself by knowing Him and loving Him (Chap. XXV). Intelligent nature therefore alone in the universe is in request for its own sake, while all other creatures are in request for the sake of it.

6. Everything is naturally made to behave as it actually does behave in the course of nature. Now we find in the actual course of nature that an intelligent subsistent being converts all other things to his own use, either to the perfection of his intellect, by contemplating truth in them, or to the execution of works of his power and development of his science, as an artist develops the conception of his art in bodily material; or again to the sustenance of his body, united as that is to an intellectual soul.

Nor is it contrary to the conclusion of the aforesaid reasons, that all the parts of the universe are subordinate to the perfection of the whole. For that subordination means that one serves another: thus there is no inconsistency in saying that unintelligent natures serve the intelligent, and at the same time serve the perfection of the universe: for if those things were wanting which subsistent intelligence requires for its perfection, the universe would not be complete. By saying that subsistent intelligences are guided by divine providence for their own sakes, we do not mean to deny that they are further referable to God and to the perfection of the universe. They are cared for their own sakes, and other things for their sake, in this sense, that the good things which are given them by divine providence are not given them for the profit of any other creature: while the gifts given to other creatures by divine ordinance make for the use of intellectual creatures.

Hence it is said: Look not on sun and moon and stars besides, to be led astray with delusion and to worship what the Lord thy God hath created for the service of all nations under heaven (Deut. iv, 19): Thou hast subjected all things under his feet, sheep and all oxen and the beasts of the field (Ps. viii, 8). Hereby is excluded the error of those who lay it down that it is a sin for man to kill dumb animals: for by the natural order of divine providence they are referred to the use of man: hence without injustice man uses them either by killing them or in any other way: wherefore God said to Noe: As green herbs have I given you all flesh (Gen. ix, 3). Wherever in Holy Scripture there are found prohibitions of cruelty to dumb animals, as in the prohibition of killing the mother-bird with the young (Deut. xxii, 6, 7), the object of such prohibition is either to turn man's mind away from practising cruelty on his fellow-men, lest from practising cruelties on dumb animals one should go on further to do the like to men, or because harm done to animals turns to the temporal loss of man, either of the author of the harm or of some other; or for some ulterior meaning, as the Apostle (1 Cor. ix, 9) expounds the precept of not muzzling the treading ox.

Saint Thomas Aquinas. 1905 [1264]. *Summa Contra Gentiles. Book Three: Providence,* Part II. Translated by Joseph Rickaby, S.J. London: Burns and Oates.

From *De Fabrica Humani Corporis*

Andreas Vesalius (1514–1564) is rightly considered to be one of the great anatomists and physiologists of the renaissance. Vesalius's writings show the growing importance that investigators were giving to the need to observe directly the animal subjects of their anatomical and physiological inquiries. The selection here comes from Vesalius's seminal work, De Fabrica

Humani Corporis (The Fabric of the Human Body), *published in 1543, while Vesalius was Professor of Anatomy at the University of Padua in what is now Italy. The early anatomists studied human cadavers and conducted comparative inquiries into the similarities and differences between humans and nonhuman animals. Gruesome as dissection of dead bodies may have been, it paled in comparison to the practice of vivisection, which involved inquiries—invariably fatal to the experimental subject—into the structure and function of the parts of living organisms. This selection has been chosen to convey to the reader something of the nature of these inquiries—and how Vesalius managed to silence his experimental subjects in the absence of techniques of anesthesia.*

Book vii, Chapter xlx. What May Be Learned by Dissection of the Dead and What of the Living

Just as the dissection of the dead teaches well the number, position and shape of each part, and most accurately the nature and composition of its material substance, thus also the dissection of a living animal clearly demonstrates at once the function itself, at another time it shows very clearly the reasons for the existence of the parts. Therefore, even though students deservedly first come to be skilled in the study of dead animals, afterward when about to investigate the action and use of the parts of the body they must become acquainted with the living animal.

On the other hand since very many small parts of the body are endowed with different uses and functions, it is fitting that no one doubt that dissections of the living present also many contradictions.

The Use of Ligaments

In the dead one sees the uses of the ligaments binding their bones together, and also that of the ligaments drawn across tendons. Thus, provided we divide the ligament placed transversely in the internal side of the foreleg and pull on the muscle flexing the second or third joints of the toes towards its origin, we discover that this tendon is so placed chiefly in order that it may hold the tendons together that they may not lift up from their bed. However it will be permitted to demonstrate this likewise in a living dog if thou shalt immediately free the skin from the elbow and paw, and shalt divide with a small knife the transverse ligament of the foreleg and the other parts which are held against the leg on the external side of the ulna and radius. Soon, forsooth, when the dog of its own volition flexes and extends its toes thou shalt see the tendons rise up from their sheaths.

The Use and Function of the Muscles

In a proper dissection thou shalt see the function of the muscles; notice during their own action they contract and become thick where they are most fleshy, and again they lengthen and become thin according as they in combination draw up a limb, either letting themselves go back and having been drawn out permit the limb to be pulled in an opposite direction by another muscle, or at other times indeed they do not put in action their own combination.

This is to be observed accurately at the elbow; indeed the skin of the same dog must be removed higher up in order that the whole leg and axilla shall be bared, and the nerves running forward into it (the leg) through the axilla may become visible. Thou shalt perceive that a chain of these nerves reaches to certain muscles and thou shalt intercept some one among them with a noose.

The Use of the Nerve in the Muscles

In truth since the number and distribution of the nerves of the dog do not correspond exactly with those same in man I would advise when thou art about to perform this dissection in a living dog that thou hast at hand a dead one also in which thou shalt have separated the series of nerves running out through the foreleg and elbow, and thus thou wilt find some one of these nerves which is distributed to particular muscles. They will be arranged almost in this manner: In man one of them is the third, and is carried into the forearm along the anterior side of the elbow joint; another, in fact, the fifth, runs to the elbow next to the posterior portion of the internal tuberosity of the humerus. For in this manner the nerves are observed also in the dog. And these nerves having been tied somewhere before they reach the elbow joint, the motion of the muscles flexing the digits and arm will be abolished, and if thou wilt intercept with a band the nerve which in man is reckoned by me the fourth and is extended along the humerus to its external tuberosity, then the motion of the muscles extending the foreleg and digits will be abolished. . . .

When thou dividest the belly of a muscle straight through thou wilt observe that the muscle draws together and contracts in one part towards its insertion, in the other portion towards its origin.

But if on the other hand thou shalt cut the tendon of any muscle thou shall see that the muscle contracts towards its origin. Likewise if thou shalt divide the origin it will contract towards its insertion. If in truth thou shalt have cut the insertion and the head at the same time the muscle will be bunched at its belly and the portion where it is most fleshy, and thus the functions of the muscles will be obvious to thee by the doing of these things.

Examination of the Uses of the Dorsal Medulla

Even so if anyone may have considered examining the function of the dorsal medulla, it will be seen when the medulla has been injured how the parts below the injury lose sensation and motion. It will be permitted anyone to fasten a dog or to bind it to a block of wood in a way that one stretches out the back and neck. Therefore some of the spines of the vertebrae can be cut in front with a large knife and then the dorsal medulla can be laid bare in its bed, when anyone will get a view of the medulla about to be cut—for nothing is easier than thus to see that movement and sensation are abolished in the parts subjected to the section.

Examination of the Uses of the Veins and Arteries

Also when inquiring into the use of the veins the work is scarcely one for the dissection of the living, since we shall become sufficiently acquainted in the case of the dead with the fact that these veins carry the blood through the whole body and that any part is not nourished in which a prominent vein has been severed in wounds.

Likewise concerning the arteries we scarcely require a dissection of the living although it will be allowable for anyone to lay bare the artery running into the groin and to obstruct it with a band, and to observe that the part of the artery cut off by the band pulsates no longer.

And thus it is observed by the easy experiment of opening an artery at any time in living animals that blood is contained in the arteries naturally.

In order that on the other hand we may be more certain that the force of pulsation does not belong to the artery or that the material contained in the arteries is not the producer of the pulsation, for in truth this force depends for its strength upon the heart. Besides, because we see that an artery bound by a cord no longer beats under the cord, it will be permitted to undertake an extensive dissection of the artery of the groin or of the thigh, and to take a small tube made of a reed of such a thickness as is the capacity of the artery and to insert it by cutting in such a way that the upper part of the tube reaches higher into the cavity of the artery than the upper part of the dissection, and in the same manner also that the lower portion of the tube is introduced downward farther than the lower part of the dissection, and thus the ligature of the artery which constricts its caliber above the canulla is passed by a circuit.

To be sure when this is done the blood and likewise the vital spirit run through the artery even as far as the foot; in fact the whole portion of the artery replaced by the canulla beats no longer. Moreover when the ligature has been cut, that part of the artery which is beyond the canulla shows no less pulsation than the portion above.

We shall see next how much force is actually carried to the brain from the heart by the arteries. Now in this demonstration thou shalt wonder greatly at a vivisection of Galen in which he advises that all things be cut off which are common to the brain and heart, always excepting the arteries which seek the head through the transverse processes of the cervical vertebrae and carry also besides a substantial portion of the vital spirit into the primary sinuses of the dura mater and also in like manner into the brain. So much so that it is not surprising that the brain performs its functions under these conditions for a long time, which Galen observed could easily be done, for the animal breathes for a long time during this dissection, and sometimes moves about. If indeed it runs, and therefore requires much breath, it falls not long afterwards although the brain will still afterwards receive the essence of the animal spirit from those arteries which I have closely observed seek the skull through the transverse processes of the cervical vertebrae. . . .

We see that the peritoneum is a wrapper for all the organs enclosed in it; that the omentum and likewise the mesentery serve in the best manner for the conduction and distribution of the blood vessels; that the stomach prepares the food and drink, and passes these through the stomach onward.

And however nothing may prevent our taking living dogs which have consumed food at less or greater intervals previously and examine them alive, and thus investigating the functions of the intestines. But on the contrary we are able to behold the functioning of the liver as also of the spleen or of the kidneys or of the bladder during the dissection of the living, scarcely better than in that of the dead;

unless someone shall wish to excise the spleen in the living dog which I have once done, and have preserved the dog (alive) for some days.

And thus also have I once excised a kidney; in truth the management of this wound is more troublesome than the pleasant knowledge which is acquired in the doing of it; unless one undertakes these dissections not so much for the sake of knowledge of the organs as in order to train his hands, and to learn to sew up wounds of the abdomen in a fitting manner; which should also be diligently practiced upon the intestines, in order to become accustomed to sew them up when wounded, and to replace them in the abdomen when they shall have slipped out.

In truth these operations, just as the dislocations and fractures of bones, which we sometimes do on brute beasts, serve more for training the hands and for determining correct treatment rather than for investigating the functions of organs. . . .

Examination of the Fetus

Quite pleasing is it in the management of the fetus to see how when the fetus touches the surrounding air it tries to breathe. And this dissection is performed opportunely in a dog or pig when the sow will soon be ready to drop her young.

To be sure if thou dividest the abdomen of such an animal down to the cavity of the peritoneum, and then thou also openest the uterus at the site of a single fetus, and when the secundines have been separated from the uterus thou shalt place the fetus on a table, thou shalt see through its coverings and its transparent membrane how the fetus attempts in vain to breathe, and dies just as if suffocated. If in fact thou shalt perforate the covering of the fetus, and shalt free its head from the coverings, thou shalt soon see that the fetus as it were comes to life again and breathes finely.

And when thou shalt have investigated this in one fetus thou shalt turn to another which thou shalt not free from the uterus but thou shalt invert the uterus opened by the same management as that of the fetuses just described, and shalt turn the edges of the dissection already made backward until the lower part or site of the coverings of another and nearest fetus shall appear, and thou shalt free this site from the uterus even up to that place where the uterus is fused in the exterior covering of the fetus, and where its abundant flesh will possess a substance similar to the spleen, which interweaves vessels stretching out from the uterus into the external coverings of the fetus.

For that network of vessels must be preserved intact during this manipulation, and the remaining external covering must be removed from the uterus in order that thou shalt see through the transparent covering of the fetus that the arteries distributed by the covering and running to the umbilicus pulsate in the rhythm of the arteries running to the uterus, and that the naked fetus attempts and struggles for respiration and thereupon when the coverings are punctured and broken thou shalt see that then the fetus breathes and that pulsations of the arteries of the fetal membranes and of the umbilicus stop. Up to this moment the arteries of the uterus are beating in unison with the rest of the arteries outside of itself.

Examination of the Function of the Heart, and Lungs and Great Vessels

And we consider many things concerning the functions of the heart and lungs; of course the motion of the lungs, and whether their rough artery takes to itself a portion of those things which are inhaled; next, the dilation and contraction of the heart, and whether the pulsation of the heart and arteries act in the same rhythm; also in what manner the venous artery is dilated and contracted, and whereby an animal continues to live after its heart has been excised. For making these investigations we particularly require an animal endowed with a wide breastbone and possessing membranes dividing the thorax separated in such a way that when the sternum has been divided lengthwise we may be able to carry the dissection between these membranes even to the heart itself without causing a perforation of the cavities of the thorax in which the lungs are contained.

Indeed, when no such animal except man or a tailless ape is to be had, these dissections must be carried out in dogs and pigs and those animals of which an abundance is furnished us, as thou hast seen all this mentioned above.

Therefore and in like manner the lung follows the movement of the thorax. It is evident from this way, when a cut is made in an intercostal space penetrating the thoracic wall, that a portion of the lung on the injured side collapses, and distends with the thorax no longer. While until then the other lung following up to this time the movement of the chest, also soon collapses if thou shalt make a cut penetrating even into the cavity of the thorax on the other side. And then the animal, even if it moves the chest for some time, will die nevertheless just as if it were suffocated.

In fact it ought to be observed in this demonstration that thou shalt make the cut as close as possible to the upper edge of any rib lest thou mayest direct the incision by chance along the lower edge and perforate the vessels stretched along there. For the blood will flow forth thence at once, which rendered frothy in consequence, of the air inhaled and exhaled through the wound, will appear to thee falsely as the lung, for this froth will appear to be the lung to those who dissect carelessly, and they will think that the lung is distended on such occasions by its own inherent force.

In order that thou mayest see the natural relationship of the lung to the thorax, thou shalt cut the cartilages of two or three median ribs from the opposite side, and when the incisions have been carried through the interspaces of these ribs thou shalt bend the separate ribs outward and break them where thou shalt decide to be the proper place through which thou seekest to observe the lung of the uninjured side, for since the membranes dividing the thorax in dogs are sufficiently translucent, it is very easy to inspect through them the portion of the lung which is following up to this time the movements of the thorax, and when these membranes have been perforated slightly, to consider in what manner that part of the lung collapses.

Before thou perforatest these membranes it will have been useful to grasp with the hands the branches of the venous artery in that part of the lung which now collapses, and in some other portion to free the substance of the lung from those vessels in order that thou mayest learn whether these vessels and the heart are moved in like manner.

For the movement of the heart is evident to thee even here especially if thou shalt divide the covering of the heart and shalt uncover the heart from it on that side where thou art carrying on this operation.

It will be fitting to try this on the left side as of course the right part of the lung may be lifted up to the point and at one time thou mayest conveniently grasp the main trunk of the venous artery easily in the hand. Also thou mayest in an operation of this kind carefully handle the base of the heart, and may cut off swiftly and at one time and with one ligature the vessels from their origins, and then thou mayest cut to excise the heart under the ligature, and when the bands have been loosened with which the animal has been tied, thou mayest allow it to run about.

Indeed we have at times seen dogs, but especially cats, treated in this manner, run around.

Besides thou wilt see the movement of the heart and also of the arteries more accurately if soon thou shalt bind the dog to a plank, and shalt carry the incision on both sides with a very sharp knife from the clavicle through the cartilages of the ribs where they are continuous with the bones, and thou shalt make a third incision even into the cavity of the peritoneum transversely from the end of one of the above incisions to the end of the other; and when the sternum has been lifted with its cartilages pressed to it, and when the transverse incision has been freed from them, thou shalt turn the sternum upward towards the head of the animal and soon when the covering of the heart has been opened thou shalt grasp the heart with one hand and with the other thou shalt take hold of the great artery extending into the back.

Examination of the Recurrent Nerves and the Loss of Voice from the Cutting of Them

And soon I begin in this wise an extended dissection in the neck with a rather sharp knife which divides the skin and muscles lying under it right to the trachea, avoiding this lest the incision may deviate to the side and wound especially the principal vein. Then by grasping the windpipe with the hands, and freeing it accurately from the overlying muscles by the use of the fingers as far as the arteries lying at its side; I seek out the nerves bordering upon the sixth pair of cerebral nerves; then I note the recurrent nerves lying on the sides of the rough artery (trachea) which I sometimes intercept with ligatures, at other times I cut. And first I do the same on the other side, in order that it may be clearly seen when one nerve has been tied or cut how half the voice disappears and is totally lost when both nerves are cut. And if I loosen the ligatures that the voice will return again. For this is carried out quickly and without an unusual loss of blood, and it is clearly proved how the animal struggles for deep breaths without its voice when the recurrent nerves have been divided with a sharp knife.

Examination of the Functions of the Diaphragm

In this wise I advise those standing near the dissection that they watch the movement of the diaphragm when it is held in the hand, and those at a distance that they observe the expansion and contraction of the stomach and liver similar to the movements in the cavity of the thorax.

Meanwhile I carry on an extensive dissection on the other side of the thorax to reach the bones of the ribs even nearly to that point where the ribs change into cartilages. And then I make transverse incisions along the bones of the ribs in order that I may free the bones in some part from the overlying muscles; and if it is seen that further painstaking work may follow I lift the intercostal muscles of two intercostal spaces from the tunic covering the ribs that I may thus tear away the intervening rib from the covering surrounding the rib with the help of my hands alone. And when that rib is broken from its cartilage and bent backward on the side, the great size of this tunic surrounding the ribs will be apparent. This transparent covering will demonstrate however the movements of the lung. After this covering has been punctured it will be seen how the lung collapses on this side although the thorax moves meanwhile just as before.

Examination of the Movements of the Lung

In fact in order that this may be made more evident I free many bones of the ribs from their cartilages, opening this side of the thorax as much as I am able in order that there may be presented through the membranes dividing the chest, another part of the lungs shows itself which being in the hitherto uninjured part of the thorax, moves well with it. And then when these membranes have been punctured this part also is soon seen to collapse from the perforation, and to fall together.

Andreas Vesalius. 1942 [1543]. *De Fabrica Corporis Humanis.* [The Fabric of the Human Body]. Translated by Dr. Samuel W. Lambert. In *Source Book of Medical History.* Compiled by Logan Clendenning. New York: Henry Schuman.

From *De motu Cordis et Sanguinis*

William Harvey (1578–1657) is without doubt one of the giants in the history of experimental investigation in the context of biomedical inquiry. Harvey was trained at the University of Padua by Fabricio di Aquapendente, who was Galileo's physician. Although Harvey did not singlehandedly uncover the pumping motions of the heart, the pulsing motions of the blood vessels and the circulatory motions of the blood, he brought together fragmentary clues about these matters from the work of other investigators, and rooted his unified account of these motions in sound experimental work in comparative anatomy and physiology. This work is important because it showed, quite early in the history of modern science, just how valuable experiments on animals could be for an understanding of human physiology in states of health and disease. The selection here is taken from his De motu Cordis et Sanguinis (Of the Motions of the Heart and Blood), *published in 1628.*

Of the Motions of Arteries, as Seen in the Dissection of Living Animals [From Chapter III]

In connection with the motions of the heart these things are further to be observed having reference to the motions and pulses of the arteries:

1. At the moment the heart contracts, and when the breast is struck, when in short the organ is in its state of systole, the arteries are dilated, yield a pulse, and are in the state of diastole. In like manner, when the right ventricle contracts and propels its charge of blood, the arterial vein (the pulmonary artery) is distended at the same time with the other arteries of the body.
2. When the left ventricle ceases to act, to contract, to pulsate, the pulse in the arteries also ceases; further, when this ventricle contracts languidly, the pulse in the arteries is scarcely perceptible. In like manner, the pulse in the right ventricle failing, the pulse in the vena arteriosa (pulmonary artery) ceases also.
3. Further, when an artery is divided or punctured, the blood is seen to be forcibly propelled from the wound at the moment the left ventricle contracts; and, again, when the pulmonary artery is wounded, the blood will be seen spouting forth with violence at the instant when the right ventricle contracts.

So also in fishes, if the vessel which leads from the heart to the gills be divided, at the moment when the heart becomes tense and contracted, at the same moment does the blood flow with force from the divided vessel. . . .

I happened upon one occasion to have a particular case under my care, which plainly satisfied me of this truth: A certain person was affected with a large pulsating tumour on the right side of the neck, called an aneurism, just at that part where the artery descends into the axilla, produced by an erosion of the artery itself, and daily increasing in size; as it received the charge of blood brought to it by the artery, with each stroke of the heart: the connexion of parts was obvious when the body of the patient came to be opened after his death. The pulse in the corresponding arm was small, in consequence of the greater portion of the blood being divided into the tumour and so intercepted.

Whence it appears that wherever the motion of the blood through the arteries is impeded, whether it be by compression or infarction, or interception, there do the remote divisions of the arteries beat less forcibly, seeing that the pulse of the arteries is nothing more than the impulse or shock of the blood in these vessels.

Of the Motion of the Heart and Its Auricles,
as Seen in the Bodies of Living Animals [From Chapter IV]

Besides the motions already spoken of, we have still to consider those that appertain to the auricles.

Casper Bauhin and John Riolan, most learned men and skilful anatomists, inform us from their observations, that if we carefully watch the movements of the heart in the vivisection of an animal, we shall perceive four motions distinct in time and in place, two of which are proper to the auricles, two to the ventricles. With all deference to such authority I say, that there are four motions distinct in point of place, but not of time; for the two auricles move together, and so also do the two ventricles, in such wise that though the places be four, the times are only two. And this occurs in the following manner:

There are, as it were, two motions going on together: one of the auricles, another of the ventricles; these by no means taking place simultaneously, but the motion of the auricles preceding, that of the heart itself following; the motion appearing to begin from the auricles and to extend to the ventricles. When all things are becoming languid, and the heart is dying, as also in fishes and the colder blooded animals, there is a short pause between these two motions, so that the heart aroused, as it were, appears to respond to the motion, now more quickly, now more tardily; and at length, and when from others; I was not surprised that Andreas Laurentius should have said that the motion of the heart was as perplexing as the flux and reflux of Euripus had appeared to Aristotle.

At length, and by using greater and daily diligence, having frequent recourse to vivisections, employing a variety of animals for the purpose, and collating numerous observations, I thought that I had attained to the truth, that I should extricate myself and escape from this labyrinth, and that I had discovered what I so much desired, both the motion and the use of the heart and arteries; since which time I have not hesitated to expose my views upon these subjects, not only in private to my friends, but also in public, in my anatomical lectures, after the manner of the Academy of old. . . .

Of the Motions of the Heart, as Seen in the Dissection of Living Animals [From Chapter II]

In the first place, then, when the chest of a living animal is laid open and the capsule that immediately surrounds the heart is slit up or removed, the organ is seen now to move, now to be at rest;—there is a time when it moves, and a time when it is motionless.

These things are more obvious in the colder animals, such as toads, frogs, serpents, small fishes, crabs, shrimps, snails, and shellfish. They also become more distinct in warm-blooded animals, such as the dog and hog, if they be attentively noted when the heart begins to flag, to move more slowly, and, as it were, to die: the movements then become slower and rarer, the pauses longer, by which it is made much more easy to perceive and unravel what the motions really are, and how they are performed. In the pause, as in death, the heart is soft, flaccid, exhausted, lying, as it were, at rest.

In the motion, and interval in which this is accomplished, three principal circumstances are to be noted:

1. That the heart is erected, and rises upwards to a point, so that at this time it strikes against the breast and the pulse is felt externally.
2. That it is everywhere contracted, but more especially towards the sides, so that it looks narrower, relatively longer, more drawn together. The heart of an eel taken out of the body of the animal and placed upon the table or the hand, shows these particulars; but the same things are manifest in the heart of small fishes and of those colder animals where the organ is more conical or elongated.
3. The heart being grasped in the hand, is felt to become harder during its

action. Now this hardness proceeds from tension, precisely as when the forearm is grasped, its tendons are perceived to become tense and resilient when the fingers are moved.

4. It may further be observed in fishes, and the colder blooded animals, such as frogs, serpents, etc., that the heart when it moves, becomes of a paler colour, when quiescent of a deeper blood-red colour.

From these particulars it appeared evident to me that the motion of the heart consists in a certain universal tension—both contraction in the line of its fibres, and constriction in every sense. It becomes erect, hard, and of diminished size during its action; the motion is plainly of the same nature as that of the muscles when they contract in the line of their sinews and fibres; for the muscles, when in action, acquire vigour and tenseness, and from soft become hard, prominent, and thickened: in the same manner the heart.

We are therefore authorized to conclude that the heart, at the moment of its action, is at once constricted on all sides, rendered thicker in its parietes and smaller in its ventricles, and so made apt to project or expel its charge of blood. This, indeed, is made sufficiently manifest by the fourth observation preceding, in which we have seen that the heart, by squeezing out the blood it contains becomes paler, and then when it sinks into repose and the ventricle is filled anew with blood, that the deeper crimson colour returns. But no one need remain in doubt of the fact, for if the ventricle be pierced the blood will be seen to be forcibly projected outwards upon each motion or pulsation when the heart is tense.

These things, therefore, happen together or at the same instant: the tension of the heart, the pulse of its apex, which is felt externally by its striking against the chest, the thickening of its parietes, and the forcible expulsion of the blood it contains by the constriction of its ventricles.

William Harvey. 1942 [1628]. *Anatomical Exercises on the Motion of the Heart and Blood in Animals.* In *Source Book of Medical History.* Compiled by Logan Clendenning. New York: Henry Schuman.

From *Apology for Raymond Sebond*

This selection is drawn from the Essais (Essays) *of Michel de Montaigne (1533–1592). The selection is a fragment of Montaigne's* Apology for Raymond Sebond, *which was written between 1575 and 1580. While Montaigne is ostensibly responding to Sebond's work* Natural Theology, *which was published in 1484, it is clear that the real business of the essay is an articulation of a skeptical philosophy in opposition to both natural theology and claims about the powers of unaided human reason. In the passage I have selected we can see that Montaigne entertains serious doubts about the claims of natural theologians and others to the effect that humans are naturally superior to nonhuman animals. In the course of his essay, Montaigne offers a very rich estimate of the cognitive possibilities for nonhuman animals, and his discussion is one in which a question is asked concerning the best way to explain human and nonhuman animal behavior. Montaigne seriously entertained the possibility that nonhuman animals may not be cognitively minimal "dumb brutes." In doing so he stood in opposition not only to influential philosophical theologians, but also to many of the new scientists who were gradually articulating mechanistic theories on the basis of discovery in renaissance anatomy and physiology. Today the spirit of Montaigne is alive and well in the works of cognitive ethologists.*

Man Is No Better than the Animals

Presumption is our natural and original malady. The most vulnerable and frail of all creatures is man, and at the same time the most arrogant. He feels and sees himself lodged here, amid the mire and dung of the world, nailed and riveted to the worst, the deadest, and the most stagnant part of the universe, on the lowest story of the house and the farthest from the vault of heaven, with the animals of the worst condition of the three [those that walk, those that fly, those that swim]; and in his imagination he goes planting himself above the circle of the moon, and bringing the sky down beneath his feet. It is by the vanity of this same imagination that he equals himself to God, attributes to himself divine characteristics, picks himself out and separates himself from the horde of other creatures, carves out their shares to his fellows and companions the animals, and distributes among them such portions of faculties and powers as he sees fit. How does he know, by the force of his intelligence, the secret internal stirrings of animals? By what comparison between them and us does he infer the stupidity that he attributes to them?

When I play with my cat, who knows if I am not a pastime to her more than she is to me? [The 1595 edition adds: "We entertain each other with reciprocal monkey tricks. If I have my time to begin or to refuse, so has she hers.] Plato, in his picture of the golden age under Saturn, counts among the principal advantages of the man of that time the communication he had with the beasts; inquiring of them and learning from them, he knew the true qualities and differences of each one of them; whereby he acquired a very perfect intelligence and prudence, and conducted his life far more happily than we could possibly do. Do we need a better proof to judge man's impudence with regard to the beasts? That great author opined that in most of the bodily form that Nature gave them, she considered solely the use of prognostications that were derived from them in his time.

This defect that hinders communication between them and us, why is it not just as much ours as theirs? It is a matter of guesswork whose fault it is that we do not understand one another; for we do not understand them any more than they do us. By this same reasoning they may consider us beasts, as we consider them. It is no great wonder if we do not understand them; neither do we understand the Basques and the Troglodytes [cave-dwellers on the western shore of the Red Sea]. However, some have boasted of understanding them, like Apollonius of Tyana, Melampus, Tiresias, Thales, and others. And since it is a fact, as the cosmographers say, that there are nations that accept a dog as their king, they must give a definite interpretation to his voice and motions. We must notice the parity there is between us. We have some mediocre understanding of their meaning; so do they of ours, in about the same degree. They flatter us, threaten us, and implore us, and we them.

Furthermore, we discover very evidently that there is full and complete communication between them and that they understand each other, not only those of the same species, but also those of different species.

> Even dumb cattle and the savage beasts
> Varied and different noises do employ
> When they feel fear or pain, or thrill with joy.
> —Lucretius

In a certain bark of the dog the horse knows there is anger; at a certain other sound of his he is not frightened. Even in the beasts that have no voice, from the mutual services we see between them we easily infer some other means of communication; their motions converse and discuss:

> Likewise in children, the tongue's speechlessness
> Leads them to gesture what they would express.
> —Lucretius

Why not; just as well as our mutes dispute, argue, and tell stories by signs? I have seen some so supple and versed in this, that in truth they lacked nothing of perfection in being able to make themselves understood. Lovers grow angry, are reconciled, entreat, thank, make assignations, and in fine say everything, with their eyes:

> And silence too records
> Our prayers and our words.
> —Tasso

What of the hands? We beg, we promise, call, dismiss, threaten, pray, entreat, deny, refuse, question, admire, count, confess, repent, fear, blush, doubt, instruct, command, incite, encourage, swear, testify, accuse, condemn, absolve, insult, despise, defy, vex, flatter, applaud, humiliate, mock, reconcile, commend, exalt, entertain, rejoice, complain, grieve, mope, despair, wonder, exclaim, are silent, and what not, with a variation and multiplication that vie with the tongue. With the head: we invite, send away, avow, disavow, give the lie, welcome, honor, venerate, disdain, demand, show out, cheer, lament, caress, scold, submit, brave, exhort, menace, assure, inquire. 'What of the eyebrows? What of the shoulders? There is no movement that does not speak both a language intelligible without instruction, and a public language; which means, seeing the variety and particular use of other languages, that this one must rather be judged the one proper to human nature. I omit what necessity teaches privately and promptly to those who need it, and the finger alphabets, and the grammars in gestures, and the sciences which are practiced and expressed only by gestures, and the nations which Pliny says have no other language.

An ambassador of the city of Abdera, after speaking at length to King Agis of

Sparta, asked him: "Well, Sire, what answer do you wish me to take back to our citizens?" "That I allowed you to say all you and as much as you wanted, without ever saying a word." Wasn't that an eloquent and thoroughly intelligible silence?

Moreover, what sort of faculty of ours do we not recognize in the actions of the animals? Is there a society regulated with more order, diversified into more charges and functions, and more consistently maintained, than that of the honeybees? Can we imagine so orderly an arrangement of actions and occupations as this to be conducted without reason and foresight?

> Some, by these signs and instances inclined,
> Have said that bees share in the divine mind
> And the ethereal spirit
>
> —Virgil

Do the swallows that we see on the return of spring ferreting in all the corners of our houses search without judgment, and choose without discrimination, out of a thousand places, the one which is most suitable for them to dwell in? And in that beautiful and admirable texture of their buildings, can birds use a square rather than a round figure, an obtuse rather than a right angle, without knowing their properties and their effects? Do they take now water, now clay, without judging that hardness is softened by moistening? Do they floor their palace with moss or with down, without foreseeing that the tender limbs of their little ones will lie softer and more comfortably on it? Do they shelter themselves from the rainy wind and face their dwelling toward the orient without knowing the different conditions of these winds and considering that one is more salutary to them than the other? Why does the spider thicken her web in one place and slacken it in another, use now this sort of knot, now that one, unless she has the power of reflection, and thought, and inference?

We recognize easily enough, in most of their works, how much superiority the animals have over us and how feeble is our skill to imitate them. We see, however, in our cruder works, the faculties that we use, and that our soul applies itself with all its power; why do we not think the same thing of them? Why do we attribute to some sort of natural and servile inclination these works which surpass all that we can do by nature and by art? Wherein, without realizing it, we grant them a very great advantage over us, by making Nature, with maternal tenderness, accompany them and guide them as by the hand in all the actions and comforts of their life; while us she abandons to chance and to fortune, and to seek by art the things necessary for our preservation, and denies us at the same time the power to attain, by any education and mental straining, the natural resourcefulness of the animals: so that their brutish stupidity surpasses in all conveniences all that our divine intelligence can do.

Truly, by this reckoning, we should be quite right to call her a very unjust stepmother. But this is not so; our organization is not so deformed and disorderly. Nature has universally embraced all her creatures; and there is none that she has not very amply furnished with all powers necessary for the preservation of its being. For these vulgar complaints that I hear men make (as the license of their opinions now raises them above the clouds, and then sinks them to the antipodes) that we are the only animal abandoned naked on the naked earth, tied, bound, having nothing to arm and cover ourselves with except the spoils of others; whereas all other creatures Nature has clothed with shells, husks, bark, hair, wool, spikes, hide, down, feathers, scales, fleece, and silk, according to the need of their being; has armed them with claws, teeth, or horns for attack and defense; and has herself instructed them in what is fit for them—to swim, to run, to fly, to sing—whereas man can neither walk, nor speak, nor eat, nor do anything but cry, without apprenticeship—

> The infant, like a sailor tossed ashore
> By raging seas, lies naked on the earth,
> Speechless, helpless for life, when at his birth
> Nature from out the womb brings him to light.
> He fills the place with wailing, as is right
> For one who through so many woes must pass.
> Yet flocks, herds, savage beasts of every class
> Grow up without the need for any rattle,
> Or for a gentle nurse's soothing prattle;
> They seek no varied clothes against the sky;
> Lastly they need no arms, no ramparts high
> To guard their own—since earth itself and nature
> Amply bring forth all things for every creature
> —Lucretius

—those complaints are false, there is a greater equality and a more uniform relationship in the organization of the world. Our skin is provided as adequately as theirs with endurance against the assaults of the weather: witness so many nations who have not yet tried the use of any clothes. Our ancient Gauls wore hardly any clothes; nor do the Irish, our neighbors, under so cold a sky. But we may judge this better by ourselves; for all the parts of the body that we see fit to expose to the wind and air are found fit to endure it: face, feet, hands, legs, shoulders, head, according as custom invites us. For if there is a part of us that is tender and that seems as though it should fear the cold, it should be the stomach, where digestion takes place; our fathers left it uncovered, and our ladies, soft and delicate as they are, sometimes go half bare down to the navel. Nor are the bindings and swaddlings of infants necessary either; and the Lacedaemonian mothers raised their children in complete freedom to move their limbs, without wrapping or binding them. Our weeping is common to most of the other animals; and there are scarcely any who are not observed to complain and wail long after their birth, since it is a demeanor most appropriate to the helplessness that they feel. As for the habit of eating, it is, in us as in them, natural and needing no instruction:

> For each one feels his powers and his needs.
> —Lucretius

Who doubts that a child, having attained the strength to feed himself, would be able to seek his food? And the earth produces and offers him enough of it for his need, with no other cultivation or artifice; and if not in all weather, neither does she for the beasts: witness the provisions we see the ants and others make for the sterile seasons of the year. These nations that we have just discovered to be so abundantly furnished with food and natural drink, without care or preparation, have now taught us that bread is not our only food, and that without plowing, our mother Nature had provided us in plenty with all we needed; indeed, as seems likely, more amply and richly than she does now that we have interpolated our artifice:

> At first and of her own accord the earth
> Brought forth sleek fruits and vintages of worth,
> Herself gave harvests sweet and pastures fair,
> Which now scarce grow, despite our toil and care,
> And we exhaust our oxen and our men;
>
> —Lucretius

the excess and unruliness of our appetite outstripping all the inventions with which we seek to satisfy it.

As for weapons, we have more that are natural than most other animals, and more varied movements of our limbs; and we get more service out of them, naturally and without lessons. Those who are trained to fight naked are seen to throw themselves into dangers like our own men. If some animals surpass us in this advantage, we surpass many others. And the skill to fortify and protect the body by acquired means, we possess by a natural instinct and precept. As proof that this is so, the elephant sharpens and whets the teeth which he uses in war (for he has special ones for this purpose, which he spares, and does not use at all for his other functions). When bulls go into combat, they spread and toss the dust around them; boars whet their tusks; and the ichneumon, when he is to come to grips with the crocodile, arms his body, coats it, and crusts it all over with mud, well pressed and well kneaded, as with a cuirass. Why shall we not say that it is just as natural to arm ourselves with wood and iron?

As for speech, it is certain that if it is not natural, it is not necessary. Nevertheless, I believe that a child who had been brought up in complete solitude, remote from all association (which would be a hard experiment to make), would have some sort of speech to express his ideas. And it is not credible that Nature has denied us this resource that she has given to many other animals: for what is it but speech, this faculty we see in them of complaining, rejoicing, calling to each other for help, inviting each other to love, as they do by the use of their voice? How could they not speak to one another? They certainly speak to us, and we to them. In how many ways do we not speak to our dogs? And they answer us. We talk to them in another language, with other names, than to birds, hogs, oxen, horses; and we change the idiom according to the species:

So ants amidst their sable-colored band
Greet one another, and inquire perchance
The road each follows, and the prize in hand.

—Dante

It seems to me that Lactantius attributes to beasts not only speech but also laughter. And the difference of language that is seen between us, according to the difference of countries, is found also in animals of the same species. Aristotle cites in this connection the various calls of partridges according to the place they are situated in,

And various birds . . .
Utter at different times far different cries . . .
And some change with the changing of the skies
Their raucous songs.

—Lucretius

But it is yet to be known what language this child would speak; and what is said about it by conjecture has not much appearance of truth. If they allege to me, against this opinion, that men naturally deaf do not speak at all, I reply that it is not only because they could not be taught speech by ear, but rather because the sense of hearing, of which they are deprived, is related to that of speech, and they hold together by a natural tie: so that what we speak we must speak first to ourselves, and make it ring on our own ears inwardly, before we send it to other ears.

I have said all this to maintain this resemblance that exists to human things, and to bring us back and join us to the majority. We are neither above nor below the rest: all that is under heaven, says the sage, incurs the same law and the same fortune,

All things are bound by their own chains of fate.

—Lucretius

There is some difference, there are orders and degrees; but it is under the aspect of one and the same nature:

And all things go their own way, nor forget
Distinctions by the law of nature set.

—Lucretius

Man must be constrained and forced into line inside the barriers of this order. The poor wretch is in no position really to step outside them; he is fettered and bound, he is subjected to the same obligation as the other creatures of his class, and in a very ordinary condition, without any real and essential prerogative or preeminence. That which he accords himself in his mind and in his fancy has neither body nor taste. And if it is true that he alone of all the animals has this freedom of imagination and this unruliness in thought that represents to him what is, what is not, what he wants, the false and the true, it is an advantage that is sold him very dear, and in which he has little cause to glory, for from it springs the principal source of the ills that oppress him: sin, disease, irresolution, confusion, despair.

So I say, to return to my subject, that there is no apparent reason to judge that the beasts do by natural and obligatory instinct the same things that we do by our choice and cleverness. We must infer from like results like faculties [The 1595 edition adds: "and from richer results, richer faculties"], and consequently confess that this same reason, this same method that we have for working, is also that of the animals. [The 1595 edition reads, instead of "is also that of the animals," "the animals have it also, or some better one."] Why do we imagine in them this compulsion of nature, we who feel no similar effect? Besides, it is more honorable, and closer to divinity, to be guided and obliged to act lawfully by a natural and inevitable condition, than to act lawfully by accidental and fortuitous liberty; and safer to leave the reins of our conduct to nature than to ourselves. The vanity of our presumption makes us prefer to owe our ability to our powers than to nature's liberality; and we enrich the other animals with natural goods and renounce them in their favor, in order to honor and ennoble ourselves with goods acquired: a very simple notion, it seems to me, for I should prize just as highly graces that were all mine and inborn as those I had gone begging and seeking from education. It is not in our power to acquire a fairer recommendation than to be favored by God and nature.

Take for example the fox, whom the inhabitants of Thrace use when they want to undertake to cross some frozen stream over the ice, turning him loose ahead of them for this reason. If we saw him at the edge of the water bring his ear very near the ice, to hear whether the water running beneath sounds near or far away, and draw back or advance according as he finds the ice too thin or thick enough, would we not have reason to suppose that there passes through his head the same reasoning that would pass through ours, and that it is a ratiocination and conclusion drawn from natural common sense: "What makes a noise, moves; what moves is not frozen; what is not frozen is liquid; and what is liquid gives way under a weight"? For to attribute this simply to a keenness of the sense of hearing, without reasoning or inference, is a chimera, and cannot enter our imagination. We must make the same supposition about the many sorts of ruses and tricks by which the animals protect themselves from the attacks we make upon them.

Michel de Montaigne. 1958 [1575–1580]. "Apologie de Raimond Sebond." Pp: 330–334 in *The Complete Essays of Montaigne.* Translated by D. M. Frame. Stanford: Stanford University Press. Reprinted courtesy Stanford University Press.

From *Discourse on Method*

René Descartes (1596–1650) is remembered today primarily as a philosopher. In fact he made important contributions not just to philosophy, but also to physics, mathematics and physiology. Descartes drew a sharp distinction between humans and nonhuman animals. The distinction is drawn in terms of cognitive abilities—humans can think and use language, whereas animals cannot. This distinction with respect to cognitive abilities itself rests on a fundamental metaphysical difference between humans and nonhuman animals: humans have souls, whereas nonhuman animals do not. The selection here is taken from his well-known work Discours de la Méthode (Discourse on Method), *first published in 1637. In this selection Descartes explains the nature of the fundamental differences he sees between humans and nonhuman animals, and he does so in a way that will shape debates about these matters down the centuries to our own time in the twenty first century.*

I had expounded all these matters with sufficient minuteness in the Treatise which I formerly thought of publishing. And after these, I had shewn what must be the fabric of the nerves and muscles of the human body to give the animal spirits contained in it the power to move the members, as when we see heads shortly after they have been struck off still move and bite the earth, although no longer animated; what changes must take place in the brain to produce waking, sleep, and dreams, how light, sounds, odours, tastes, heat, and all the other qualities of external objects impress it with different ideas by means of the senses; how hunger, thirst, and the other internal affections can likewise impress upon it divers ideas; what must be understood by the common sense *(sensus communis)* in which these ideas are received, by the memory which retains them, by the fantasy which can change them in various ways, and out of them compose new ideas, and which, by the same means, distributing the animal spirits through the muscles, can cause the members of such a body to move in as many different ways, and in a manner as suited, whether to the objects that are presented to its senses or to its internal affections, as can take place in our own case apart from the guidance of the will. Nor will this appear at all strange to those who are acquainted with the variety of movements performed by the different automata, or moving machines fabricated by human industry, and that with help of but few pieces compared with the great multitude of bones, muscles, nerves, arteries, veins, and other parts that are found in the body of each animal. Such persons will look upon this body as a machine made by the hands of God, which is incomparably better arranged, and adequate to movements more admirable than is any machine of human invention. And here I specially stayed to show that, were there such machines exactly resembling in organs and outward form an ape or any other irrational animal, we could have no means of knowing that they were in any respect of a different nature from these animals; but if there were machines bearing the image of our bodies, and capable of imitating our actions as far as it is morally possible, there would still remain two most certain tests whereby to know that they were not therefore really men. Of these the first is that they could never use words or other signs arranged in such a manner as is competent to us in order to declare our thoughts to others: for we may easily conceive a machine to be so constructed that it emits vocables, and even that it emits some correspondent to the action upon it of external objects which cause a change in its organs; for example, if touched in a particular place it may demand what we wish to say to it; if in another it may cry out that it is hurt, and such like; but not that it should arrange them variously so as appositely to reply to what is said in its presence, as men of the lowest grade of intellect can do. The second test is, that although such machines might execute many things with equal or perhaps greater perfection than any of us, they would, without doubt, fail in certain others from which it could be discovered that they did not act from knowledge, but solely from the disposition of their organs: for while Reason is an universal instrument that is alike available on every occasion, these organs, on the contrary, need a particular arrangement for each particular action; whence it must be morally impossible that there should exist in any machine a diversity of organs sufficient to enable it to act in all the occurrences of life, in the way in which our reason enables us to act. Again, by means of these two tests we may likewise know the difference between men and brutes. For it is highly deserving of remark, that there are no men so dull and stupid, not even idiots, as to be incapable of joining together different words, and thereby constructing a declaration by which to make their thoughts understood; and that on the other hand, there is no other animal, however perfect or happily circumstanced, which can do the like. Nor does this inability arise from want of organs: for we observe that magpies and parrots can utter words like ourselves, and are yet unable to speak as we do, that is, so as to show that they understand what they say; in place of which men born deaf and dumb, and thus not less, but rather more than the brutes, destitute of the organs which others use in speaking, are in the habit of spontaneously inventing certain signs by which they discover their thoughts to those who, being usually in their company, have leisure to learn their language. And this proves not only that the brutes have less Reason than man, but that they have none at all: for we see that very little is required to enable a person to speak; and since a certain inequality of capacity is observable among animals of the same species, as well as among men, and since some are more capable of being instructed than others, it is incredible that the most perfect ape or parrot of its species, should not in this be equal to the most stupid infant of its kind, or at least to one that was crack-brained, unless the soul of brutes were of a nature wholly different from ours. And we ought not to confound speech with the natural movements which indicate the passions, and can be imitated by machines as well as manifested by animals; nor must it be thought with certain of the ancients, that the brutes speak, although we do not understand their language. For if such were the case, since they are endowed with many organs analogous to ours, they could as easily communicate their thoughts to us as to their fellows. It is also very worthy of remark, that, though there are many animals which manifest more industry than we in certain of their actions, the same animals are yet observed to show none at all in many others: so that the circumstance that they do better than we does not prove that they are endowed with mind, for it would thence follow that they possessed greater Reason than any of us, and could surpass us in all things; on the contrary, it rather proves that they are destitute of Reason, and that it is Nature which acts in them according to the disposition of their organs: thus it is seen, that a clock composed only of wheels and weights can number the hours and measure time more exactly than we with all our skill.

René Descartes. 1899 [1637]. *Discourse on Method.* Translated by John Veitch. Chicago: Open Court.

From *Introduction to the Principles of Morals and Legislation*

Jeremy Bentham (1748–1852) is one of the founders of the utilitarian school of moral, social and political philosophy. The selection here comes from his Introduction to the Principles of Morals and Legislation, *which was published in 1789. Bentham articulates a moral philosophy deeply rooted in the moral significance of pleasure and pain, happiness and suffering. It was Bentham who articulated the claim that the morality of actions should be evaluated by their tendency to promote or hinder the greatest happiness of the greatest number. Bentham saw no reason to exclude any beings capable of enjoying pleasure and suffering pain. This included persons of color, women and nonhuman animals. Although Bentham's philosophy has been used to focus our attention on the pain and suffering endured by nonhuman animals, the cost-benefit analysis implicit in his dictum concerning the greatest happiness of the greatest number has set up the context for a modern debate concerning the great benefits to humans from experimentation that inflicts pain on animals. It is against the utilitarian background of the moral relevance of pleasure and pain that the contemporary debates about animal consciousness take place.*

Chapter 1. Of the Principle of Utility

I. Nature has placed mankind under the governance of two sovereign masters, *pain* and *pleasure*. It is for them alone to point out what we ought to do, as well as to determine what we shall do. On the one hand the standard of right and wrong, on the other the chain of causes and effects, are fastened to their throne. They govern us in all we do, in all we say, in all we think: every effort we can make to throw off our subjection, will serve but to demonstrate and confirm it. In words a man may pretend to abjure their empire: but in reality he will remain subject to it all the while. The *principle of utility* recognises this subjection, and assumes it for the foundation of that system, the object of which is to rear the fabric of felicity by the hands of reason and of law. Systems which attempt to question it, deal in sounds instead of sense, in caprice instead of reason, in darkness instead of light.

But enough of metaphor and declamation: it is not by such means that moral science is to be improved.

II. The principle of utility is the foundation of the present work: it will be proper therefore at the outset to give an explicit and determinate account of what is meant by it. By the principle of utility is meant that principle which approves or disapproves of every action whatsoever, according to the tendency which it appears to have to augment or diminish the happiness of the party whose interest is in question: or, what is the same thing in other words, to promote or to oppose that happiness. I say of every action whatsoever; and therefore not only of every action of a private individual, but of every measure government.

III. By utility is meant that property in any object, whereby it tends to produce benefit, advantage, pleasure, good, or happiness, (all this in the present case comes to the same thing) or (what comes again to the same thing) to prevent the happening of mischief, pain, evil, or unhappiness to the party whose interest is considered: if that party be the community in general, then the happiness of the community: if a particular individual, thou the happiness of that individual.

IV. The interest of the community is one of the most general expressions that can occur in the phraseology of morals: no wonder that the meaning of it is often lost. When it has a meaning, it is this. The community is a fictitious *body,* composed of the individual persons who are considered as constituting as it were its *members.* The interest of the community then is, what?—the sum of the interests of the several members who compose it.

V. It is in vain to talk of the interest of the community, without understanding what is the interest of the individual. A thing is said to promote the interest, or to be for the interest, of an individual, when it tends to add to the sum total of his pleasures: or, what comes to the same thing, to diminish the sum total of his pains.

VI. An action then may be said to be conformable to the principle of utility, or, for shortness sake, to utility, (meaning with respect to the community at large) when the tendency it has to augment the happiness of the community is greater than any it has to diminish it.

VII. A measure of government (which is but a particular kind of action, performed by a particular person or persons) may be said to be conformable to or dictated by the principle of utility, when in like manner the tendency which it has to augment the happiness of the community is greater than any which it has to diminish it.

VIII. When an action, or in particular a measure of government, is supposed by a man to be conformable to the principle of utility, it may be convenient, for the purposes of discourse, to imagine a kind of law or dictate, called a law or dictate of utility: and to speak of the action in question, as being conformable to such law or dictate.

IX. A man may be said to be a partizan of the principle of utility, when the approbation or disapprobation he annexes to of any action, or to any measure, is determined by and proportioned to the tendency which he conceives it to have to augment or to diminish the happiness of the community: or in other words, to its conformity or nonconformity to the laws or dictates of utility.

X. Of an action that is conformable to the principle of utility one may always say either that it is one that ought to be done, or at least that it is not one that ought not to be done. One may say also, that it is right it should be done; at least that it is not wrong it should be done: that it is a right action; at least that it is not a wrong action. When thus interpreted, the words *ought,* and *right* and *wrong,* and others of that stamp, have a meaning: when otherwise, they have none.

XI. Has the rectitude of this principle been ever formally contested? It should seem that it had, by those who have not known what they have been meaning. Is it susceptible of any direct proof? it should seem not: for that which is used to prove every thing else, cannot itself be proved: a chain of proofs must have their commencement somewhere. To give such proof is as impossible as it is needless.

XII. Not that there is or ever has been that human creature breathing, however stupid or perverse, who has not on many, perhaps on most occasions of his life, deferred to it. By the natural constitution of the human frame, on most occasions of their lives men in general embrace this principle, without thinking of it: if not for the ordering of their own actions, yet for the trying of their own actions, as well as of those of other men. There have been, at the same time, not many, perhaps, even of the most intelligent, who hare been disposed to embrace it purely and without reserve. There are even few who have not taken some occasion or other to quarrel with it, either on account of their not understanding always how to apply if, or on account of some prejudice or other which they were afraid to examine into or could not bear to part with. For such is the stuff that man is made of: in principle and in practice, in a right track and in a wrong one, the rarest of all human qualities is consistency.

XIII. When a man attempts to combat the principle of utility, it is with reasons drawn, without his being aware of it, from that very principle itself. His arguments, if they prove any thing, prove not that the principle is *wrong,* but that, according to the applications he supposes to be made of it, it is *misapplied.* Is it possible for a man to move the earth? Yes; but he must first find out another earth to stand upon.

XIV. To disprove the propriety of it by arguments is impossible; but, from the causes that have been mentioned, or from some confused or partial view of it, a man may happen to be disposed not to relish it. Where this is the case, if he thinks the settling of his opinions on such a subject worth the trouble, let him take the following steps, and at length, perhaps, he may come to reconcile himself to it.

1. Let him settle with himself, whether he would wish to discard this principle altogether; if so, let him consider what it is that all his reasonings (in matters of politics especially) can amount to?
2. If he would, let him settle with himself, whether he would judge and act without any principle, or whether there is any other he would judge and act by?
3. If there be, let him examine and satisfy himself whether the principle he thinks he has found is really any separate intelligible principle; or whether it be not a mere principle in words, a kind of phrase, which at bottom expresses neither more nor less than the mere averment of his own unfounded sentiments; that is, what in another person he might be apt to call caprice?
4. If he is inclined to think that his own approbation or disapprobation, annexed to the idea of an act, without any regard to its consequences, is a sufficient foundation for him to judge and act upon, let him ask himself whether his sentiment is to be a standard of right and wrong, with respect to every other man, or whether every man's sentiment has the same privilege of being a standard to itself?
5. In the first case, let him ask himself whether his principle is not despotical, and hostile to all the rest of human race?
6. In the second case, whether it is not anarchial, and whether at this rate there are not as many different standards of right and wrong as there are men? and whether even to the same man, the same thing, which is right to-day, may not (without the least change in its nature) be wrong

to-morrow? and whether the same thing is not right and wrong in the same place at the same time? and in either case, whether all argument is not at an end? and whether, when two men have said, ' I like this,' and ' I don't like it,' they can (upon such a principle) have any thing more to say?

7. If he should have said to himself, No: for that the sentiment which he proposes as a standard must be grounded on reflection, let him say on what particulars the reflection is to turn? if on particulars having relation to the utility of the act, then let him say whether this is not deserting his own principle, and borrowing assistance from that very one in opposition to which he sets it up: or if not on those particulars, on what other particulars?
8. If he should be for compounding the matter, and adopting his own principle in part, and the principle of utility in part, let him say how far he will adopt it?
9. When he has settled with himself where he will stop, then let him ask himself how he justifies to himself the adopting it so far? and why he will not adopt it any farther?
10. Admitting any other principle than the principle of utility to be a right principle, a principle that it is right for a man to pursue; admitting (what is not true) that the word *right* can have a meaning without reference to utility, let him say whether there is any such thing as a *motive* that a man can have to pursue the dictates of it: if there *is,* let him say what that motive is, and how it is to be distinguished from those which enforce the dictates of utility: if not, then lastly let him say what it is this other principle can be good for?

Jeremy Bentham. 1879. *An Introduction to the Principles of Morals and Legislation.* Reprint. Oxford: Clarendon.

From *Animal Liberation*

Peter Singer (b. 1946) is currently De Camp Professor of Bioethics at Princeton University. He is arguably one of the most controversial philosophers of his generation. His book Animal Liberation *was first published in 1975 and has since then become both a classic in moral philosophy and a rallying point for activists in various animal rights and animal liberation movements. Singer's arguments take place against a background of utilitarian moral philosophy. He defends the claim that all animals are equal, and compares speciesism (discrimination on the basis of species membership) to both racism and sexism. The selection here is from the introduction to* Animal Liberation. *It should suffice to give the flavor of the enterprise.*

ALL ANIMALS ARE EQUAL . . . OR WHY THE ETHICAL PRINCIPLE ON WHICH HUMAN EQUALITY RESTS REQUIRES US TO EXTEND EQUAL CONSIDERATION TO ANIMALS TOO

"Animal Liberation" may sound more like a parody of other liberation movements than a serious objective. The idea of "The Rights of Animals" actually was once used to parody the case for women's rights. When Mary Wollstonecraft, a forerunner of today's feminists, published her *Vindication of the Rights of Woman* in 1792, her views were widely regarded as absurd, and before long an anonymous publication appeared entitled *A Vindication of the Rights of Brutes.* The author of this satirical work (now known to have been Thomas Taylor, a distinguished Cambridge philosopher) tried to refute Mary Wollstonecraft's arguments by showing that they could be carried one stage further. If the argument for equality was sound when applied to women, why should it not be applied to dogs, cats, and horses? The reasoning seemed to hold for these "brutes" too; yet to hold that brutes had rights was manifestly absurd. Therefore the reasoning by which this conclusion had been reached must be unsound, and if unsound when applied to brutes, it must also be unsound when applied to women, since the very same arguments had been used in each case.

In order to explain the basis of the case for the equality of animals, it will be helpful to start with an examination of the case for the equality of women. Let us assume that we wish to defend the case for women's rights against the attack by Thomas Taylor. How should we reply?

One way in which we might reply is by saying that the case for equality between men and women cannot validly be extended to nonhuman animals. Women have a right to vote, for instance, because they are just as capable of making rational decisions about the future as men are; dogs, on the other hand, are incapable of understanding the significance of voting, so they cannot have the right to vote. There are many other obvious ways in which men and women resemble each other closely, while humans and animals differ greatly. So, it might be said, men and women are similar beings and should have similar rights, while humans and nonhumans are different and should not have equal rights.

The reasoning behind this reply to Taylor's analogy is correct up to a point, but it does not go far enough. There are obviously important differences between humans and other animals, and these differences must give rise to some differences in the rights that each have. Recognizing this evident fact, however, is no barrier to the case for extending the basic principle of equality to nonhuman animals. The differences that exist between men and women are equally undeniable, and the supporters of Women's Liberation are aware that these differences may give rise to different rights. Many feminists hold that women have the right to an abortion on request. It does not follow that since these same feminists are campaigning for equality between men and women they must support the right of men to have abortions too. Since a man cannot have an abortion, it is meaningless to talk of his right to have one. Since dogs can't vote, it is meaningless to talk of their right to vote. There is no reason why either Women's Liberation or Animal Liberation should get involved in such nonsense. The extension of the basic principle of equality from one group to another does not imply that we must treat both groups in exactly the same way, or grant exactly the same rights to both groups. Whether we should do so will depend on the nature of the members of the two groups. The basic principle of equality does not require equal or identical *treatment;* it requires equal consideration. Equal consideration for different beings may lead to different treatment and different rights.

So there is a different way of replying to Taylor's attempt to parody the case for women's rights, a way that does not deny the obvious differences between human beings and nonhumans but goes more deeply into the question of equality and concludes by finding nothing absurd in the idea that the basic principle of equality applies to so-called brutes. At this point such a conclusion may appear odd; but if we examine more deeply the basis on which our opposition to discrimination on grounds of race or sex ultimately rests, we will see that we would be on shaky ground if we were to demand equality for blacks, women, and other groups of oppressed humans while denying equal consideration to nonhumans. To make this clear we need to see, first, exactly why racism and sexism are wrong. When we say that all human beings, whatever their race, creed, or sex, are equal, what is it that we are asserting? Those who wish to defend hierarchical, inegalitarian societies have often pointed out that by whatever test we choose it simply is not true that all humans are equal. Like it or not we must face the fact that humans come in different shapes and sizes; they come with different moral capacities, different intellectual abilities, different amounts of benevolent feeling and sensitivity to the needs of others, different abilities to communicate effectively, and different capacities to experience pleasure and pain. In short, if the demand for equality were based on the actual equality of all human beings, we would have to stop demanding equality.

Still, one might cling to the view that the demand for equality among human beings is based on the actual equality of the different races and sexes. Although, it may be said, humans differ as individuals, there are no differences between the races and sexes as such. From the mere fact that a person is black or a woman we cannot infer anything about that person's intellectual or moral capacities. This, it may be said, is why racism and sexism are wrong. The white racist claims that whites are superior to blacks, but this is false; although there are differences among individuals, some blacks are superior to some whites in all of the capacities and abilities that could conceivably be relevant. The opponent of sexism would say the same: a person's sex is no guide to his or her abilities, and this is why it is unjustifiable to discriminate on the basis of sex.

The existence of individual variations that cut across the lines of race or sex, however, provides us with no defense at all against a more sophisticated opponent of equality, one who proposes that, say, the interests of all those with IQ scores below 100 be given less consideration than the interests of those with ratings over 100. Perhaps those scoring below the mark would, in this society, be made the slaves of those scoring higher. Would a hierarchical society of this sort really be so much better than one based on race or sex? I think not. But if we tie the moral principle of equality to the factual equality of the different races or sexes, taken as a whole, our opposition to racism and sexism does not provide us with any basis for objecting to this kind of inegalitarianism.

There is a second important reason why we ought not to base our opposition to racism and sexism on any kind of factual equality, even the limited kind that asserts that variations in capacities and abilities are spread evenly among the different races and between the sexes: we can have no absolute guarantee that these capacities and abilities really are distributed evenly, without regard to race or sex, among human beings. So far as actual abilities are concerned there do seem to be certain measurable differences both among races and between sexes. These differences do not, of course, appear in every case, but only when averages are taken. More important still, we do not yet know how many of these differences are really due to the different genetic endowments of the different races and sexes, and how many are due to poor schools, poor housing, and other factors that are the result of past and continuing discrimination. Perhaps all of the important differences will eventually prove to be environmental rather than genetic. Anyone opposed to racism and sexism will certainly hope that this will be so, for it will make the task of ending discrimination a lot easier; nevertheless, it would be dangerous to rest the case against racism and sexism on the belief that all significant differences are environmental in origin. The opponent of, say, racism who takes this line will be unable to avoid conceding that if differences in ability did after all prove to have some genetic connection with race, racism would in some way be defensible.

Fortunately there is no need to pin the case for equality to one particular outcome of a scientific investigation. The appropriate response to those who claim to have found evidence of genetically based differences in ability among the races or between the sexes is not to stick to the belief that the genetic explanation must be wrong, whatever evidence to the contrary may turn up; instead we should make it quite clear that the claim to equality does not depend on intelligence, moral capacity, physical strength, or similar matters of fact. Equality is a moral idea, not an assertion of fact. There is no logically compelling reason for assuming that a factual difference in ability between two people justifies any difference in the amount of consideration we give to their needs and interests. *The principle of the equality of human beings is not a description of an alleged actual equality among humans: it is a prescription of how we should treat human beings.*

Jeremy Bentham, the founder of the reforming utilitarian school of moral philosophy, incorporated the essential basis of moral equality into his system of ethics by means of the formula: "Each to count for one and none for more than one." In other words, the interests of every being affected by an action are to be taken into account and given the same weight as the like interests of any other being. A later utilitarian, Henry Sidgwick, put the point in this way: "The good of any one individual is of no more importance, from the point of view (if I may say so) of the Universe, than the good of any other." More recently the leading figures in contemporary moral philosophy have shown a great deal of agreement in specifying as a fundamental presupposition of their moral theories some similar requirement that works to give everyone's interests equal consideration—although these writers generally cannot agree on how this requirement is best formulated.

It is an implication of this principle of equality that our concern for others and our readiness to consider their interests ought not to depend on what they are like or on what abilities they may possess. Precisely what our concern or consideration requires us to do may vary according to the characteristics of those affected by what we do: concern for the well-being of children growing up in America would require that we teach them to read; concern for the well-being of pigs may require no more than that we leave them with other pigs in a place where there is adequate food and room to run freely. But the basic element—the taking into account of the interests of the being, whatever those interests may be—must, according to the principle of equality, be extended to all beings, black or white, masculine or feminine, human or nonhuman.

Thomas Jefferson, who was responsible for writing the principle of the equality of men into the American Declaration of Independence, saw this point. It led him to oppose slavery even though he was unable to free himself fully from his slaveholding background. He wrote in a letter to the author of a book that emphasized the notable intellectual achievements of Negroes in order to refute the then common view that they had limited intellectual capacities:

> Be assured that no person living wishes more sincerely than I do, to see a complete refutation of the doubts I myself have entertained and expressed on the grade of understanding allotted to them by nature, and to find that they are on a par with ourselves . . . but whatever be their degree of talent it is no measure of their rights. Because Sir Isaac Newton was superior to others in understanding, he was not therefore lord of the property or persons of others.

Similarly, when in the 1850s the call for women's rights was raised in the United States, a remarkable black feminist named Sojourner Truth made the same point in more robust terms at a feminist convention:

> They talk about this thing in the head; what do they call it? ["Intellect," whispered someone nearby.] That's it. What's that got to do with women's rights or Negroes' rights? If my cup won't hold but a pint and yours holds a quart, wouldn't you be mean not to let me have my little half-measure full?

It is on this basis that the case against racism and the case against sexism must both ultimately rest; and it is in accordance with this principle that the attitude that we may call "speciesism," by analogy with racism, must also be condemned. Speciesism—the word is not an attractive one, but I can think of no better term—is a prejudice or attitude of bias in favor of the interests of members of one's own species and against those of members of other species. It should be obvious that the fundamental objections to racism and sexism made by Thomas Jefferson and Sojourner Truth apply equally to speciesism. If possessing a higher degree of intelligence does not entitle one human to use another for his or her own ends, how can it entitle humans to exploit nonhumans for the same purpose?

Many philosophers and other writers have proposed the principle of equal consideration of interests, in some form or other, as a basic moral principle; but not many of them have recognized that this principle applies to members of other species as well as to our own. Jeremy Bentham was one of the few who did realize this. In a forward-looking passage written at a time when black slaves had been freed by the French but in the British dominions were still being treated in the way we now treat animals, Bentham wrote:

> The day *may* come when the rest of the animal creation may acquire those rights which never could have been withholden from them but by the hand of tyranny. The French have already discovered that the blackness of the skin is no reason why a human being should be abandoned without redress to the caprice of a tormentor. It may one day come to be recognized that the number of the legs, the villosity of the skin, or the termination of the *os sacrum* are reasons equally insufficient for abandoning a sensitive being to the same fate. What else is it that should trace the insuperable line? Is it the faculty of reason, or perhaps the faculty of discourse? But a full-grown horse or dog is beyond comparison a more rational, as well as a more conversable animal, than an infant of a day or a week or even a month, old. But suppose they were otherwise, what would it avail? The question is not, Can they *reason?* nor Can they *talk?* but, Can they *suffer?*

In this passage Bentham points to the capacity for suffering as the vital characteristic that gives a being the right to equal consideration. The capacity for suffering—or more strictly, for suffering and/or enjoyment or happiness—is not just another characteristic like the capacity for language or higher mathematics. Bentham is not saying that those who try to mark "the insuperable line" that determines whether the interests of a being should be considered happen to have chosen the wrong characteristic. By saying that we must consider the interests of all beings with the capacity for suffering or enjoyment Bentham does not arbitrarily exclude from consideration any interests at all—as those who draw the line with reference to the possession of reason or language do. The capacity for suffering and enjoyment is *a prerequisite for having interests at all,* a condition that must be satisfied before we can speak of interests in a meaningful way. It would be nonsense to say that it was not in the interests of a stone to be kicked along the road by a schoolboy. A stone does not have interests because it cannot suffer. Nothing that we can do to it could possibly make any difference to its welfare. The capacity for suffering and enjoyment is, however, not only necessary, but also sufficient for us to say that a being has interests—at an absolute minimum, an interest in not suffering. A mouse, for example, does have an interest in not being kicked along the road, because it will suffer if it is.

Although Bentham speaks of "rights" in the passage I have quoted, the argument is really about equality rather than about rights. Indeed, in a different passage, Bentham famously described "natural rights" as "nonsense" and "natural and imprescriptable rights" as "nonsense upon stilts." He talked of moral rights as a shorthand way of referring to protections that people and animals morally ought to have; but the real weight of the moral argument does not rest on the assertion of the existence of the right, for this in turn has to be justified on the basis of the possibilities for suffering and happiness. In this way we can argue for equality for animals without getting embroiled in philosophical controversies about the ultimate nature of rights.

In misguided attempts to refute the arguments of this book, some philosophers have gone to much trouble developing arguments to show that animals do not have rights. They have claimed that to have rights a being must be autonomous, or must be a member of a community, or must have the ability to respect the rights of others, or must possess a sense of justice. These claims are irrelevant to the case for Animal Liberation. The language of rights is a convenient political shorthand. It is even more valuable in the era of thirty-second TV news clips than it was in Bentham's day; but in the argument for a radical change in our attitude to animals, it is in no way necessary.

If a being suffers there can be no moral justification for refusing to take that suffering into consideration. No matter what the nature of the being, the principle of equality requires that its suffering be counted equally with the like suffering—insofar as rough comparisons can be made—of any other being. If a being is not capable of suffering, or of experiencing enjoyment or happiness, there is nothing to be taken into account. So the limit of sentience (using the term as a convenient if not strictly accurate shorthand for the capacity to suffer and/or experience enjoyment) is the only defensible boundary of concern for the interests of others. To mark this boundary by some other characteristic like intelligence or rationality would be to mark it in an arbitrary manner. Why not choose some other characteristic, like skin color?

Racists violate the principle of equality by giving greater weight to the interests of members of their own race when there is a clash between their interests and the interests of those of another race. Sexists violate the principle of equality by favoring the interests of their own sex. Similarly, speciesists allow the interests of their own species to override the greater interests of members of other species. The pattern is identical in each case.

Peter Singer. 1990. *Animal Liberation*. 2d ed. New York: New York Review of Books. Pp. 1–9. Reprinted courtesy of New York Review of Books.

From *Introduction à l' étude de la Médicine Expérimentale*

Claude Bernard (1813–1878) was one of the great biomedical experimenters of the nineteenth century. He made significant contributions to anatomy, physiology, toxicology and early biochemistry. But more than this, he was one of the great philosophers of science of his century. And this meant that Bernard had reflected long and hard about the methods to be employed by biomedical investigators. Bernard had a particular horror of the haphazard methods employed by physicians making unsystematic observations of their patients. He believed that the advance of biomedical inquiry must be deeply rooted in carefully controlled experiments on nonhuman animals. The selection here comes from Bernard's Introduction à l' étude de la Médicine Expérimentale (An Introduction to the Study of Experimental Medicine), *published in 1865, where he offers a sustained defense of the practice of vivisection.*

III. Vivisection

We have succeeded in discovering the laws of inorganic matter only by penetrating into inanimate bodies and machines; similarly we shall succeed in learning the laws and properties of living matter only by displacing living organs in order to get into their inner environment. After dissecting cadavers, then, we must necessarily dissect living beings, to uncover the inner or hidden parts of the organisms and see them work; to this sort of operation we give the name of vivisection, and without this mode of investigation, neither physiology nor scientific medicine is possible; to learn how man and animals live, we cannot avoid seeing great numbers of them die, because the mechanisms of life can be unveiled and proved only by knowledge of the mechanisms of death.

Men have felt this truth in all ages; and in medicine, from the earliest times, men have performed not only therapeutic experiments but even vivisection. We are told that the kings of Persia delivered men condemned to death to their physicians, so that they might perform on them vivisections useful to science. According to Galen, Attalus III (Philometor), who reigned at Pergamum, one hundred thirty-seven years before Jesus Christ, experimented with poisons and antidotes on criminals condemned to death. Celsus recalls and approves the vivisection which Herophilus and Erasistratus performed on criminals with the Ptolemies' consent. It is not cruel, he says, to inflict on a few criminals, sufferings which may benefit multitudes of innocent people throughout all centuries. The Grand Duke of Tuscany had a criminal given over to the professor of anatomy, Fallopius, at Pisa, with permission to kill or dissect him at pleasure. As the criminal had a quartan fever, Fallopius wished to investigate the effects of opium on the paroxysms. He administered two drams of opium during an intermission; death occurred after the second experiment. Similar instances have occasionally recurred, and the story is well known of the archer of Meudon who was pardoned because a nephrotomy was successfully performed on him. Vivisection of animals also goes very far back. Galen may be considered its founder. He performed his experiments especially on monkeys and on young pigs and described the instruments and methods used in experimenting. Galen performed almost no other kind of experiment than that which we call disturbing experiments, which consist in wounding, destroying or removing a part, so as to judge its function by the disturbance caused by its removal. He summarized earlier experiments and studied for himself the effects of destroying the spinal cord at different heights, of perforating the chest on one side or both sides at once; the effects of section of the nerves leading to the intercostal muscles and of section of the recurrent nerve. He tied arteries and performed experiments on the mechanism of deglutition. Since Galen, at long intervals in the midst of medical systems, eminent vivisectors have always appeared. As such, the names of Graaf, Harvey, Aselli, Pecquet, Hailer, etc., have been handed down to us. In our time, and especially under the influence of Magendie, vivisection has entered physiology and medicine once for all, as an habitual or indispensable method of study.

The prejudices clinging to respect for corpses long halted the progress of anatomy. In the same way, vivisection in all ages has met with prejudices and detractors. We cannot aspire to destroy all the prejudice in the world; neither shall we allow ourselves here to answer the arguments of detractors of vivisection; since they thereby deny experimental medicine, i.e., scientific medicine. However, we shall consider a few general questions, and then we shall set up the scientific goal which vivisection has in view.

First, have we a right to perform experiments and vivisections on man? Physicians make therapeutic experiments daily on their patients, and surgeons perform vivisections daily on their subject. Experiments, then, maybe performed on man, but within what limits? It is our duty and our right to perform an experiment on man whenever it can save his life, cure him or gain him some personal benefit. The principle of medical and surgical morality, therefore, consists in never performing on man an experiment which might be harmful to him to any extent, even though the result might be highly advantageous to science, i.e., to the health of others. But performing experiments and operations exclusively from the point of view of the patient's own advantage does not prevent their turning out profitably to science. It cannot indeed be otherwise; an old physician who has often administered drugs and treated many patients is more experienced, that is, he will experiment better on new patients, because he has learned from experiments made on others. A surgeon who has performed operations on different kinds of patients learns and perfects himself experimentally. Instruction comes only through experience; and that fits perfectly into the definitions given at the beginning of this introduction.

May we make experiments on men condemned to death or vivisect them? Instances have been cited, analogous to the one recalled above, in which men have permitted themselves to perform dangerous operations on condemned criminals, granting them pardon in exchange. Modern ideas of morals condemn such actions; I completely agree with these ideas; I consider it wholly permissible, however, and useful to science, to make investigations on the properties of tissues immediately after the decapitations of criminals. A helminthologist had a condemned woman without her knowledge swallow larvae of intestinal worms, so as to see whether the worms developed in the intestines after her death. Others have made analogous experiments on patients with phthisis doomed to an early death; some men have made experiments on themselves. As experiments of this kind are of great interest to science and can be conclusive only on man, they seem to be wholly permissible when they involve no suffering or harm to the subject of the experiment. For we must not deceive ourselves, morals do not forbid making experiments on one's neighbor or on one's self; in everyday life men do nothing but experiment on one another. Christian morals forbid only one thing, doing ill to one's neighbor. So, among the experiments that may be tried on man, those that can only harm are forbidden, those that are innocent are permissible, and those that may do good are obligatory.

Another question presents itself. Have we the right to make experiments on animals and vivisect them? As for me, I think we have this right, wholly and absolutely. It would be strange indeed if we recognized man's right to make use of animals in every walk of life, for domestic service, for food, and then forbade him to make use of them for his own instruction in one of the sciences most useful to humanity. No hesitation is possible; the science of life can be established only through experiment, and we can save living beings from death only after sacrificing others. Experiments must be made either on man or on animals. Now I think that physicians already make too many dangerous experiments on man, before carefully studying them on animals. I do not admit that it is moral to try more or less dangerous or active remedies on patients in hospitals, without first experimenting with them on dogs; for I shall prove, further on, that results obtained on animals may all be conclusive for man when we know how to experiment properly. If it is immoral, then, to make an experiment on man when it is dangerous to him, even though the result may be useful to others, it is essentially moral to make experiments on an animal, even though painful and dangerous to him, if they may be useful to man.

After all this, should we let ourselves be moved by the sensitive cries of people of fashion or by the objections of men unfamiliar with scientific ideas? All feelings deserve respect, and I shall be very careful never to offend anyone's. I easily explain them to myself, and that is why they cannot stop me. I understand perfectly how physicians under the influence of false ideas, and lacking the scientific sense, fail to appreciate the necessity of experiment and vivisection in establishing biological science. I also understand perfectly how people of fashion, moved by ideas wholly different from those that animate physiologists, judge vivisection quite differently. It cannot be otherwise. Somewhere in this introduction we said that, in science, ideas are what give facts their value and meaning. It is the same in morals, it is everywhere the same. Facts materially alike may have opposite scientific meaning, according to the ideas with which they are connected. A cowardly assassin, a hero and a warrior each plunges a dagger into the breast of his fellow. What differentiates them, unless it be the ideas which guide their hands? A surgeon, a physiologist and Nero give themselves up alike to mutilation of living beings. What differentiates them also, if not ideas? I therefore shall not follow the example of LeGallois, in trying to justify physiologists in the eyes of strangers to science who reproach them with cruelty; the difference in ideas explains everything. A physiologist is not a man of fashion, he is a man of science, absorbed by the scientific idea which he pursues: he no longer hears the cry of animals, he no longer sees the blood that flows, he sees only his idea and perceives only organisms concealing problems which he intends to solve. Similarly, no surgeon is stopped by the most moving cries and sobs, because he sees only his idea and the purpose of his operation. Similarly again, no anatomist feels himself in a horrible slaughter house; under the influence of a scientific idea, he delightedly follows a nervous filament through stinking livid flesh, which to any other man would be an object of disgust and horror. After what has gone before we shall deem all discussion of vivisection futile or absurd. It is impossible for men, judging facts by such different ideas, ever to agree; and as it is impossible to satisfy everybody, a man of science should attend only to the opinion of men of science who understand him, and should derive rules of conduct only from his own conscience.

The scientific principle of vivisection is easy, moreover, to grasp. It is always a question of separating or altering certain parts of the living machine, so as to study them and thus to decide how they function and for what. Vivisection, considered as an analytic method of investigation of the living, includes many successive steps, for we may need to act either on organic apparatus, or on organs, or on tissue, or on the histological units themselves In extemporized and other vivisections, we produce mutilations whose results we study by preserving the animals At other times, vivisection is only an autopsy on the living, or a study of properties of tissues immediately after death. The various processes of analytic study of the mechanisms of life in living animals are indispensable, as we shall see, to physiology, to pathology and to therapeutics. However, it would not do to believe that vivisection in itself can constitute the whole experimental method as applied to the study of vital phenomena. Vivisection is only anatomical dissection of the living; it is necessarily combined with all the other physico-chemical means of investigation which must be carried into the organism. Reduced to itself, vivisection would have only a limited range and in certain cases must even mislead us as to the actual role of organs, By these reservations I do not deny the usefulness or even the necessity of vivisection in the study of vital phenomena. I merely declare it insufficient. Our instruments for vivisection are indeed so coarse and our senses so imperfect that we can reach only the coarse and complex parts of an organism. Vivisection under the microscope would make much finer analysis possible, but it presents much greater difficulties and is applicable only to very small animals.

But when we reach the limits of vivisection we have other means of going deeper and dealing with the elementary parts of organisms where the elementary properties of vital phenomena have their seat We may introduce poisons into the circulation, which carry their specific action to one or another histological unit. Localized poisonings, as Fontana and J. Muller have already used them, are valuable means of physiological analysis. Poisons are veritable reagents of life, extremely delicate instruments which dissect vital units. I believe myself the first to consider the study of poisons from this point of view, for I am of the opinion that studious attention to agents which alter histological units should form the common foundation of general physiology, pathology and therapeutics. We must always, indeed, go back to the organs to find the simplest explanations of life.

To sum up, dissection is a displacing of a living organism by means of instruments and methods capable of isolating its different parts. It is easy to understand that such dissection of the living presupposes dissection of the dead.

Claude Bernard. 1927 [1865]. "Bernard's Introduction à l' étude de la Médicine Expérimentale." Translated by Henry Copley Greene. In *An Introduction to the Study of Experimental Medicine.* New York: The Macmillan Company. Pp: 99–105.

From *The Descent of Man, and Selection in Relation to Sex*

Charles Darwin (1809–1882) is perhaps best known today for the theory of evolution that he expounded in The Origin of Species, *published in 1859. Darwin was a subtle thinker who saw not only how adaptations could be brought about through the mechanism of natural selection; he also saw clearly the implications of evolution for the relatedness of all species. All species descend from common ancestor species, accumulating evolutionary modifications in the process. That all animal species around today are related to each other through descent with modification from common ancestors is a consequence of the evolutionary principle of phylogenetic continuity. Even in Darwin's day there had been an accumulation of much significant evidence concerning this continuity—especially from the standpoint of anatomy and physiology. Were there further continuities to be inferred from the comparative study of animal behavior? In particular were there cognitive continuities that undercut the traditional differences between humans and nonhuman animals that theologians had sought to draw? Darwin believed so, and the selection here comes from his book* The Descent of Man, *published in 1871. As will be seen, very much in the spirit of Montaigne, Darwin makes a case for a richer estimate of the cognitive abilities of nonhuman animals than had been found in the works of both theologians and mechanistically minded physiologists working in the spirit of Cartesianism.*

It has, I think, now been shown that man and the higher animals, especially the Primates, have some few instincts in common. All have the same senses, intuitions, and sensations—similar passions, affections, and emotions, even the more complex ones; they feel wonder and curiosity; they possess the same faculties of imitation, attention, memory, imagination, and reason, though in very different degrees. Nevertheless many authors have insisted that man is separated through his mental faculties by an impassable barrier from all the lower animals. I formerly made a collection of above a score of such aphorisms, but they are not worth giving, as their wide difference and number prove the difficulty, if not the impossibility, of the attempt. It has been asserted that man alone is capable of progressive improvement; that he alone makes use of tools or fire, domesticates other animals, possesses property, or employs language; that no other animal is self-conscious, comprehends itself, has the power of abstraction, or possesses general ideas; that man alone has a sense of beauty, is liable to caprice, has the feeling of gratitude, mystery, etc.; believes in God, or is endowed with a conscience. I will hazard a few remarks on the more important and interesting of these points.

Archbishop Sumner formerly maintained that man alone is capable of progressive improvement. With animals, looking first to the individual, every one who has had any experience in setting traps knows that young animals can be caught much more easily than old ones; and they can be much more easily approached by an enemy. Even with respect to old animals, it is impossible to catch many in the same place and in the same kind of trap, or to destroy them by the same kind of poison; yet it is improbable that all should have partaken of the poison, and impossible that all should have been caught in the trap. They must learn caution by seeing their brethren caught or poisoned. In North America, where the fur-bearing animals have long been pursued, they exhibit, according to the unanimous testimony of all observers, an almost incredible amount of sagacity, caution, and cunning; but trapping has been there so long carried on that inheritance may have come into play.

If we look to successive generations, or to the race, there is no doubt that birds and other animals gradually both acquire and lose caution in relation to man or other enemies; and this caution is certainly in chief part an inherited habit or instinct, but in part the result of individual experience. A good observer, Leroy, states that in districts where foxes are much hunted, the young when they first leave their burrows are incontestably much more wary than the old ones in districts where they are not much disturbed.

Our domestic dogs are descended from wolves and jackals, and though they may not have gained in cunning, and may have lost in wariness and suspicion, yet they have progressed in certain moral qualities, such as in affection, trustworthiness, temper, and probably in general intelligence. The common rat has conquered and beaten several other species throughout Europe, in parts of North America, New Zealand, and recently in Formosa, as well as on the mainland of China. Mr. Swinhoe, who describes these latter cases, attributes the victory of the common rat over the large *Mus coninga* to its superior cunning; and this latter quality may be attributed to the habitual exercise of all its faculties in avoiding extirpation by man, as well as to nearly all the less cunning or weak-minded rats having been successively destroyed by him. To maintain, independently of any direct evidence, that no animal during the course of ages has progressed in intellect or other mental faculties, is to beg the question of the evolution of species. Hereafter we shall see that, according to Lartet, existing mammals belonging to several orders have larger brains than their ancient tertiary prototypes.

It has often been said that no animal uses any tool; but the chimpanzee in a state of nature cracks a native fruit, somewhat like a walnut, with a stone. Rengger easily taught an American monkey thus to break open hard palm-nuts, and afterward of its own accord it used stones to open other kinds of nuts, as well as boxes. It thus also removed the soft rind of fruit that had a disagreeable flavor. Another monkey was taught to open the lid of a large box with a stick, and afterward it used the stick as a lever to move heavy bodies; and I have myself seen a young orang put a stick into a crevice, slip his hand to the other end, and use it in the proper manner as a lever. In the cases just mentioned stones and sticks were employed as implements; but they are likewise used as weapons. Brehm states, on the authority of the well-known traveller Schimper, that in Abyssinia when the baboons belonging to one species (*C. gelada)* descend in troops from the mountains to plunder the fields, they sometimes encounter troops of another species (*C. hamadryas),* and then a fight ensues. The Geladas roll down great stones, which the Hamadryas try to avoid, and then both species, making a great uproar, rush furiously against each other. Brehm, when accompanying the Duke of Coburg-Gotha, aided in an attack with fire-arms on a troop of baboons in the pass of Mensa in Abyssinia. The baboons in return rolled so many stones down the mountain, some as large as a man's head, that the attackers had to beat a hasty retreat; and the pass was actually for a time closed against the caravan. It deserves notice that these baboons thus acted in concert. Mr. Wallace on three occasions saw female orangs, accompanied by their young, "breaking off branches and the great spiny fruit of the Durian-tree, with every appearance of rage; causing such a shower of missiles as effectually kept us from approaching too near the tree."

In the Zoological Gardens a monkey which had weak teeth used to break open nuts with a stone; and I was assured by the keepers that this animal, after using the stone, hid it in the straw, and would not let any other monkey touch it. Here, then, we have the idea of property; but this idea is common to every dog with a bone, and to most or all birds with their nests.

The Duke of Argyll remarks, that the fashioning of an implement for a special purpose is absolutely peculiar to man; and he considers that this forms an immeasurable gulf between him and the brutes. It is no doubt a very important distinction, but there appears to me much truth in Sir J. Lubbock's suggestion, that when primeval man first used flint-stones for any purpose, he would have accidentally splintered them, and would then have used the sharp fragments. From this step it would be a small one to intentionally break the flints, and not a very wide step to rudely fashion them. This latter advance, however, may have taken long ages, if we may judge by the immense interval of time which elapsed before the men of the neolithic period took to grinding and polishing their stone tools. In breaking the flints, as Sir J. Lubbock likewise remarks, sparks would have been emitted, and in grinding them heat would have been evolved: "thus the two usual methods of obtaining fire may have originated." The nature of fire would have been known in the many volcanic regions where lava occasionally flows through forests. The anthropomorphous apes, guided probably by instinct, build for themselves temporary platforms; but as many instincts are largely controlled by reason, the simpler ones, such as this of building a platform, might readily pass into a voluntary and conscious act. The orang is known to cover itself at night with the leaves of the Pandanus; and Brehm states that one of his baboons used to protect itself from the heat of the sun by throwing a straw mat over its head. In these latter habits, we probably see the first steps toward some of the simpler arts; namely, rude architecture and dress, as they arose among the early progenitors of man.

Language.—This faculty has justly been considered as one of the chief distinctions between man and the lower animals. But man, as a highly competent judge, Archbishop Whately remarks, "is not the only animal that can make use of language to express what is passing in his mind, and can understand, more or less, what is so expressed by another." In Paraguay the *Cebus azaroe* when excited utters at least six distinct sounds, which excite in other monkeys similar emotions. The movements of the features and gestures of monkeys are understood by us, and they partly understand ours, as Rengger and others declare. It is a more remarkable fact that the dog, since being domesticated, has learned to bark in at least four or five distinct tones. Although barking is a new art, no doubt the wild species, the parents of the dog, expressed their feelings by cries of various kinds. With the domesticated dog we have the bark of eagerness as in the chase; that of anger; the yelping or howling bark of despair, as when shut up; that of joy, as when starting on a walk with his master; and the very distinct one of demand or supplication, as when wishing for a door or window to be opened.

Articulate language is, however, peculiar to man; but he uses in common with the lower animals inarticulate cries to express his meaning, aided by gestures and the movements of the muscles of the face. This especially holds good with the more simple and vivid feelings, which are but little connected with our higher intelligence. Our cries of pain, fear, surprise, anger, together with their appropriate actions, and the murmur of a mother to her beloved child, are more expressive than any words. It is not the mere power of articulation that distinguishes man from other animals, for, as every one knows, parrots can talk; but it is his large power of connecting definite sounds with definite ideas; and this obviously depends on the development of the mental faculties.

As Horne Tooke, one of the founders of the noble science of philology, observes, language is an art, like brewing or baking; but writing would have been a much more appropriate simile. It certainly is not a true instinct, as every language has to be learned. It differs, however, widely from all ordinary arts, for man has an instinctive tendency to speak, as we see in the babble of our young children; while no child has an instinctive tendency to brew, bake, or write. Moreover, no philologist now supposes that any language has been deliberately invented; each has been slowly and unconsciously developed by many steps. The sounds uttered by birds offer in several respects the nearest analogy to language, for all the members of the same species utter the same instinctive cries expressive of their emotions; and all the kinds that have the power of singing exert this power instinctively; but the actual song, and even the call-notes, are learned from their parents or foster-parents. These sounds, as Daines Barrington has proved, "are no more innate than language is in man." The first attempt to sing "may be compared to the imperfect endeavor in a child to babble."

Charles Darwin. 1871. *The Descent of Man, and Selection in Relation to Sex.* New York: D. Appleton. Pp. 47–53.

Sigma Xi Statement on the Use of Animals in Research

Sigma Xi: The Scientific Research Society is an international organization whose membership is mainly composed of research scientists. Many of these researchers are engaged in biomedical research, and a lot of this research involves the use of nonhuman animal subjects. In 1992 the society's journal, American Scientist, *produced a statement explaining and defending the use of nonhuman animals in scientific research. The selection below is a reproduction of that statement. It is clear that the authors of the statement believed that there had been significant benefits to human health and well-being from the practice of animal experimentation. In formulating and considering utilitarian defenses and critiques of animal experimentation, it will be important to assess very carefully the benefits to humans and the cost to nonhuman animals, of the conduct of such research.*

APPROVED UNANIMOUSLY BY THE SIGMA XI
BOARD OF DIRECTORS ON NOVEMBER 16, 1991

Early in 1990, the Executive Committee of Sigma Xi requested that the Society's Committee on Science and Society develop a general mechanism for responding to controversial issues affecting the scientific research community. Dr. John Ahearne, Executive Director of Sigma Xi, suggested that the Committee first consider the issue of the use of animals in research. He noted that a number of scientists who do not use animals in their own research, as well as some who do, were concerned that resistance to the use of animals in biomedical research reflected a general questioning of all scientific investigation.

In response to this request, Alan McGowan, Chair of the Committee on Science and Society, convened a workshop on the use of animals in research at Sigma Xi's 1990 Annual Meeting. At the Annual Meeting, Dr. Colin Blakemore, Waynflete Professor of Physiology at

Oxford University, received Sigma Xi's John P. McGovern Award in Science and Society, in recognition of his contributions to the public understanding of this issue. Professor Blakemore participated in the workshop on the use of animals, which was attended by more than 200 Society members. Participants in the workshop overwhelmingly recommended that Sigma Xi become involved in this issue.

Based on this recommendation, the Board of Directors of Sigma Xi requested commentary from all chapters and clubs of the Society, as well as from individual Society members. More than 90 chapters and clubs and 100 individual members responded, and the response was overwhelming: Sigma Xi should take a stand supporting the responsible use of animals in research. In April 1991, having reviewed a summary of responses from chapters and clubs and from individual members, the Board of Directors requested that a draft statement of Sigma Xi's position be developed for discussion during the 1991 Annual Meeting.

A small group was convened in July 1991 to provide more specific comments and to discuss what issues such a statement should include. That group included members and non-members of Sigma Xi; scientists who use animals in research and those who do not; representatives of other professional organizations; and individuals who have long been involved in the national discussion of the use of animals in research. Input from this group and from Sigma Xi members, responding via chapters and clubs and individually, was used to prepare the following statement outlining Sigma Xi's position on the use of animals in research. The statement is one step in series of activities planned to address this issue.

Sigma Xi, The Scientific Research Society, advocates sound research. The Society recognizes the importance and value of animals in scientific research and science education and it supports responsible use of animals in research and teaching. Sigma Xi opposes unnecessary restrictions on the use of animals in these endeavors, and it encourages public education on the importance of continuing animal research to support advances in scientific knowledge and medical applications. Freedom of opinion and discussion concerning the use of animals in research must be safeguarded. However, attacks on life or property, hostile campaigns against individuals, and the use of distorted, inaccurate, or misleading evidence should be publicly condemned.

Issues Associated with the Use of Animals in Research

Sigma Xi's strong support for the use of animals in research follows from a balanced and thorough consideration of three separate, but related, aspects of animal research: its importance for science, its value, and its conduct. These three issues are considered first in a general way, and subsequently within the context of the research process.

Importance of Animal Research. The use of animals in research and teaching is important for science. Much research is performed primarily to advance basic knowledge. This basic research is vital to the success of the research process, often in ways that are not fully evident at the time the research is being performed.

For centuries, thoughtful research with animals also has advanced understanding of chemical, biological, and behavioral processes that provide a direct application to medical treatment. For example, Pavlov won a Nobel prize for studies of digestion in dogs, even before his discovery of classical conditioning in the same subjects. Most recently, research in neuroscience, using rats as subjects, has provided basic knowledge for the development of diverse pharmacological agents and for the investigation of Alzheimer's disease and other human disorders. These examples emphasize the importance of major advances resulting from animal research. As important as these striking contributions are, however, most progess in research does not come from such major advances, but from the slow accumulation of results from many studies.

Theories, methods, and concepts derived from the study of animals stimulate hypotheses, not only within closely related fields, but also in distant ones. Darwin's theory of evolution by natural selection is an exemplar case. Thus, to some extent, restrictions imposed on animal research are restrictions on the entire research enterprise and on the substantial advances it fosters. These advances are perhaps most obvious in medicine, but are also present in many other areas of research.

Value of Research Using Animals. Well-conducted research with animals has provided, and continues to provide, information, ideas, and applications that can be obtained in no other way. Much medical research produces clear benefit for human health care: Medical advances have contributed substantially to decreased infant mortality and increased life expectancy. (In addition, medical and related research also contribute to the quality and length of life for many animals—pets, zoo animals and wildlife, including endangered species.)

Results from work with animals have led to understanding mechanisms of bodily function in humans, with substantial and tangible applications to medicine and surgery (e.g., antibiotics, imaging technologies, coronary bypass surgery, anticancer therapies), public health (e.g., nutrition, agriculture, immunization, toxicology, and product safety), and also propagation of endangered species (e.g., via captive breeding). As the Surgeon General has stated, research with animals has made possible most of the advances in medicine that we today take for granted. An end to animal research would mean an end to our best hope for finding treatments that still elude us. Animal research has also contributed substantially to the understanding of behavior patterns and ecological principles and to developing the means for responding to environmental problems.

Research with animals has been remarkably successful in generating both basic and applied knowledge. Without such research, many of us would not have survived diseases that were once common. Without further research with animals, there will be no vaccine for AIDS and dramatically fewer advances for treating and preventing heart disease, cancer, and other serious health problems. For this reason, the overwhelming majority of Sigma Xi members providing input on this matter believes that the benefits accrued from this increase in knowledge justifies the responsible use of animals in research.

Conduct of Animal Research. Significant issues regarding the conduct of animal research include the treatment, number, and appropriate use of animals; the efficiency of experimental designs; the use of alternatives to animal research; and the duplication of results.

State and federal agencies have established numerous regulations addressing housing, handling, and experimental procedures for research animals. Disciplinary professional societies representing practicing scientists in North America and abroad have also adopted guidelines for the use of animals in research. These regulations were instituted to ensure humane treatment of animal subjects, and they have become increasingly detailed and comprehensive. Currently, before federally funded research with animal subjects can be undertaken in the United States, the experimental protocol and treatment of subjects must first be approved by an institutional animal care and use committee of prescribed membership (including a veterinarian, a lay person, and a person from outside the institution). Scientists recognize that the health and well-being of animal subjects is essential to good research: Healthy animals are needed to ensure valid and reliable results. However, the level of detail of some mandated procedures far exceeds what is needed to ensure humane treatment of subjects and may, in fact, impede even well-designed animal research. New regulations for personnel, equipment, procedures, and facilities mandated by recent regulations have made research with animals increasingly difficult and expensive. For some researchers, these regulations have limited the scope of research. For others, particularly those at smaller institutions (such as liberal arts colleges) that cannot institute mandated structural and personnel changes, these regulations have interfered more substantially with research and research training.

The use of statistical models and the application of proper experimental designs can help determine the number of animals needed to test hypotheses. With knowledge of certain parameters, one can determine the minimum number of subjects needed to produce results of required statistical power. This approach to research addresses both the naive and unnecessary use of too many subjects, and the equally wasteful use of fewer subjects than are needed to produce valid and statistically reliable results. Sigma Xi advocates the use of sound statistical methodology and recognizes that following good scientific practice (e.g., replication, control groups, and adequate sample size) may actually increase the total number of animals used. Courses providing formal instruction in statistics and experimental design have long been a standard part of the undergraduate curriculum in the social sciences and in some areas of biology. Supporting and strengthening such courses, providing for all of the life sciences, and re-educating current investigators who do not use these methods would result in more efficient use of animal subjects.

The use of alternatives to animals in research is at an early stage. Animals have been used for many routine tests because they represent the best methods currently available. Biochemical, bioenzymatic, or radioimmunologic procedures have replaced *in vivo* tests in some instances (e.g., pregnancy assessment). The goal of developing such procedures was to provide the fastest, cheapest, most reliable and simplest test. For many purposes, tests using animals are rarely as cheap, reliable, or as sensitive as desired. Therefore, we can expect that more biochemical tests will be developed and instituted, replacing tests using whole animals for these specific purposes. Yet, the biochemical tests for pregnancy could not have been developed without using animals to reveal the basic details of reproductive biology underlying these procedures. Indeed, even when the newest procedures (using antibodies produced in cell cultures) are used, animals remain the original source of antibodies (and tumor cell lines) for these procedures. Thus, instituting biochemical methods for specific procedures may reduce the number of animals used for that purpose, but will not necessarily eliminate the use of animals.

The development and validation of computer simulations and cell and tissue culture techniques may in the future diminish the number of animals used in some routine procedures. However, these developments will not entirely replace the use of animals. Indeed, the number of animals used in research may actually increase, for several reasons: First, virtually all of these alternative methods are now adjuncts to the use of animal subjects in research, not replacements for such subjects. Second, because of complex interactions between organ systems, some physiological processes cannot be studied in isolation, but require entire animals. Third, new lines of animal research (e.g., transgenic animals) will be needed to reap the benefits of recent progress in fields such as molecular biology and genetics. Finally, results of computer simulations may raise questions that can be addressed only by the use of animal subjects.

Some scientists do object to the use of animals in product—especially cosmetic—safety testing (which represents less than 1 percent of all animal testing). However, if products are to be marketed, they should be tested by some means to ensure reasonable safety. The use of animals is critical in many instances for the testing of products, especially pharmaceuticals. The challenge is to develop, validate, and institute reliable and cost-effective procedures that require a minimal number of animal subjects to achieve satisfactory test performance. As in the case of *in vitro* pregnancy testing, alternative methods of assuring product safety may involve the use of fewer animals.

Replication of results is a necessary and beneficial part of the scientific research enterprise. Computerized data bases and literature searches provide a mechanism for scientists to determine the extent of published replication, preventing unnecessary repetition of investigations. Grant and publication review processes also contribute to preventing excess replication of results.

Using animals for teaching may appear to represent unnecessary duplication of results. However, the purpose of teaching is not to get a result, but rather to provide a learning experience. As an example, textbooks, lectures, videotapes, models, and simulations can teach some aspects of anatomy. Yet, they cannot provide the training for a surgeon that a real specimen can. In some cases, using alternatives may meet educational goals more effectively than using animals. For example, a number of computer programs simulate the effects of specified selective forces on successive generations of animals, or the effects of altering ion concentrations on neural membrane potentials. These simulations do not entirely replace the use of animals. However, they can be effective adjuncts to animal use by illustrating for students phenomena that cannot be easily demonstrated in the classroom. By contrast, careful use of animal material with clear educational goals (e.g., teaching surgery to veterinary or medical students or teaching morphology to biology majors) is an essential part of professional training.

Ethical Basis for the Use of Animals in Research. The importance of animals for scientific study and the value of such investigations for the public constitute valid reasons for using animals in research. Additionally, establishing and enforcing standards for the care and use of animal subjects ensures that animals so used are well treated. However, the more complex issue, from an ethical perspective, is to determine the conditions under which humans should use members of other species in research. This issue has been raised particularly in the case of research that produces discomfort or pain. To be sure, discomfort or pain should not be produced when a method exists to alleviate pain or discomfort without affecting the results of the study. Beyond this point, however, the issue becomes more difficult.

Do non-human animals have rights? A useful distinction is made between animal rights and animal welfare. In the discussion of ethics, the term "right" is generally reserved for a legitimate claim to a particular treatment or resource, a claim that carries concomitant responsibilities. We do not attribute responsibilities to nonhuman animals, and we do not attribute rights. By contrast, the position for animal welfare asserts that animals should be treated with respect, that animals should be used only for legitimate purposes, and that (within the limitations of an experiment) every reasonable effort should be made to minimize or reduce pain or discomfort. We conclude that, although nonhuman animals intrinsically cannot have rights in the sense that humans do, researchers who enjoy these rights assume with them the strong responsibility to provide for animal welfare.

What, then, are appropriate ethical criteria for using animals in research? The above discussion suggests that: 1) a reasoned judgment must be made that the benefit derived from the research is sufficient to justify the use of animals in the experiment, and 2) when animals are used, reasonable means be employed to provide for the welfare of subjects. How these criteria are applied is best addressed within the context of the research process.

Use of Animals and the Research Process

Scientific research proceeds within the framework of designing and completing systematic studies to test well-defined research hypotheses. These hypotheses make specific, often highly quantitative, predictions based upon the results of previous studies. The research hypothesis often defines a small set of highly probable—but contradictory—results. Thus, an experiment or series of experiments must be performed to determine which results occur under particular conditions. For example, the chemical structure of a newly synthesized compound may resemble a class of antibiotics that work in a particular way. Based on this information, one might hypothesize similar activity for the new compound. However, the extent of antibiotic activity, and the conditions under which the compound shows such activity, must be determined by experiments based on results from similar, related compounds. Moreover, experiments must be conducted to determine effective dosage and possible side effects of the compound. These characteristics may be predictable from structural similarities with other compounds, but they must nonetheless be empirically verified, particularly if the antibiotic is to be marketed for use in humans or animals.

This process of scientific hypothesis testing has several implications for the use of animals in research. First, although carefully planned studies most often yield results falling within a set of predicted outcomes, the precise results are rarely known before the experiments are performed. Therefore, even if one could state that the use of animal subjects would be justified by a given result, such logic is difficult to employ. It is often impossible to know in advance what the result will be and, therefore, to weigh its importance.

Second, on occasion, a study yields an entirely unpredicted result. Such results may reflect the serendipitous discovery of the effect of an important—but previously unrecognized—variable. Thus, despite their rarity and unpredictability, unexpected results may make valuable contributions to the direction and progress of research. Yet, given that one cannot predict when such a result may occur, one cannot weigh its importance before the experiment is performed.

Third, a low yield is intrinsic to the research process. Experiments often suggest unexpected questions that must be addressed before a clear-cut answer can be obtained to the original question. Therefore, many experiments must often be performed to provide an unequivocal advance in knowledge. To be sure, scientists attempt to design and perform experiments in order to maximize the information obtained from each experiment. Yet, when dealing with complex systems about which little is known, multiple experiments are often necessary to isolate important variables and to determine with a high degree of certainty the relations of cause and effect. Consequently, although a line of research using animal subjects very well may yield substantial useful information, any one experiment in that line of research may appear unimportant.

Fourth, the body of scientific data generally increases by painstaking research that advances knowledge in small, incremental steps. Many such advances are usually needed to produce significant breakthroughs, and the value and importance of individual experiments are difficult to assess until the entire process has been completed. Therefore, it often is impossible to estimate the value of such experiments soon after they are finished, and thus to consider their worth in relation to any animals that may be used in the work.

Fifth, not only is it difficult to predict the value of results before an experiment is performed, or even immediately afterward, but the ultimate value may be unrecognized for some time. In advance of contributions to a line of research or other applications, we can not determine with certainty which results will have applications, what these applications may be, or when that application will arise. Long before AIDS appeared, veterinary scientists investigated retroviral infections of livestock. Their knowledge of how to work with these viruses provided a basis for initial work on AIDS. The common delay in application of research findings further complicates attempts to justify the use of animals in research in terms of the benefits of an experiment or line of research.

Sixth, repeating studies, often under slightly different conditions, are necessary to validate results in all fields of science. Duplication may appear to contribute no new information. However, both replication (under similar conditions) and systematic replication (in which parameters are systematically altered) are necessary to document the reliability of phenomena, to address the extent to which results may be generalized, and to isolate important (but otherwise undetected) variables. Concern over apparently needless duplication of experiments should be tempered by considering the possible contributions of studies completed under both similar and dissimilar conditions.

Scientists are concerned about the humane treatment of animals. Because of their work, investigators using animals in research have a heightened awareness concerning the issues surrounding animal welfare. In addition to their own personal feelings, those who work with animals recognize that to achieve valid, reliable results, one must have healthy subjects that are well treated. Under laboratory conditions, animals are not exposed to competition, parasitism, predation, or the level of disease present under natural conditions.

Conclusion

Sigma Xi supports the responsible use of animals in scientific investigations and in science education. The use of animals in research has been essential for advances in the life sciences and medicine, resulting in enormous benefit to human health and welfare. Their use will continue to be necessary for future progress. Given the world's health and other problems, it seems unwise to curtail research that is likely to have a major impact on these problems. Sigma Xi recognizes that the use of animals in research carries serious responsibilities for the welfare of the animals. Therefore, mechanisms must be in place to ensure that unnecessary suffering is avoided and the number of animals used is not excessive. Sigma Xi also supports the development of alternatives to animal experimentation when such alternatives can meet the scientific objectives of a study. Finally, Sigma Xi encourages public education on the importance of animal research in the production of scientific advances and medical treatment and the education, early in their careers, of scientists on the proper use of animals.

"Sigma Xi Statement on the Use of Animals in Research," *American Scientist* 80 (Jan.–Feb. 1992): 73–76. Reprinted with permission of *American Scientist,* magazine of Sigma Xi, the Scientific Research Society.

From *About Behaviorism*

Burrhus Frederic Skinner (1904–1990) was one of the architects of the radical behaviorist school of psychological theory. Skinner was concerned with an articulation of the best way to explain human and nonhuman animal behavior. What would such explanation look like? What sorts of objects and states would such explanations refer to? What evidence could be marshaled for and against such explanations? Skinner's work has been enormously influential in both human psychology and the study of animal behavior, yet few theorists in twentieth century psychology have been maligned as badly as Skinner on the basis of misrepresentations of their ideas. Skinner's book About Behaviorism, *first published in 1974, provides a very readable account of his theories and his understanding of their implications. The selection below is drawn from this book.*

1. The Causes of Behavior

Why do people behave as they do? It was probably first a practical question: How could a person anticipate and hence prepare for what another person would do?

Later it would become practical in another sense: How could another person be induced to behave in a given way? Eventually it became a matter of understanding and explaining behavior. It could always be reduced to a question about causes.

We tend to say, often rashly, that if one thing follows another, it was probably caused by it—following the ancient principle of *post hoc, ergo propter hoc* (after this, therefore because of this). Of many examples to be found in the explanation of human behavior, one is especially important here. The person with whom we are most familiar is ourself; many of the things we observe just before we behave occur within our body, and it is easy to take them as the causes of our behavior. If we are asked why we have spoken sharply to a friend, we may reply, "Because I felt angry."

It is true that we felt angry before, or as, we spoke, and so we take our anger to be the cause of our remark. Asked why we are not eating our dinner, we may say, "Because I do not feel hungry." We often feel hungry when we eat and hence conclude that we eat because we feel hungry. Asked why we are going swimming, we may reply, "Because I feel like swimming." We seem to be saying, "When I have felt this before, I have behaved in such and such a way." Feelings occur at just the right time to serve as causes of behavior, and they have been cited as such for centuries. We assume that other people feel as we feel when they behavae as we behave.

But where are these feelings and states of mind? Of what stuff are they made? The traditional answer is that they are located in a world of nonphysical dimensions called the mind and that they are mental. But another question then arises: How can a mental event cause or be caused by a physical one? If we want to predict what a person will do, how can we discover the mental causes of his behavior, and how can we produce the feelings and states of mind which will induce him to behave in a given way? Suppose, for example, that we want to get a child to eat a nutritious but not very palatable food. We simply make sure that no other food is available, and eventually he eats. It appears that in depriving him of food (a physical event) we have made him feel hungry (a mental event), and that because he has felt hungry, he has eaten the nutritious food (a physical event). But how did the physical act of deprivation lead to the feeling of hunger, and how did the feeling move the muscles involved in ingestion? There are many other puzzling questions of this sort. What is to be done about them?

The commonest practice is, I think, simply to ignore them. It is possible to believe that behavior expresses feelings, to anticipate what a person will do by guessing or asking him how he feels, and to change the environment in the hope of changing feelings while paying little if any attention to theoretical problems. Those who are not quite comfortable about such a strategy sometimes take refuge in physiology. Mind, it is said, will eventually be found to have a physical basis. As one neurologist recently put it, "Everyone now accepts the fact that the brain provides the physical basis of human thought." Freud believed that his very complicated mental apparatus would eventually be found to be physiological, and early introspective psychologists called their discipline Physiological Psychology. The theory of knowledge called Physicalism holds that when we introspect or have feelings we are looking at states or activities of our brains. But the major difficulties are practical: we cannot anticipate what a person will do by looking directly at his feelings *or* his nervous system, nor can we change his behavior by changing his mind *or* his brain. But in any case we seem to be no worse off for ignoring philosophical problems.

Structuralism

A more explicit strategy is to abandon the search for cause and simply describe what people do. Anthropologists can report customs and manners, political scientists can take the line of "behavioralism" and record political action, economists can amass statistics about what people buy and sell, rent and hire, save and spend, and make and consume, and psychologists can sample attitudes and opinions. All this may be done through direct observation, possibly with the help of recording systems, and with interviews, questionnaires, tests, and polls. The study of literature, art, and music is often confined to the forms of these products of human behavior, and linguists may confine themselves to phonetics, semantics, and syntax. A kind of prediction is possible on the principle that what people have often done they are likely to do again; they follow customs because it is customary to follow them, they exhibit voting or buying habits, and so on. The discovery of organizing principles in the structure of behavior—such as "universals" in cultures or languages, archetypal patterns in literature, or psychological types—may make it possible to predict instances of behavior that have not previously occurred.

The structure or organization of behavior can also be studied as a function of time or age, as in the development of a child's verbal behavior or his problem-solving strategies or in the sequence of stages through which a person passes on his way from infancy to maturity, or in the stages through which a culture evolves. History emphasizes changes occurring in time, and if patterns of development or growth can be discovered, they may also prove helpful in predicting future events.

Control is another matter. Avoiding mentalism (or "psychologism") by refusing to look at causes exacts its price. Structuralism and developmentalism do not tell us why customs are followed, why people vote as they do or display attitudes or traits of character, or why different languages have common features. Time or age cannot be manipulated; we can only wait for a person or a culture to pass through a developmental period. In practice the systematic neglect of useful information has usually meant that the data supplied by the structuralist are acted upon by others—for example, by decision-makers who in some way manage to take the causes of behavior into account. In theory it has meant the survival of mentalistic concepts. When explanations are demanded, primitive cultural practices are attributed to "the mind of the savage," the acquisition of language to "innate rules of grammar," the development of problem-solving strategies to the "growth of mind," and so on. In short, structuralism tells us how people behave but throws very little light on why they behave as they do. It has no answer to the question with which we began.

Methodological Behaviorism

The mentalistic problem can be avoided by going directly to the prior physical causes while bypassing intermediate feelings or states of mind. The quickest way to do this is to confine oneself to what an early behaviorist, Max Meyer, called the "psychology of the other one": consider only those facts which can be objectively observed in the behavior of one person in its relation to his prior environmental history. If all linkages are lawful, nothing is lost by neglecting a supposed nonphysical link. Thus, if we know that a child has not eaten for a long time, and if we know that he therefore feels hungry and that because he feels hungry he then eats, then we know that if he has not eaten for a long time, he will eat. And if by making other food inaccessible, we make him feel hungry, and if because he feels hungry he then eats a special food, then it must follow that by making other food inaccessible, we induce him to eat the special food.

Similarly, if certain ways of teaching a person lead him to notice very small differences in his "sensations," and if because he sees these differences he can classify colored objects correctly, then it should follow that we can use these ways of teaching him to classify objects correctly. Or, to take still another example, if circumstances in a white person's history generate feelings of aggression toward blacks, and if those feelings make him behave aggressively, then we may deal simply with the relation between the circumstances in his history and his aggressive behavior.

There is, of course, nothing new in trying to predict or control behavior by observing or manipulating prior public events. Structuralists and developmentalists have not entirely ignored the histories of their subjects, and historians and biographers have explored the influences of climate, culture, persons, and incidents. People have used practical techniques of predicting and controlling behavior with little thought to mental states. Nevertheless, for many centuries there was very little systematic inquiry into the role of the physical environment, although hundreds of highly technical volumes were written about human understanding and the life of the mind. A program of methodological behaviorism became plausible only when progress began to be made in the scientific observation of behavior, because only then was it possible to override the powerful effect of mentalism in diverting inquiry away from the role of the environment.

Mentalistic explanations allay curiosity and bring inquiry to a stop. It is so easy to observe feelings and states of mind at a time and in a place which make them seem like causes that we are not inclined to inquire further. Once the environment begins to be studied, however, its significance cannot be denied.

Methodological behaviorism might be thought of as a psychological version of logical positivism or operationism, but they are concerned with different issues. Logical positivism or operationism holds that since no two observers can agree on what happens in the world of the mind, then from the point of view of physical science mental events are "unobservables"; there can be no truth by agreement, and we must abandon the examination of mental events and turn instead to how they are studied. We cannot measure sensations and perceptions as such, but we can measure a person's capacity to discriminate among stimuli, and the *concept* of sensation or perception can then be reduced to the *operation* of discrimination.

The logical positivists had their version of "the other one." They argued that a robot which behaved precisely like a person, responding in the same way to stimuli, changing its behavior as a result of the same operations, would be indistinguishable from a real person, even though it would not have feelings, sensations, or ideas. If such a robot could be built, it would prove that none of the supposed manifestations of mental life demanded a mentalistic explanation.

With respect to its own goals, methodological behaviorism was successful. It disposed of many of the problems raised by mentalism and freed itself to work on its own projects without philosophical digressions. By directing attention to genetic and environmental antecedents, it offset an unwarranted concentration on an inner life. It freed us to study the behavior of lower species, where introspection (then regarded as exclusively human) was not feasible, and to explore similarities and differences between man and other species. Some concepts previously associated with private events were formulated in other ways.

But problems remained. Most methodological behaviorists granted the existence of mental events while ruling them out of consideration. Did they really mean to say that they did not matter, that the middle stage in that three-stage sequence of physical-mental-physical contributed nothing—in other words, that feelings and states of mind were merely epiphenomena? It was not the first time that anyone had said so. The view that a purely physical world could be self-sufficient had been suggested centuries before, in the doctrine of psychophysical parallelism, which held that there were two worlds—one of mind and one of matter—and that neither had any effect on the other. Freud's demonstration of the unconscious, in which an awareness of feelings or states of mind seemed unnecessary, pointed in the same direction.

But what about other evidence? Is the traditional *post hoc, ergo propter hoc* argument entirely wrong? Are the feelings we experience just before we behave wholly unrelated to our behavior? What about the power of mind over matter in psychosomatic medicine? What about psychophysics and the mathematical relation between the magnitudes of stimuli and sensations? What about the stream of consciousness? What about the intrapsychic processes of psychiatry, in which feelings produce or suppress other feelings and memories evoke or mask other memories? What about the cognitive processes said to explain perception, thinking, the construction of sentences, and artistic creation? Must all this be ignored because it cannot be studied objectively?

Radical Behaviorism

The statement that behaviorists deny the existence of feelings, sensations, ideas, and other features of mental life needs a good deal of clarification. Methodological behaviorism and some versions of logical positivism ruled private events out of bounds because there could be no public agreement about their validity. Introspection could not be accepted as a scientific practice, and the psychology of people like Wilhelm Wundt and Edward B. Titchener was attacked accordingly. Radical behaviorism, however, takes a different line. It does not deny the possibility of self-observation or self-knowledge or its possible usefulness, but it questions the nature of what is felt or observed and hence known. It restores introspection but not what philosophers and introspective psychologists had believed they were "specting," and it raises the question of how much of one's body one can actually observe.

Mentalism kept attention away from the external antecedent events which might have explained behavior, by seeming to supply an alternative explanation. Methodological behaviorism did just the reverse: by dealing exclusively with external antecedent events it turned attention away from self-observation and self-knowledge. Radical behaviorism restores some kind of balance. It does not insist upon truth by agreement and can therefore consider events taking place in the private world within the skin. It does not call these events unobservable, and it does not dismiss them as subjective. It simply questions the nature of the object observed and the reliability of the observations.

The position can be stated as follows: what is felt or introspectively observed is not some nonphysical world of consciousness, mind, or mental life but the observer's own body. This does not mean, as I shall show later, that retrospection is a kind of physiological research, nor does it mean (and this is the heart of the argument) that what are felt or introspectively observed are the causes of behavior. An organism behaves as it does because of its current structure, but most of this is out of reach of introspection. At the moment we must content ourselves, as the methodological behaviorist insists, with a person's genetic and environmental histories. What are introspectively observed are certain collateral products of those histories.

The environment made its first great contribution during the evolution of the species, but it exerts a different kind of effect during the lifetime of the individual, and the combination of the two effects is the behavior we observe at any given time. Any available information about either contribution helps in the prediction and control of human behavior and in its interpretation in daily life. To the extent that either can be changed, behavior can be changed.

Our increasing knowledge of the control exerted by environment makes it possible to examine the effect of the world within the skin and the nature of self-knowledge. It also makes it possible to interpret a wide range of mentalistic expressions. For example, we can look at those features of behavior which have led people to speak of an act of will, of a sense of purpose, of experience as distinct from reality, of innate or acquired ideas, of memories, meanings, and the personal knowledge of the scientist, and of hundreds of other mentalistic things or events. Some can be "translated into behavior," others discarded as unnecessary or meaningless.

In this way we repair the major damage wrought by mentalism. When what a person does is attributed to what is going on inside him, investigation is brought to an end. Why explain the explanation? For twenty-five hundred years people have been preoccupied with feelings and mental life, but only recently has any interest been shown in a more precise analysis of the role of the environment. Ignorance of that role led in the first place to mental fictions, and it has been perpetuated by the explanatory practices to which they gave rise.

A Few Words of Caution

As I noted in the Introduction, I am not speaking as *the* behaviorist. I believe I have written a consistent, coherent account, but it reflects my own environmental history. Bertrand Russell once pointed out that the experimental animals studied by

American behaviorists behaved like Americans, running about in an almost random fashion, while those of Germans behaved like Germans, sitting and thinking. The remark may have been apt at the time, although it is meaningless today. Nevertheless, he was right in insisting that we are all culture-bound and that we approach the study of behavior with preconceptions. (And so, of course, do philosophers. Russell's account of how people think is very British, very Russellian. Mao Tse-tung's thoughts on the same subject are very Chinese. How could it be otherwise?)

I have not presupposed any technical knowledge on the part of the reader. A few facts and principles will, I hope, become familiar enough to be useful, since the discussion cannot proceed in a vacuum, but the book is not about a science of behavior but about its philosophy, and I have kept the scientific material to a bare minimum. Some terms appear many times, but it does not follow that the text is very repetitious. In later chapters, for example, the expression "contingencies of reinforcement" appears on almost every page, but contingencies are what the chapters are about. If they were about mushrooms, the word "mushroom" would be repeated as often.

Much of the argument goes beyond the established facts. I am concerned with interpretation rather than prediction and control. Every scientific field has a boundary beyond which discussion, though necessary, cannot be as precise as one would wish. One writer has recently said that "mere speculation which cannot be put to the test of experimental verification does not form part of science," but if that were true, a great deal of astronomy, for example, or atomic physics would not be science. Speculation is necessary, in fact, to devise methods which will bring a subject matter under better control.

I consider scores, if not hundreds, of examples of mentalistic usage. They are taken from current writing, but I have not cited the sources. I am not arguing with the authors but with the practices their terms or passages exemplify. I make the same use of examples as is made in a handbook of English usage. (I express my regrets if the authors would have preferred to be given credit, but I have applied the Golden Rule and have done unto others what I should have wished to have done if I had used such expressions.) Many of these expressions I "translate into behavior." I do so while acknowledging that *Traduttori traditori*—Translators are traitors—and that there are perhaps no exact behavioral equivalents, certainly none with the overtones and contexts of the originals. To spend much time on exact redefinitions of consciousness, will, wishes, sublimation, and so on would be as unwise as for physicists to do the same for ether, phlogiston, or *vis viva*.

Finally, a word about my own verbal behavior. The English language is heavy-laden with mentalism. Feelings and states of mind have enjoyed a commanding lead in the explanation of human behavior; and literature, preoccupied as it is with how and what people feel, offers continuing support. As a result, it is impossible to engage in casual discourse without raising the ghosts of mentalistic theories. The role of the environment was discovered very late, and no popular vocabulary has yet emerged.

For purposes of casual discourse I see no reason to avoid such an expression as "I have chosen to discuss . . ." (though I question the possibility of free choice), or "I have in mind . . ." (though I question the existence of a mind), or "I am aware of the fact . . ." (though I put a very special interpretation on awareness). The neophyte behaviorist is sometimes embarrassed when he finds himself using mentalistic terms, but the punishment of which his embarrassment is one effect is justified only when the terms are used in a technical discussion. When it is important to be clear about an issue, nothing but a technical vocabulary will suffice. It will often seem forced or roundabout. Old ways of speaking are abandoned with regret, and new ones are awkward and uncomfortable, but the change must be made.

This is not the first time a science has suffered from such a transition. There were periods when it was difficult for the astronomer not to sound like an astrologer (or to be an astrologer at heart) and when the chemist had by no means freed himself from alchemy. We are in a similar stage in a science of behavior, and the sooner the transition is completed the better. The practical consequences are easily demonstrated: education, politics, psychotherapy, penology, and many other fields of human affairs are suffering from the eclectic use of a lay vocabulary. The theoretical consequences are harder to demonstrate but, as I hope to show in what follows, equally important.

From *The Question of Animal Awareness*

Donald Griffin (b. 1915) has been one of the major figures behind a scientific reassessment of the cognitive abilities of nonhuman animals. Working under the aegis of the theory of evolution, Griffin rejects many of the philosophical and scientific presuppositions that lie behind traditional, minimalist estimates of cognition in nonhuman species. In particular, Griffin is one of the architects of cognitive ethology—a new way of thinking about animal behavior that is prepared to ask about the best way to explain animal behavior, and is willing to countenance, in contrast to some strands of the behaviorist schools of thinking that dominated twentieth century science of animal behavior, the possibility that the best explanations of animal behavior might be couched in mentalistic terms not that different from the sorts of explanations that are routinely offered for human behavior. Griffin's work is very much in the cognitively generous tradition of thought extending back from Darwin to Montaigne and the dawn of modern science in the renaissance. The selections here are taken from his book The Question of Animal Awareness, *first published in 1976.*

The Adaptive Value of Conscious Awareness

That social communication is adaptively valuable to some species of animals is demonstrated by the "cost" of anatomically growing, physiologically maintaining, and behaviorally displaying structures that have communication as their principal or, sometimes, their only known function. An extreme, but not unique, case is the one greatly enlarged claw of male fiddler crabs. The claw constitutes a third or more of the body weight in certain species. The structure is very rarely, if ever, used for anything but social communication—chiefly ritualized aggression between males, and courtship (Grane, 1975). Fiddler crabs and many other animals need more food and are more vulnerable to predation than would otherwise be the case, because of conspicuous structures or conspicuous behavior involved in social communication.

Even more commonly, larger amounts of time and energy are consumed in intermale aggression or courtship behavior than would seem at all necessary for the simple requirement of bringing together males and females ready to mate. These costs seem large compared to those proposed by evolutionary biologists to account for the evolution of morphological characters. For example, the trend for many species of birds and mammals to be larger and to have shorter extremities at higher latitudes is usually explained by postulating that the relatively slight decrease in surface-to-volume ratio reduces heat loss and thus conserves metabolic energy. While such things are difficult to measure, it seems likely that the added cost of competitive group displays at "leks" of grouse and other birds far exceed these differences in heat loss as a function of body surface. Hence, according to the basic axioms of evolutionary biology, such social displays must have been favored by some selective advantage great enough to outweigh their cost. Wilson (1975) discusses these questions with many specific examples.

Communication behavior is probably most likely to resemble human language in species whose social behavior involves a high degree of interdependence, so that it is adaptively advantageous to have an efficient means of communication between individuals to coordinate their activities. But, as social communication is by no means limited to men and honeybees, versatile signaling systems should be advantageous to many species.

Humphrey (1979) and Crook (1980) have elaborated the argument that in human evolution the development of interdependent societies made it adaptively advantageous to recognize other members of one's group as individuals, and to react appropriately to their individual attributes and behavioral idiosyncrasies. They believe that this, in turn, led to the development of symbolic language and the kind of thinking which they and others feel to be possible only with the aid of language. These considerations have so far been applied only to the evolutionary history by which our own species developed from other primates. But the general argument is equally applicable to other social animals.

However vulnerable modern civilized men might seem if forced to live in isolation, under favorable climatic conditions and where food supplies are relatively abundant, some of us would succeed in surviving and reproducing without any of the artifacts to which we have become so accustomed. Yet such a possibility is almost unthinkable in the case of the highly evolved social insects. Their coordinated group activities are so dependent upon effective communication that the adaptiveness of such communicative behavior is even more overwhelmingly event than in the case of our own ancestors. Individual recognition is felt to be impossible for social insects, but it is not clear whether the kinds of data so far available would reveal individual recognition if it did occur. One can (and perhaps Humphrey and Crook would) accept this argument, but insist that the social communication of insects results entirely from evolutionary selection operating on unthinking automata. But, as discussed in Chapter 5, this belief must rest on other arguments than the adaptive value of symbolic communication for highly interdependent social creatures.

In a wide-ranging and stimulating discussion of the evolution of mental processes, Julian Huxley and Nikolaas Tinbergen expressed a fundamental disagreement concerning the likelihood than animals have subjective mental experiences (Tax and Callender, 1960, pp. 175–206 and 267). Huxley held that they probably do, and that the question is a valid one, open to scientific investigation. Tinbergen argued the contrary position that we have no basis for inferring subjective experiences in other species. During this discussion, Huxley was asked whether conscious awareness is adaptive in the sense that this term is used by evolutionary biologists; that is, whether it has a survival value and hence has been favored by natural selection. Huxley was sure that it does, but his reasons were not stated in any detail in that symposium. Because of its importance, I should like to take up this question where Huxley left off and present arguments that awareness is indeed adaptive.

In strictly operational terms, awareness can be considered as readiness to respond to certain patterns of stimulation. Because responsiveness and awareness are not the same thing, behavioral evidence is indirect and may be unduly limited, or even misleading. For instance, an animal with only one operative sensory channel—an electric fish in muddy water, for example—could be subjectively aware of a simple, but important, communication signal, such as threat of attack conveyed by a change in the frequency of the electric discharges from another electric fish (Hopkins, 1974). Yet, in the absence of other information, such a fish would be operationally indistinguishable from an electronic frequency meter. Must we therefore define awareness in terms of the numbers and complexity of the signal patterns to which an animal is ready to respond, or is the criterion necessarily a subjective one?

An alternate approach is to consider awareness as the existence of internal images available for comparison with current sensory input. This recalls the cybernetic concept of a "Sollwert," the value of a sensory input which the animal tends to keep constant by adjustments of its behavior (Mittelstaedt, 1972). But this concept would have to be extended to include more than one sensory channel. Also related is the notion of a neural template (Marler, 1969). A sufficiently versatile template-matching machine, again in principle, could fulfill the *behavioral* criteria involved here. The psychological concept of a Gestalt is also applicable, at least in part. It is usually defined as a moderately complex pattern recognized from any of several viewpoints or when any of several redundant, overlapping stimulus patterns are perceived. One example is provided by searching images, postulated internal images of something for which the animal searches (Croze, 1970).

The possession of mental images could well confer an important adaptive advantage on an animal by providing a reference pattern against which stimulus patterns can be compared; and it may well be an efficient form of pattern recognition. It is characteristic of much animal, as well as human, behavior that patterns are recognized not as templates so rigid that slight deviations cause the pattern to be rejected, but as multidimensional entities that can be matched by new and slightly different stimulus patterns, as when a familiar object is recognized from a novel angle of view. This ability to abstract the essential qualities of an important object and recognize it, despite various kinds of distortion, is obviously adaptive. Even greater adaptive advantage results when such a mental image also includes time as one of its dimensions, that is, the relationships to past and future events. Mental images with a time dimension would be far more useful than static searching images, because they would allow the animal to adapt its behavior to the probable flow of events, rather than limiting it to separate reactions as successive perceptual pictures of the animal's surroundings present themselves one at a time. Anticipation of future enjoyment of food and mating or fear of injury could certainly be adaptive, by leading to behavior that increases the likelihood of positive reinforcement and decreases the probability of pain or injury.

All these attributes can also be postulated in nonconscious systems, but conscious awareness may be an efficient, and hence adaptive, way in which complex animals cope with changing situations. The matrix of concepts needed to encompass the variety of spatial and temporal patterns successfully dealt with by many animals tends to approach a working definition of conscious awareness. For instance, the image of food within reach might well be coupled with an image of the act of grasping the food, another of swallowing it, or even the image of its pleasant taste. Thus, if the existence of mental images in animals can be accepted as plausible, one need only postulate an appropriate linkage between them to sketch out a working definition of conscious awareness. It may be helpful, and even parsimonious, to assume some limited degree of conscious awareness in animals, rather than postulating cumbersome chains of interacting reflexes and internal states of motivation. Conscious attention to the performance of new and challenging tasks ordinarily improves our performance; perhaps this principle also applies to other species.

Behavior patterns that are adaptive in the evolutionary biologist's sense may be reinforcing in the psychologist's terms, as well. Perhaps natural selection has also favored the mental experiences that accompany adaptive behavior. It thus becomes almost a truism, once one reflects upon the question, that conscious awareness could have great adaptive value in the sense that this term is used by evolutionary biologists. The better an animal understands its physical, biological, and social environment, the better it can adjust its behavior to accomplish whatever goals may be important in its life, including those that contribute to its evolutionary fitness. The basic assumption of contemporary behavioral ecology and sociobiology, as the latter term is used by Wilson (1975) and many others, is that behavior is acted upon by natural selection along with morphological and physiological attributes. From this plausible assumption it follows that—insofar as any mental experiences animals have are significantly interrelated with their behavior—they, too, must feel the impact of natural selection. To the extent that they convey an adaptive advantage on animals, they will be reinforced by natural selection.

Arguments of this kind, which appeal to a presumed selective advantage, suffer from the limitation that a sufficiently fertile imagination can almost always find a plausible adaptive advantage for any observed trait. The very success of such arguments tends to undermine their strength, because if one can make up an equally plausible case for alternate explanations, there is little basis for preferring one explanation over any of the others. A stronger form of evolutionary argument is the converse position that any attribute with a selective *dis*advantage will almost certainly be eliminated unless it is genetically coupled with some compensating advantage. On this basis, we can at least argue that no selective drawbacks to conscious awareness have been demonstrated. Indeed, I am not aware that any have even been suggested.

In recent years, ethologists and ecologists have analyzed in great detail the behavioral strategies and tactics which animals employ in their daily affairs. These include how they distribute their time and energy in searching for food, in seeking mates, and even in managing their reproductive affairs in terms of the times selected for producing young, how many young to rear, and in general how to maximize the individual's contribution to the future gene pool of its species. As a result of these investigations, it is becoming clear that most animals behave in a relatively efficient manner, doing those things that tend to enhance their individual fitness, in the evolutionary biologist's sense of maximizing the proportion of future generations consisting of their offspring or collateral descendants (nephews, cousins, etc.). In such discussions, ethologists scarcely ever consider to what extent the animals exhibiting these effective tactics are consciously aware that they are doing so. Do the male lions or monkeys that displace previously dominant males from a group of females kill their predecessor's cubs with any understanding that they will thereby have a better chance of producing their own offspring at an earlier time or in greater numbers (Bertram, 1976; Hrdy, 1978)? Because similar sociobiological arguments can be applied to algae and to chimpanzees, the likelihood of awareness on the part of the actors cannot readily be judged from the nature and effectiveness of the tactics themselves, and other evidence, such as the sorts of complex and communicative behavior discussed in this book, must be relied on. But it may well be that animals which are consciously aware of their socio-biological goals can achieve them more effectively than would otherwise be the case.

The Nature and Nurture of Mental Experiences

One consequence of the view that awareness is simply one aspect of neurophysiological processes is to raise the nature-nurture question with regard to mental experiences themselves. To the extent that they are dealt with at all by scientists, it seems to be tacitly assumed that mental experiences result solely from individual experience and, in particular, from learning. This implication is clear in the statements of Pollio (1974), Maritain (1957), and Adler (1967), quoted in Chapters 5 and 6.

Whether or not we accept the behaviorists' axiom of psycho-neural identity (discussed from several viewpoints in the volume edited by Feyerabend and Maxwell, 1966), we should face up to the possibility that a nervous system might attain those properties leading to mental experiences primarily on the basis of genetic information. The development of mental experiences might depend on environmental influences only in the general and unspecific sense that DNA cannot lead to a complete animal without an environment that provides the necessary nourishment and other conditions. We might therefore conclude that the assumption of psychoneural identity leads to the likelihood of something akin to "innate ideas" in the philosophical sense.

The nature-nurture issue with respect to behavior has aroused some of the most violent passions and heated debates known among scientists. As is usual in such cases, balanced consideration strongly suggests that both sides are partly correct and that both individual experience and genetic heritage have significant effects on behavior (reviewed by J. L. Brown, 1975, and Marler et al., 1980). It is also obvious that the relative importance of these two major factors varies widely among behavior patterns and groups of animals. This means that the two extreme, all-or-nothing positions are clearly and equally untenable. Furthermore, interactions between genetic and environmental factors are of considerable importance, and this, together with the enormous difficulty of controlled experiments, makes it almost impossible to estimate their relative importance with anything approaching adequate accuracy. This does not mean that either can safely be assumed *a priori* to be all-important (or unimportant) in the absence of direct evidence of a sort that is rarely available at present.

Applying the same balanced approach to mental experience leads to a cautiously open mind concerning the possibility that both genetic and environmental influences, and interactions between them, may be important in the causation of mental processes, including conscious awareness. Because we know so little about mental experiences in other species, we can scarcely begin to attack the nature-nurture question, despite its potential importance. But we should not overlook the reality of the question, any more than it seems sensible to ignore the possible existence of mental expressions in more than one species. . . .

Summary and Conclusions

The communication behavior of certain animals is complex, versatile, and, to a limited degree, symbolic. The best-analyzed examples are the dances of honeybees and the signing of captive chimpanzees. These and other animal communication systems share many of the basic properties of human language, although in very much simpler form.

Language has generally been regarded as a unique attribute of human beings, different in kind from animal communication. But on close examination of this view, as it has been expressed by linguists, psychologists, and philosophers, it becomes evident that one of the major criteria on which this distinction has been based is the assumption that animals lack any conscious intent to communicate, whereas men know what they are doing. The available evidence concerning communication behavior in animals suggests that there may be no qualitative dichotomy, but rather a large quantitative difference in complexity of signals and range of intentions that separates animal communication from human language.

Human thinking has generally been held to be closely linked to language, and some philosophers have argued that the two are inseparable or even identical. To the extent that this assertion is accepted, and insofar as animal communication shares basic properties of human language, the employment of versatile communication systems by animals becomes evidence that they have mental experiences and communicate with conscious intent. The contrary view is supported only by negative evidence, which justifies, at the most, an agnostic position.

According to the strict behaviorists, it is more parsimonious to explain animal behavior without postulating that animals have any mental experiences. But mental experiences are also held by behaviorists to be identical with neurophysiological processes. Neurophysiologists have so far discovered no fundamental differences between the structure or function of neurons and synapses in man and other animals. Hence, unless one denies the reality of human mental experiences, it is actually parsimonious to assume that mental experiences are as similar from species to species as are the neurophysiological processes with which they are held to be identical. This, in turn, implies qualitative evolutionary continuity (though not identity) of mental experiences among multicellular animals.

The possibility that animals have mental experiences is often dismissed as anthropomorphic because it is held to imply that other species have the same mental experiences a man might have under comparable circumstances. But this widespread view itself contains the questionable assumption that human mental experiences are the only kind that can conceivably exist. This belief that mental experiences are a unique attribute of a single species is not only unparsimonious; it is conceited. It seems more likely than not that mental experiences, like many other characters, are widespread, at least among multicellular animals, but differ greatly in nature and complexity.

Awareness probably confers a significant adaptive advantage by enabling animals to react appropriately to physical, biological, and social events and signals from the surrounding world with which their behavior interacts.

Opening our eyes to the theoretical possibility that animals have significant mental experiences is only a first step toward the more difficult procedure of investigating their actual nature and importance to the animals concerned. Great caution is necessary until adequate methods have been developed to gather independently verifiable data about the properties and significance of any mental experiences animals may prove to have.

It has long been argued that human mental experiences can only be detected and analyzed through the use of language and introspective reports, and that this avenue is totally lacking in other species. Recent discoveries about the versatility of some animal communication systems suggest that this radical dichotomy may also be unsound. It seems possible, at least in principle, to detect and examine any mental experiences or conscious intentions that animals may have through the experimental use of the animal's capabilities for communication. Such communication channels might be learned, as in recent studies of captive apes, or it might be possible, through the use of models or by other methods, to take advantage of communication behavior which animals already use.

Recognizing the hazards of both positive and negative dogmatism in our present state of ignorance, how can ethologists handle the unsettled (and to some, unsettling) questions of animal awareness and consciousness? Open-minded agnosticism is clearly a necessary first step. Then, when the behavior of an animal suggests awareness, conscious intention, or simple forms of knowledge and belief, a second step might be to entertain the hypothesis that the particular animal under the given conditions may be aware of a certain fact or relationship or may be experiencing some feeling or perception. Granting that such hypotheses are difficult to test by currently available procedures, the tentative consideration of their plausibility might pave the way for thoughtful ethologists to devise improved methods to study when and where animal consciousness may occur and what its content may be. The future extension and refinement of two-way communication between ethologists and the animals they study offer the prospect of developing in due course a truly experimental science of cognitive ethology.

From D. R. Griffin. 1981. *The Question of Animal Awareness: Evolutionary Continuity of Mental Experience.* New York: Rockefeller University Press. Pp. 142–148; 169–171. Reproduced by copyright permission of The Rockefeller University Press.

References

Alcock, J. 1984. *Animal Behavior: An Evolutionary Approach.* Sunderland, MA: Sinauer Associates.

Allen, C., and M. Bekoff. 1997. *Species of Mind: The Philosophy and Biology of Cognitive Ethology.* Cambridge, MA: MIT Press.

Amdur, A., J. Doull, and C. Klaassen, eds. 1993. *Casarett and Doull's Toxicology.* New York: McGraw-Hill.

American Medical Association. 1992. *Statement on the Use of Animals in Biomedical Research: The Challenge and Response.* Chicago: American Medical Association.

Anscombe, E., and P. T. Geach., eds. 1976. *Descartes' Philosophical Writings.* London: Nelson University Paperbacks.

Behe, M. J. 1996. *Darwin's Black Box: The Biochemical Challenge to Evolution.* New York: The Free Press.

Bekoff, M. 2000. "Animal Emotions: Exploring Passionate Natures." *BioScience* 50: 861–869.

Bekoff, M., ed. 1998. *Encyclopedia of Animal Rights and Animal Welfare.* Westport, CT: Greenwood Press.

Bekoff, M., and C. Allen. 1997. "Cognitive Ethology: Slayers, Sceptics and Proponents." In *Anthropomorphism, Anecdote and Animals: The Emperor's New Clothes,* edited by R. W. Mitchell, N. Thompson, and L. Miles, pp. 313–334. Albany: SUNY Press.

Bernard, C. [1865] 1949. *An Introduction to the Study of Experimental Medicine.* Paris: Henry Schuman.

Bickerton, D. 1990. *Species and Language.* Chicago: University of Chicago Press.

Biddle, W. 1995. *A Field Guide to Germs.* New York: Doubleday.

Bonner, J. T. 1980. *The Evolution of Culture in Animals.* Princeton, NJ: Princeton University Press.

Brambell Committee. 1965. *Report of Technical Committee of Inquiry into the Welfare of Livestock Husbandry Systems.* London: Her Majesty's Stationary Office.

Burggren, W. W., and W. E. Bemis. 1990. "Studying Physiological Evolution: Paradigms and Pitfalls." In *Evolutionary Innovations,* edited by M. H. Nitecki, pp. 191–228. Chicago: University of Chicago Press.

Burnet, F. M., and D. O. White. 1972. *Natural History of Infectious Disease.* Cambridge: Cambridge University Press.

Burnet, T. [1691] 1965. *The Sacred Theory of the Earth.* Carbondale: SIU Press.

Butterfield, H. 1957. *The Origins of Modern Science.* New York: The Free Press.

Butts, R. E., ed. 1989. *William Whewell: Theory of Scientific Method.* Indianapolis: Hackett.

Cairns-Smith, A. G. 1985. *Seven Clues to the Origins of Life: A Scientific Detective Story.* Cambridge: Cambridge University Press.

Caldwell, J. 1980. "Comparative Aspects of Detoxification in Mammals." In *Basis of Detoxification,* edited by W. Jacoby, vol. 1, pp.85–113. New York: Academic Press.

Caldwell, J. 1992. "Species Differences in Metabolism and Their Toxicological Significance." *Toxicology Letters* 64/65: 651–659.

Campbell, N. A. 1996. *Biology:* Menlo Park, CA: Benjamin Cummings Publishing.

Carroll, S. B., J. K. Grenier, and S. D. Weatherbee. 2001. *From DNA to Diversity: Molecular Genetics and the Evolution of Animal Design.* Malden, MA: Blackwell Science.

Carruthers, P. 1992. *The Animals Issue.* Cambridge: Cambridge University Press.

Castiglioni, A. 1961. "The Medical School at Padua and the Renaissance of Medicine." In *Toward Modern Science,* edited by R. M. Palter, vol. 2, pp. 48–66. New York: Farrar, Strauss and Cudahy.

Changeaux, J-P. 1985. *Neuronal Man: The Biology of Mind.* Oxford: Oxford University Press.

Chomsky, N. 1957. *Syntactic Structures.* The Hague: Mouton.

Chomsky, N. 1986. *Knowledge of Language: Its Nature, Origin and Use.* New York: Praeger.

Church, F. J., trans. 1987. *The Phaedo.* New York: Macmillan.

Clendening, L., ed. 1960. *Sourcebook of Medical History.* New York: Dover.

Cohen, C. 1986. "The Case for the Use of Animals in Biomedical Research." *New England Journal of Medicine* 315: 865–870.

Cohen, I. B. 1985. *The Birth of a New Physics.* New York: W. W. Norton.

Connor, R. C., and K. S. Norris. 1982. "Are Dolphins Reciprocal Altruists?" *American Naturalist* 119: 358–374.

Copi, I. M. 1982. *Introduction to Logic.* New York: Macmillan.

Copleston, F. C. 1961. *Medieval Philosophy.* New York: Harper Torchbooks.

Cottingham, J. 1986. *Descartes.* Oxford: Basil Blackwell.

Cottingham, J. 1988. *The Rationalists.* Oxford: Oxford University Press.

Crawley, J., M. Sutton, and D. Pickar. 1985. "Animal Models of Self-Destructive Behavior and Suicide." *Psychiatric Clinical, North American* 8: 299–310.

Crombie, A. C. 1959a. *Medieval and Early Modern Science.* Vol. 1. New York: Doubleday Anchor Book.

Crombie, A. C. 1959b. *Medieval and Early Modern Science.* Volume 2. New York: Doubleday Anchor Books.

Crossley, J. N., C. J. Ash, C. J. Brickhill, J. C. Stillwell, and N. H. Williams. 1979. *What Is Mathematical Logic?* Oxford: Oxford University Press.

Cziko, G. 1995. *Without Miracles: Universal Selection Theory and the Second Darwinian Revolution.* Cambridge, MA: MIT Press.

Darwin, C. [1871] 1896. *The Descent of Man, and Selection in Relation to Sex.* New York: D. Appleton and Co.

Darwin, C. [1872] 1965. *The Expression of the Emotions in Man and Animals.* Chicago: University of Chicago Press.

Darwin, C. [1859] 1970. *The Origin of Species.* Edited by P. Appleman. New York: W. W. Norton.

Davidson, E. H. 2001. *Genomic Regulatory Systems: Development and Evolution.* San Diego, CA: Academic Press

Dawkins, M. S. 1998. *Through Our Eyes Only? The Search for Animal Consciousness.* Oxford: Oxford University Press.

Dawkins, R. 1989. *The Selfish Gene.* Oxford: Oxford University Press.

Day, M. H. 1999. "Historical Overview." In *Evolutionary Medicine,* edited by W. R. Trevathan, E. O. Smith, and J. J. McKenna, pp. vii–ix. Oxford: Oxford University Press.

Dennett, D. C. 1987. *The Intentional Stance.* Cambridge, MA: MIT Press.

Depew, D. J., and B. H. Weber. 1995. *Darwinism Evolving: Systems Dynamics and the Genealogy of Natural Selection.* Cambridge, MA: MIT Press.

Dick, O. L., ed. 1978. *Aubrey's Brief Lives.* London: Penguin.

Dobzhansky, T. 1973. "Nothing in Biology Makes Sense Except in the Light of Evolution." *American Biology Teacher* 35: 125–129.

Dugatkin, L. A. 1997. *Cooperation among Animals: An Evolutionary Perspective.* Oxford: Oxford University Press.

Edey, M. A., and D. C. Johanson. 1989. *Blueprints: Solving the Mystery of Evolution.* New York: Penguin.

Elliot, P. 1987. "Vivisection and the Emergence of Experimental Medicine in Nineteenth-Century France." In *Vivisection in Historical Perspective,* edited by N. Rupke, pp. 48–77. New York: Croom Helm.

Ewald, P. W. 1994. *The Evolution of Infectious Disease.* Oxford: Oxford University Press.

Finsen, S., and L. Finsen. 1998. "Animal Rights Movement." In *Encyclopedia of Animal Rights and Animal Welfare,* edited by M. Bekoff, pp. 50–53. Westport, CT: Greenwood Press.

Fouts, R. 1997. *Next of Kin: My Conversations with Chimpanzees.* New York: Avon Books.

Fox, M. 1986. *The Case for Animal Experimentation.* Berkeley: University of California Press.

Frame, D. M., trans. 1968. *The Complete Essays of Montaigne.* Stanford, CA: Stanford University Press.

Frey, R. 1994. "The Ethics of the Search for Benefits: Experimentation in Medicine." In *Principles of Health Care Ethics,* edited by R. Gillon, pp. 1067–1075. Chichester, UK: John Wiley.

Frost, R. 1972. *Selected Prose of Robert Frost.* Edited by H. Cox and E. C. Lathem. New York: Collier Books.

Futuyma, D. 1986. *Evolution.* Sunderland, MA: Sinauer Associates.

Futuyma, D. 1998. *Evolutionary Biology.* Sunderland, MA: Sinauer Associates.

Gallup, G . G. 1970. "Chimpanzees: Self-Recognition." *Science* 167: 86–87.

Giere, R. N. 1991. *Understanding Scientific Reasoning.* New York: Harcourt Brace Jovanovich.

Goodwin, B. 1996. *How the Leopard Changed Its Spots: The Evolution of Complexity.* New York: Harper Torchbooks.

Gorbman, A., W. Dickhoff, S. Vigna, N. Clark, and C. Ralph. 1983. *Comparative Endocrinology.* New York: John Wiley and Sons.

Gordon, R. 1993. *The Alarming History of Medicine: Amusing Anecdotes from Hippocrates to Heart Transplants.* New York: St. Martin's Press.

Gould, S. J. 1977. *Ontogeny and Phylogeny.* Cambridge: Harvard University Press.

Gould, S. J. 1987. *Time's Arrow, Time's Cycle: Myth and Metaphor in the Discovery of Geological Time.* Cambridge: Harvard University Press.

Gould, S. J. 1993. *Eight Little Piggies: Reflections in Natural History.* New York: W. W. Norton.

Greaves, M. 2000. *Cancer: The Evolutionary Legacy.* Oxford: Oxford University Press.

Greenfield, S. A. 1997. *The Human Brain: A Guided Tour.* New York: Basic Books.

Griffin, D. R. 1981. *The Question of Animal Awareness: Evolutionary Continuity of Mental Experience.* Los Altos, CA: William Kaufmann.

Griffin, D. R. 1992. *Animal Minds.* Chicago: University of Chicago Press.

Griffin, D. R. 2001. "Animals Know More than We Used to Think." *Proceedings of the National Academy of Sciences* 98: 4833–4834.

Hall, A. R. 1956. *The Scientific Revolution 1500–1800: The Formation of the Modern Scientific Attitude.* Boston, MA: The Beacon Press.

Hamilton, W. D. 1964a. The Genetical Evolution of Social Behavior, I. *Journal of Theoretical Biology* 7: 1–16.

Hamilton, W. D. 1964b. The Genetical Evolution of Social Behavior, II. *Journal of Theoretical Biology* 7: 17–52.

Hart, J. 1990. "Endocrine Pathology of Estrogens: Species Differences." *Pharmacological Therapy* 47: 203–218.

Hauser, M. D. 1997. *The Evolution of Communication.* Cambridge, MA: MIT Press.

Hawkins, D. J. B. 1946. *A Sketch of Mediaeval Philosophy.* London: Sheed and Ward.

Herzog, H. A. 1998. Sociology of the Animal Rights Movement. In *Encyclopedia of Animal Rights and Animal Welfare,* edited by M. Bekoff, pp. 53–54. Westport, CT: Greenwood Press.

Hooke, R. 1996.. "Preserving Animals Alive by Blowing through Their Lungs with Bellows." In *Introduction to the Philosophy of Science,* edited by A. Zucker, pp. 21–22. Upper Saddle River, NJ: Prentice Hall.

Hoover, H. C., and L. Hoover, eds. [1556] 1950. *George Agricola: De Re Metallica.* New York: Dover.

Irwin, T. 1989. *Classical Thought.* Oxford: Oxford University Press.

Jacquette, D. 1994. *Philosophy of Mind.* Englewood Cliffs, NJ: Prentice Hall.

Johnson, P. E. 1997. *Defeating Darwinism by Opening Minds.* Downers Grove, IL: InterVarsity Press.

Kauffman, S. A. 1993. *The Origins of Order: Self-Organization and Selection in Evolution.* Oxford: Oxford University Press.

Kauffman, S. A. 1995. *At Home in the Universe: The Search for the Laws of Self-Organization and Complexity.* Oxford: Oxford University Press.

Kearney, H. F., ed. 1966. *Origins of the Scientific Revolution.* London: Longmans.

Keegan, J. 1976. *The Face of Battle.* New York: Penguin.

Keller, E. F., and E. A. Lloyd., eds. 1992. *Keywords in Evolutionary Biology.* Cambridge: Harvard University.

King, M. C., and A . C. Wilson. 1975. "Evolution at Two Levels in Humans and Chimpanzees." *Science* 188: 107–116.

Kirby, R., S. Withington, A. B. Darling, and F. Kilgour. 1990. *Engineering in History.* New York: Dover.

Klaassen, C., and D. M. Eaton. 1993. "Principles of Toxicology." In *Casarett and Doull's Toxicology,* edited by A. Amdur, J. Doull, and C. Klaassen, pp.12–49. New York: McGraw-Hill.

Kline, M. 1953. *Mathematics in Western Culture.* London: Penguin.

Lafleur, L. J., ed. 1948. *Bentham: An Introduction to the Principles of Morals and Legislation.* New York: Hafner Publishing.

Lafleur, L. J., ed. 1987. *Descartes: Discourse on Method and Meditations.* New York: Macmillan Publishing.

LaFollette, H. L., and N. Shanks. 1996. *Brute Science: The Dilemmas of Animal Experimentation.* London: Routledge.

Lahav, R., and N. Shanks. 1992. "How to be a Scientifically Respectable Property-Dualist." *The Journal of Mind and Behavior* 13: 211–232.

Lehninger, A., D. Nelson, and M. Cox. 1993. *Principles of Biochemistry.* New York: Worth.

Lewontin, R. C. 1992. "Genotype and Phenotype." In *Keywords in Evolutionary Biology,* edited by E. F. Keller and E. A. Lloyd, 137–144. Cambridge: Harvard University Press.

Lewontin, R. C. 1995. "Primate Models of Human Traits." *Perspectives on Medical Research* 5: 5–19.

Lewontin, R. C. 2000. *The Triple Helix: Gene, Organism and Environment.* Cambridge: Harvard University Press.

Lindeboom, G. A. 1979. *Descartes and Medicine.* Amsterdam: Rodopi.

Linzey, A. 1998. "Saints." In *Encyclopedia of Animal Rights and Animal Welfare,* edited by M. Bekoff, pp. 38–39. Westport, CT: Greenwood Press.

Lyell, C. 1861a. *Principles of Geology or the Modern Change of the Earth and Its Inhabitants Considered as Illustrative of Geology.* Vol. 1. London: John Murray.

Lyell, C. 1861b. *Principles of Geology or the Modern Change of the Earth and Its Inhabitants Considered as Illustrative of Geology.* Vol. 1. London: John Murray.

Macphail, E. M. 1998. *The Evolution of Consciousness.* Oxford: Oxford University Press.

Malthus, T. R. [1798] 1966. *First Essay on Population.* New York: St. Martin's Press.

Manson, J., and L. D. Wise. 1993. "Teratogens." In *Casarett and Doull's Toxicology,* edited by A. Amdur, J. Doull, and C. Klaassen, pp.12–49. New York: McGraw-Hill.

Mason, J. 1998. "Animal Presence." In *Encyclopedia of Animal Rights and Animal Welfare,* edited by M. Bekoff, pp. 38–39. Westport, CT: Greenwood Press.

Mason, S. F. 1970. *A History of the Sciences.* New York: Collier Books.

Maynard Smith, J. 1988. *Did Darwin Get It Right: Essays on Games, Sex and Evolution.* New York: Chapman and Hall.

Maynard Smith, J., and E. Szathmáry. 1999. *The Origins of Life: From the Birth of Life to the Origins of Language.* Oxford: Oxford University Press.

Mayr, E. 1988. *Toward a New Philosophy of Biology: Observations of an Evolutionist.* Cambridge: Harvard University Press.

Mayr, E. 1991. *One Long Argument: Charles Darwin and the Genesis of Modern Evolutionary Thought.* Cambridge: Harvard University Press.

Mayr, E. 1997. *This Is Biology: The Science of the Living World.* Cambridge: Harvard University Press.

McClain, M. 1992. "Thyroid Gland Neoplasia: Non-Genotoxic Mechanisms." *Toxicology Letters* 64/65: 397–408.

Medawar, P. 1984. *Pluto's Republic.* Oxford: Oxford University Press.

Miller, G. 2001. "No Pain, No Brains?" *ScienceNow* 129: 1.

Miller, K. R. 1999. *Finding Darwin's God: A Scientist's Search for Common Ground between God and Evolution.* New York: HarperCollins.

Minogue, K. R., ed. 1973. *Hobbes' Leviathan.* New York: Dutton.

Mitruka, B. M., H. M. Rawnsley, and D. V. Vadhera. 1976. *Animals for Medical Research: Models for the Study of Human Disease.* New York: Wiley.

Molnar, S. 1998. *Human Variation: Races, Types, and Ethnic Groups.* Upper Saddle River, NJ: Prentice Hall.

Morley, H. 1961. "Andreas Vesalius." In *Toward Modern Science,* edited by R. M. Palter. Vol. 2, pp. 67–89. New York: Farrar, Strauss and Cudahy.

Murphy, J. G. 1970. *Kant: The Philosophy of Right.* London: Macmillan.

Nesse, R. M. 1999. "What Darwinian Medicine Offers Psychiatry." In *Evolutionary Medicine,* edited by W. R. Trevathan, E. O. Smith, and J. J. McKenna, pp. 351–373. Oxford: Oxford University Press.

Nesse, R. M., and G . C. Williams. 1995. *Why We Get Sick: The New Science of Darwinian Medicine.* New York: Vintage Books.

Nitecki, M. H., ed. 1990. *Evolutionary Innovations.* Chicago: University of Chicago Press.

Orlans, F. B. 1998. "Invasiveness Scales." In *Encyclopedia of Animal Rights and Animal Welfare,* edited by M. Bekoff, pp. 267–269. Westport, CT: Greenwood Press.

Paley, W. [1801] 1850. *Natural Theology, or Evidence of the Existence and Attributes of the Deity, Collected from the Appearances of Nature.* New York: American Tract Society.

Palmer, A. 1978. "Design of Subprimate Animal Studies." In *Handbook of Teratology,* edited by J. Wilson and F. C. Fraser, pp. 215–253. New York: Plenum Press.

Palter, R. M., ed. 1961a. *Toward Modern Science.* Vol. 1. New York: Farrar, Strauss and Cudahy.

Palter, R. M., ed. 1961b. *Toward Modern Science.* Vol. 2. New York: Farrar, Strauss and Cudahy.

Parascandola, J. 1995. "The History of Animal Use in the Life Sciences." In *Alternative Methods in Toxicology and the Life Sciences,* edited by A. M. Goldberg and L. F. M. Zutphen, vol. 1, pp. 11–21. Larchmont, NY: Mary Ann Liebert.

Park, M. A. 1996. *Biological Anthropology.* Mountain View, CA: Mayfield Publishing.

Parkinson, G. H. R., ed. 1977. *Leibniz: Philosophical Writings.* London: J. M. Dent and Sons.

Patton, W. 1993. *Mouse and Man.* Oxford: Oxford University Press.

Petroski, H. 1994. *The Evolution of Useful Things.* New York: Vintage.

Pinker, S. 1994. *The Language Instinct.* New York: Morrow.

Plamenatz, J. 1963a. *Man and Society.* Vol. 1. London: Longman.

Plamenatz, J. 1963b. *Man and Society.* Vol. 2. London: Longman.

Pledge, H. T. 1959. *Science since 1500: A Short History of Mathematics, Physics, Chemistry and Biology.* New York: Harper Torchbooks.

Poole, J. 1998. "An Exploration of a Commonality between Ourselves and Elephants." *Etica et Animali* 9: 85–110.

Popham, A. E., ed. 1952. *The Drawings of Leonardo Da Vinci.* London: The Reprint Society.

Price, P. W. 1996. *Biological Evolution.* New York: Saunders College Publishing.

Quine, W. V. 1963. *From a Logical Point of View: Logico-Philosophical Essays.* New York: Harper and Row.

Rachels, J. 1991. *Created from Animals: The Moral Implications of Darwinism.* Oxford: Oxford University Press.

Reé, J. 1974. *Descartes.* London: Penguin.

Regan, T. 1987. *The Case for Animal Rights.* Berkeley: The University of California Press.

Regan, T. 1998. "Animal Rights." In *Encyclopedia of Animal Rights and Animal Welfare,* edited by M. Bekoff, pp. 42–43. Westport, CT: Greenwood Press.

Regan, T., and P. Singer., eds. 1989. *Animal Rights and Human Obligations.* Englewood Cliffs, NJ: Prentice Hall.

Reiss, D., and L. Marino. 2001. "Mirror Self-Recognition in the Bottlenose Dolphin: A Case of Cognitive Convergence." *Proceedings of the National Academy of Sciences* 98: 5937–5942.

Restak, R. 1994. *The Modular Brain.* New York: Simon and Schuster.

Rhodes, R. 1997. *Deadly Feasts: Tracking the Secrets of a Terrifying New Plague.* New York: Simon and Schuster.

Rossum, G. D-v. 1996. *History of the Hour: Clocks and Modern Temporal Orders.* Chicago: University of Chicago Press.

Rowan, A. N. 1984. *Of Mice, Models, and Men: A Critical Examination of Animal Research.* Albany: SUNY Press.

Rowan, A. N., F. M. Loew, and J. C. Weer. 1995. *The Animal Research Controversy: Protest, Process and Public Policy.* Medford, MA: Center for Animals and Public Policy, Tufts University School of Veterinary Medicine.

Rupke, N., ed. 1987. *Vivisection in Historical Perspective.* New York: Croom Helm.

Ruse, M. 2000. *The Evolution Wars: A Guide to the Debates.* Santa Barbara, CA: ABC-CLIO.

Russell, B. 1972. *A History of Western Philosophy.* New York: Simon and Schuster.

Schardein, J. L. 1976. *Drugs as Teratogens.* Cleveland, OH: CRC Press.

Schardein, J. L. 1985. *Chemically Induced Birth Defects.* New York: Marcel Dekker.

Shanks, N. 2000. "Creationism, Evolution and Baloney." *Metascience* 9: 86–93.

Shanks, N. 2001. "Modeling Biological Systems: The Belousov-Zhabotinski Reaction." *Foundations of Chemistry* 3: 33–53.

Shanks, N., and K. H. Joplin. 1999. "Redundant Complexity: A Critical Analysis of Intelligent Design in Biochemistry." *Philosophy of Science* 66: 268–282.

Shanks, N., and K. H. Joplin. 2000. "Of Mousetraps and Men: Behe on Biochemistry." *Reports of the National Center for Science Education* 20: 25–30.

Shanks, N., and K. H. Joplin. 2001. "Behe, Biochemistry and the Invisible Hand." *Philo* 4: 54–67.

Shanks, N., and Lahav, R. 1992. How to Be a Scientifically Respectable Property-Dualist." *Journal of Mind and Behavior*13: 211–232.

Sheets-Johnstone, M. 1998. "Consciousness: A Natural History." *Journal of Consciousness Studies* 5: 260–294.

Sherry, D. F., and B. G. Galef. 1990. "Social Learning without Imitation: More about Milk Bottle Opening by Birds." *Animal Behavior* 32: 987–989.

Shettleworth, S. J. 1998. *Cognition, Evolution and Behavior.* Oxford: Oxford University Press.

Sigma Xi. 1992. "Sigma Xi Statement of the Use of Animals in Research." *American Scientist* 80: 73–76.

Simon, H. J. 1960. *Attenuated Infection: The Germ Theory in Contemporary Perspective.* Philadelphia: J. B. Lippincott.

Singer, P. 1990. *Animal Liberation.* New York: Avon Books.

Sipes, I. G., and A. J. Gandolfi. 1993. "Biotransformation of Toxicants." In *Casarett and Doull's Toxicology,* edited by A. Amdur, J. Doull, and C. Klaassen, pp.12–49. New York: McGraw-Hill.

Skinner, B. F. 1976. *About Behaviorism.* New York: Vintage Books.

Sober, E. 2000. *Philosophy of Biology.* Boulder, CO: Westview Press.

Stumpf, S. E. 1982. *Socrates to Sartre: A History of Philosophy.* New York: McGraw-Hill.

Szathmáry, E., F. Jordan, and C. Pál. 2001. "Can Genes Explain Biological Complexity?" *Science* 292: 1315–1320.

Thayer, H. S., ed. 1953. *Newton's Philosophy of Nature: Selections from His Writings.* New York: Hafner Publishing.

Tolman, E. C. 1948. "Cognitive Maps in Rats and Men." *Psychological Review* 55: 189–208.

Trevathan, W. R., Smith, E. O. and J. J. McKenna, eds. 1999. *Evolutionary Medicine.* Oxford: Oxford University Press.

von Neuman, J. 1966. *Theory of Self-Reproducing Automata.* Edited by A. W. Burks. Urbana: University of Illinois Press.

Ward, P. 1994. *The End of Evolution: A Journey in Search of Clues to the Third Mass Extinction Facing Planet Earth.* New York: Bantam.

Weiner, J. 1994. *The Beak of the Finch.* New York: Vintage.

Werkmeister, W. H. 1980. *Kant: The Architectonic and Development of His Philosophy.* La Salle, IL: Open Court.

Wilson, E. O. 1978. *On Human Nature.* Cambridge: Harvard University Press

Wilson, E . O. 1992. *The Diversity of Life.* New York: W. W. Norton.

Winny, J. 1965. *The General Prologue to the Canterbury Tales.* Cambridge: Cambridge University Press.

Woolhouse, R. S. 1988. *The Empiricists.* Oxford: Oxford University Press.

Wray, G. A. 2001. "Development: Resolving the *Hox* Paradox." *Science* 292: 2256–2257.

Wynne, C. D. L. 2001. "The Soul of the Ape." *American Scientist* 89: 120–122.

Zucker, A., ed. 1996. *Introduction to the Philosophy of Science.* Upper Saddle River, NJ: Prentice Hall.

Zusne, L., and W. H. Jones. 1982. *Anomalistic Psychology.* Hillsdale, NJ: Lawrence Ehrlauer and Associates.

Index

About the Author

Niall Shanks is professor of Philosophy, adjunct professor of Biological Sciences, and adjunct professor of Physics and Astronomy at East Tennessee State University, where he is also president of the Institute for Mathematical and Physical Sciences. He is the author of numerous articles on the theory of evolution and the history and philosophy of science. He is also coauthor, with Hugh LaFollette, of *Brute Science: The Dilemmas of Animal Experimentation*.